Audio Education

Audio Education: Theory, Culture, and Practice is a groundbreaking volume of 16 chapters exploring the historical perspectives, methodologies, and theoretical underpinnings that shape audio in educational settings. Bringing together insights from a roster of international contributors, this book presents perspectives from researchers, practitioners, educators, and historians. *Audio Education* highlights a range of timely topics, including environmental sustainability, inclusivity, interaction with audio industries, critical listening, and student engagement, making it recommended reading for teachers, researchers, and practitioners engaging with the field of audio education.

Daniel Walzer is an assistant professor of music at the University of Massachusetts Lowell.

Mariana Lopez is a senior lecturer in sound production and postproduction in the Department of Theatre, Film, Television and Interactive Media at the University of York.

Audio Education

Theory, Culture, and Practice

Edited by
Daniel Walzer and Mariana Lopez

Routledge
Taylor & Francis Group
NEW YORK AND LONDON

First published 2021
by Routledge
605 Third Avenue, New York, NY 10017

and by Routledge
2 Park Square, Milton Park, Abingdon, Oxon, OX14 4RN

Routledge is an imprint of the Taylor & Francis Group, an informa business

Library of Congress Cataloging-in-Publication Data
A catalog record for this book has been requested

ISBN: 978-0-367-07443-2 (hbk)
ISBN: 978-0-367-07444-9 (pbk)
ISBN: 978-0-429-02078-0 (ebk)

DOI: 10.4324/9780429020780

Typeset in Times New Roman
by Apex CoVantage, LLC

Contents

Contributors

Mark Bassett is an internationally experienced academic director, lecturer, mastering engineer, and acoustician. He is the Director of Academic Quality Assurance for the SAE Institute's international licensed campuses and his professional audio experience includes managing a successful acoustic consultancy business and working as a mastering engineer for nearly a decade in Sydney. He holds a PhD from the University of Sydney's Spatial Audio and Acoustics Lab and his research in technical ear training has been published and presented internationally, including on several expert panels. In 2018, he founded Zipper Noise, a company specializing in developing audio engineering education technology.

Doug Bielmeier is an associate teaching professor at Northeastern University with a doctorate in education and over 15 years of experience teaching music/audio engineering at graduate and undergraduate levels. Bielmeier, formerly a freelance engineer in Nashville, Tennessee, has 15 years proven success as a studio and live-sound engineer. His live-sound work has included working at the Kennedy Center, in Washington D.C., doing sound for Vice President Joe Biden. Bielmeier was the Designer/Studio Manager of the C.L.E.A.R. Lab recording studio at the Purdue School of Engineering. His papers on audio education, circular design, and music have been published internationally.

Jude Brereton is a senior lecturer in music technology, University of York. She was the founding program leader for the MSc in audio and music technology in the Department of Electronic Engineering. She currently teaches in the areas of virtual acoustics and auralization, music performance analysis, and voice analysis/synthesis. She is dedicated to improving diversity in audio engineering, through innovative, creative approaches to teaching. She serves as a panel member for the UK Athena SWAN charter, which promotes gender equality in higher education. With a background in arts and music administration, she is still active in promoting research-inspired live performance events combining art and science for public engagement.

George Brock-Nannestad graduated in signal processing in 1971, focusing on musical acoustics. With public funding, from 1981 to 1986 he carried out the project The Establishment of Objective Criteria for Correct Reproduction of

Historical Sound Recordings. From 1991 to 1998 he was responsible for the media at the School of Conservation in Copenhagen. Subsequently, he has researched and consulted on sound restoration and AV technology history. He was a regular contributor to the AHRC Research Centre for the History and Analysis of Recorded Music, 2004–09, and contributed chapters to *The Cambridge Companion to Recorded Music* (2009) and *The Art of Record Production* (2012).

Helena Daffern is currently a senior lecturer in music technology in the Department of Electronic Engineering at the University of York, UK. She received a BA (honors) degree in music, an MA degree in music, and PhD in music technology, all from the University of York, UK, in 2004, 2005, and 2009, respectively, before completing postgraduate training as a classical singer at Trinity College of Music, London. Her research utilizes interdisciplinary approaches and virtual reality technology to investigate voice science and acoustics, particularly singing performance, vocal pedagogy, choral singing, and singing for health and well-being.

Marc Green received a BA degree in music production in 2013 from the University of Central Lancashire followed by an MSc in audio and music technology in 2015 from the University of York, UK. Presently a PhD candidate at the AudioLab, University of York, UK, following a successful year as an associate lecturer, his research interests include spatial audio, environmental soundscapes, and machine listening.

Brian Heller is an artist and technician who works with sound using a unique blend of skills. Over 20 years, he has built a practice integrating composition, audio engineering, and education. This diverse range of experience has taken him everywhere—recording in a grain silo, collaborating at a Bamboo festival, composing in an abandoned Manhattan storefront, and writing for *Electronic Musician* magazine. His positions at the Banff Centre, the Tanglewood Music Center, and the Aspen Music Festival have allowed him the privilege of recording concerts featuring some of the great classical performing artists of our time. Brian graduated from The Hartt School, and is currently the director of the Sound Arts program at Minneapolis College.

Clare Hibberd graduated in 2004 from the Royal Central School of Speech and Drama, London, with first class BA (honors) in theater practice–sound design. She worked extensively in musical theater sound departments across the West End and on national and international tours for nearly ten years before completing a PGCE at Canterbury Christchurch University, Kent, and then teaching science at a large comprehensive school in the South of England. In 2015 she briefly relocated to Oman to work at the Royal Opera House Muscat as senior sound technician. Clare is currently the sound lecturer on the BA production, technology, management course at the Royal Conservatoire of Scotland.

Andy Hunt is a professor of teaching and scholarship in the Department of Electronic Engineering at the University of York, UK. He has served as the deputy head for teaching and has been heavily involved in the implementation of the department's strategy for effective education. He teaches audio computer programming, app development, multimedia technology, and project management and development. His book *Managing Your Project: Achieving Success with Minimal Stress* is aimed at all university students having to undertake a large-scale solo project.

Elissa Johnson-Green is Assistant Professor of Music and Music Education and Coordinator of the Program in Music Education at the University of Massachusetts Lowell's Department of Music. She is the inventor and project leader of the EcoSonic Playground Project, an open-access, immersive music program for all children. She also directs the Creative Sound Play Children's Chorus of the Lowell String Project, which is a program that fosters and strengthens bonds between university and community. Her research interests focus on children's musical thinking through composition, immersive and differentiated learning in music education, and on music as a transitional property in children's lives.

Elsa Lankford is a full professor at Towson University in the US. She teaches audio production and sound creation in the Department of Electronic Media and Film. She is an interdisciplinary artist and composer whose individual creative work typically touches on the overlaps between sound, cities, and social action. Other pedagogical and research interests include designing sounds and soundtracks for visual media, student radio, audio documentaries and podcasts, and inclusive film and media festivals. She lives in Baltimore City with her wife, cats, and chickens.

M. Nyssim Lefford facilitates creativity in a variety of settings: academic and professional, artistic and scientific. Cognition and creative processes are the focus of her research. She studied music production and engineering and film scoring at Berklee College of Music, received her MS from the Massachusetts Institute of Technology for work on network music collaboration, and her PhD from the same institution for investigations into the perceptions of music creators in situ. Currently, she is a researcher and teacher at Luleå University of Technology in Sweden.

Mariana Lopez is a senior lecturer in sound production and postproduction in the Department of Theatre, Film, Television and Interactive Media at the University of York. In addition to lecturing in the field of sound, Mariana also leads the BSc in film and television production. She has a background in music and sound design, having been awarded a BA degree in arts with specialization in music and an MA degree in postproduction with specialization in sound design. In 2013 she completed her PhD at the University of York on the importance of virtual acoustics to further our understanding of the York Mystery Plays. Before joining the University of York, Mariana worked at Anglia Ruskin University as a senior research fellow.

Ben Mosley is a senior lecturer at Leeds Beckett University. He is course leader for BSc (honors) audio engineering course and teaches across a range of undergraduate and postgraduate audio and acoustics modules. As an installation engineer, he has designed and installed recording studios and music technology suites across the UK. Ben also continues to work as a recording engineer and acoustician and his research is focused on both audio education and sustainability within the audio industries. He has served as chair of the Audio Engineering Society UK Section, is a member of the Institute of Acoustics, and a senior fellow of the Higher Education Academy.

Adam Scott Neal is a composer working in a variety of genres, from toy piano solos to modern dance scores. His music has received nearly 300 performances in 12 countries. He has also presented and published research on experimental computer music, improvisation, and the agency of musical instruments. Adam earned a PhD at the University of Florida and previous degrees at Queen's University Belfast and Georgia State University. Originally from Atlanta, he currently lives in Dallas, Texas.

Frank Stevens received the MEng (honors) degree in electronic engineering with music technology systems and the PhD in music technology, both from the University of York, UK in 2014 and 2019, respectively. His research interests include environmental sound measurement, modeling, and evaluation. He is an associate lecturer in music technology in the Department of Electronic Engineering at the University of York, UK, where he teaches, among other things, digital signal processing in MATLAB, acoustics and psychoacoustics, synthesiser design, and foundation year electronics.

Paul Thompson is a professional recording engineer who has worked in the music industry for more than 15 years. He is currently a reader at Leeds Beckett University in the School of Arts and his research is centered on record production, audio education, popular music learning practices, creativity, and cultural production in popular music. His book *Creativity in the Recording Studio: Alternative Takes* was published in early 2019 by Palgrave Macmillan.

Daniel Walzer is an assistant professor of music at the University of Massachusetts Lowell. Walzer received his PhD from the University of the Cumberlands, MFA from Academy of Art University, MM from the University of Cincinnati, and BM from Bowling Green State University. Walzer draws on nearly 20 years of experience as an audio and media professional, session and touring musician, academic administrator, and faculty member at colleges and universities in the US.

An experienced music technology researcher and practitioner, Walzer maintains an active career as a composer and has published widely on issues relating to audio and music technology education.

Jeremy J. Wells is a senior lecturer in the Department of Music at the University of York, having previously worked as a teaching fellow, lecturer and research associate in the Department of Electronic Engineering. A graduate of the University of Surrey's Tonmeister program, Jez then went on to work for Digital Audio Research and Fairlight before undertaking an MSc, and then a PhD (completed in 2006), at York. From 2011 to 2012 he held an *Ingenious* Public Engagement Fellowship with the Royal Academy of Engineering. He is currently the program leader of his department's MA in music production. His research interests are audio design and sound recording practice.

Duncan Williams is a lecturer (assistant professor) in acoustics and audio engineering at the University of Salford, UK. Prior to this he worked in academia at the universities of York, Plymouth, and Oxford, and has taught in the US and China. Duncan holds a PhD in psychoacoustics from the University of Surrey, UK, and worked extensively in industry as a sound engineer and a psychoacoustic consultant, in particular working with automotive noise evaluations. He is a qualified teacher and a fellow of the Higher Education Academy.

Victor Zottmann moved from Brazil to Canada in 2014 to study sound design for visual media at Vancouver Film School. Shortly thereafter he went to Melbourne, Australia, to pursue a bachelor's degree in audio at SAE Institute. Recently, he completed his master's degree in architectural science (audio and acoustics) at the University of Sydney. With a focus on helping beginners improve both their creative and professional skills, Victor wrote for *DesigningSound.org* from 2017 to 2019. Among the factors discussed in this chapter, the experience he acquired inspired him to pursue technical and academic careers.

Affiliations

Mark Bassett—SAE Institute
Doug Bielmeier—Northeastern University
Jude Brereton—AudioLab, University of York
George Brock-Nannestad—Patent Tactics
Helena Daffern—AudioLab, University of York
Marc Green—AudioLab, University of York
Brian Heller—Sound Arts Program, Minneapolis College
Clare Hibberd—Royal Conservatoire of Scotland
Andy Hunt—AudioLab, University of York
Elissa Johnson-Green—Department of Music, University of Massachusetts Lowell
Elsa Lankford—Electronic Media and Film Department, Towson University

M. Nyssim Lefford—Luleå University of Technology
Mariana Lopez—Department of Theatre, Film, Television and Interactive Media, University of York
Ben Mosley—Leeds Beckett University
Adam Scott Neal—Southern Methodist University
Frank Stevens—AudioLab, University of York
Paul Thompson—Leeds Beckett University
Daniel Walzer—Department of Music, University of Massachusetts Lowell
Jeremy J. Wells—Department of Music, University of York
Duncan Williams—Acoustics Research Centre, University of Salford
Victor Zottmann—SAE Institute Melbourne

Acknowledgments

The editors wish to thank the staff at Routledge/Focal Press for their support and assistance throughout the project. Hannah Rowe and Shannon Neill, in particular, were helpful in guiding the book to completion. We are grateful to the two referees, who offered helpful feedback on the initial proposal for the volume. Finally, we extend our thanks to the volume's contributors for sharing their critical work in furthering audio education globally.

Introduction

Audio Education and the Makings of Creative and Inclusive Scholarship

Daniel Walzer and Mariana Lopez

As the editors of *Audio Education: Theory, Culture, and Practice*, we are pleased to share the first compendium exploring audio and its influence in educational settings. The volume's 16 chapters include a broad coalition of researchers, practitioners, educators, and historians from the United States, the United Kingdom, Australia, Brazil, and across Europe. From detailing the origins of audio specializations in education to assessing future sustainability efforts through an eco-literate lens, the chapters in *Audio Education: Theory, Culture, and Practice* aim to provide a practical and engaging resource for all with a keen interest in audio's rich potentialities.

The book's authors tackle what "audio" and "education" signify, both together and as distinctly separate entities in learning environments. Intentionally conceived, the book's title invites readers to reflect on a central premise of the volume itself—that defining education cannot happen without considering the social and cultural spaces where teaching and learning occur. Moreover, if the book intends to provide a tool for educators to utilize, then the content ought to reflect a similar openness. Theory must be well researched and balanced with thoughtful application in ways that educators can use in their work.

What *Audio Education: Theory, Culture, and Practice* cannot do, however, is provide a unified answer as to what the definition of audio education is in full. We believe it is neither wise nor feasible to "name and claim" a medium (audio) and a discipline (education) in a single, solitary fashion. What we do see, as is evidenced across the 16 chapters in this book, is that audio and its parts vary widely based on the perspectives of the authors and their students. The contributors in this volume come from backgrounds in audio engineering, television and film, acoustics, music education, sound design, and theater. What *is* clear, nonetheless, is that audio *informs* their pedagogy, *inspires* their practice, and *promotes* new and applied forms of knowledge exchange. As Boehm et al. (2018) remind us, music technology and its related specializations now have six generations of researchers and practitioners in the UK and globally.

In considering the volume's scope, we encouraged authors to exercise care when presenting ideas on specific technologies. Indeed, technology plays a critical role in teaching and learning. However, if one is to assume that audio functions in a broader media landscape, then prioritizing specific techniques becomes risky, as not everyone has the same access to particular tools. We are also aware that scholarly publication cycles vary. A project of this depth takes a few years to materialize. What might be considered cutting-edge technology at the time of writing may prove redundant once the volume is eventually published, as software platforms become obsolete—replaced by newer technologies. While these are commonly found risks in academic publishing, our hope here is that pedagogical discussions offer flexible options for educators to use in the ways they see fit.

Considering the book's philosophical aims, as editors, we deemed it essential to provide a supportive platform for emerging scholarly voices writing on topics of inclusion, social justice, and equality to be heard and eventually cited. Indeed, such issues find a home in music education and social science literature. Our aim here is not to present a volume solely on educational policy. On the contrary, we believe there is more room for reflective discourse on how to expand the audio-related canon beyond techno-centric content. That several chapters in this volume expand on such ideas is intentional and, in our view, vital for improving audio education's pathway for inclusion.

Chapter 1 begins with a historical look into how audio education evolved over two decades in the early 20th century. George Brock-Nannestad analyzes early training literature and hardware used in audio, along with the emergence of professional organizations dedicated to audio-related matters. In tracing audio's rapid evolution over two decades, Brock-Nannestad questions what it means to be an "audio person" in modern times. By understanding the history of the audio field, educators might better discern where the profession will be in the future.

Chapter 2 introduces the concept of primary school children as musical affectors—those with a capacity to influence others through creative means. Here, Elissa Johnson-Green elucidates how children's artistic expression provides a more in-depth understanding of their thinking patterns and ways of being. Drawing on a rich literature review, Johnson-Green provides evidence that audio education begins early in a child's development. Her chapter reports on a four-year study analyzing student compositions through scores, spectrographic analyses, videos, and field observations. In connecting audio education with music composition and creativity, Johnson-Green provides teachers with suggestions on how to cultivate similar methods with young children in the classroom.

Chapter 3 puts forward a more comprehensive and topical view of materials used in an electronic music history course. Brian Heller argues that by broadening the syllabus to include nontraditional voices and perspectives on sound technology, student learning is more abundant and the pedagogy more expansive. Heller's lens, informed by his experience working in a Sound Arts pathway at a two-year college in the United States, advance the nascent discourse on how to bridge academic topics and audio-focused curriculum in a fashion that embraces musicality, technology, popular culture, and sociocultural relevance.

Chapter 4 offers a holistic alternative to overreliance on digital audio workstations in audio education. Mark Bassett argues that to truly guide students to a deeper understanding of sound engineering concepts, they must develop problem-solving and autonomous learning competencies. Using Outcomes-Based Teaching and Learning and Max/MSP as theoretical models, Bassett writes on how emphasizing process over technology requires students to design a personalized mixing interface to meet course requirements. In so doing, Bassett maintains that students learn more than just a particular software—they solve problems and acquire a stronger foundation in audio engineering.

Chapter 5 details Mariana Lopez's research on incorporating gamification techniques to supplement lectures in an introductory sound design module for undergraduate film and television students. Lopez writes on how practical game theory enhances student learning by reinforcing concepts, improving class participation and motivation, and incorporating real-world examples to help articulate class topics. Lopez's research sheds light on future research in game theory and pedagogical best practices by using a logical analysis of student surveys, course learning outcomes, and her experience working with film students. Similarly, data gleaned from Lopez's preliminary research suggests that with thoughtful implementation, gamification techniques can support student learning, improve class participation, and facilitate deeper reflection by those working in audio education and related fields.

Chapter 6 examines how popular music serves to reinforce a broader, multifaceted foundation of critical listening competencies in audio and music technology education. In this chapter, Adam Scott Neal describes how using particular recordings, supportive teaching, and positive socialization help discern a deeper understanding of musical form, text painting, sound characteristics, and genre across grade levels. Neal takes care to consider both the audio-specific aspects of critical listening, and its broader merits as a gateway to a more in-depth student agency, interpersonal communication style, and understanding of course content. Neal offers suggestions on how to create similar units for music teachers across grade levels, and the practical considerations of including such content among more stringent learning rubrics in elementary and secondary schools in the United States.

Chapter 7 builds on the theme of critical listening and social contexts. Nyssim Lefford considers the importance of group work and dynamics, and how thoughtfully mediated listening sessions facilitate more meaningful peer feedback and cultivate a sense of aesthetics among audio engineering students. Lefford asserts that listening and reflection, both essential for success in audio engineering professions, are intertwined and connected; their success in practice requires the learner to be self-directed and aware. Drawing on a series of constructivist and educational theories, Lefford offers a method whereby listening session participants verbalize their reactions to sounds, ask questions, and discuss ways to improve recording and mixing. The sum of these efforts is a context where students and teachers have space to reflect on their social interactions without the pressures of "the profession" and "the academy" combined. Lefford's research reminds those of us in audio education that respect, autonomy, and attentiveness form the basis of a more nuanced language designed to improve teaching and learning in Sweden and beyond.

Chapter 8 introduces the reader to the different connotations of what an engineer is in this day and age. Jeremy J. Wells interrogates the limitations in placing too narrow a view on what engineering is, particularly as the definition relates to music production and recording. Using survey data and group discussions with varied stakeholders in the audio industry, Wells presents the epistemological possibilities and ramifications of what an engineer does, who that person might be, their diverse skill sets, and what their influence will be on the audio profession in the future. While a single description of the engineer remains elusive, Wells takes a humanistic view of audio engineering, is careful to detail the many artistic and scientific aspects of the field through synthesized learning, and urges educators to take an active role in addressing inequality in the profession at large.

Chapter 9 addresses the issue of sustainability in audio education. Ben Mosley's contribution is both timely and appropriate when considering the realities of climate change and environmental sustainability. Mosley's chapter is both a clarion call for educators to consider their ethical responsibilities to train eco-literate graduates and a case study in how to foster sustainability initiatives in audio education coursework. Equally crucial to Mosley's research is the overarching mission to prepare future graduates to be informed citizens and to use their audio-related knowledge and skills for the betterment of the world. Equally inspiring and firmly grounded in solid practice, Mosley's sustainability curriculum provides audio educators with points for reflection and practical implementation in related courses in the UK and elsewhere.

Chapter 10 focuses on interdisciplinarity in audio education. Elsa Lankford unpacks the plethora of technical and artistic possibilities associated with audio while providing a focused set of methods that audio educators can use to maximize departmental and institutional collaboration. Lankford discusses service and experiential learning opportunities for students through audio documentaries and multimedia production and guides the reader to a clearer understanding of the breadth and depth of disciplinary options and the possible implications for those working in audio education in the United States. By providing a thorough analysis of what interdisciplinary curricula is, along with related audio education programs in North America, Lankford's chapter reveals much about current trends in audio education and the likely ramifications for those working in colleges and universities individually.

Chapter 11 places audio education within the fluid policy-making framework of the UK higher education sector. Paul Thompson and Ben Mosley summarize audio education's space and place in the UK Creative Industries area, and clarify the essential aspects of audio education and its functions in a time when rising tuition costs and unpredictable job markets remain constant. Incorporating recent literature reviews, email correspondence with industry professionals, semi-structured interviews, and focus groups from the authors' institution, Thompson and Mosley's work contributes to the expanding discourse on how audio education can best serve vested stakeholders in the UK. Thompson and Mosley surmise that audio education, while ever improving and expanding, must be realistic and sensitive when assessing its abilities to balance vocational and sociocultural aims in the audio profession at large.

Chapter 12 presents a case study on how audio education functions within the realm of distance learning. Duncan Williams discusses the design and implementation of a course on sound recording technology through an online delivery format at a public university in the United States. While e-learning is commonly accepted in colleges and universities, its role in audio education (and audio engineering coursework specifically) is less clear. Through personal reflection and a synthesized and pertinent literature review, Williams offers a balanced perspective on the merits and drawbacks of delivering audio-related courses online. Additionally, through a comparative analysis of e-learning theories in the UK and abroad, Williams provides valuable considerations on assessment, course design, and how to meet audio-specific course outcomes for institutions offering similar course offerings online.

Chapter 13, written by Victor Zottmann, analyzes the state of audio education in Brazil and how online tutorials on audio-related topics provide a critical resource to non-English speakers. Zottmann's case study is the first of its kind to examine not only how the internet (and social media) functions as a dissemination tool for audio education topics, but also how the sites of learning extend beyond colleges and universities. Zottmann documents his process in generating 40 tutorials on audio-related topics suitable for a general audience, both in English and in Portuguese. In his case study, Zottmann interviews an audio education expert from Brazil and guides the reader to a sobering understanding of how challenging it can be to pursue audio education training in certain parts of the world. Zottmann's quantitative data reveals a deeper understanding of the potential audience interacting with the video tutorials, their learning preferences, and the role of social media in delivering audio-related content.

Chapter 14 places effective communication in audio education at the fore. Doug Bielmeier builds on his considerable body of scholarship in audio education both reflectively and by synthesizing the existing literature to lead the reader to a clearer understanding of the types of effective communication required from audio-related stakeholders. Bielmeier reinforces the social and relational aspects of audio education and the audio industry. Such partnerships, when and where appropriate, enhance why communication matters, and how educational stakeholders can inform their decision making accordingly to meet the needs of students. Ultimately, it is in the local and regional partnerships where the most actionable parts of communication can benefit audio education. Bielmeier's chapter reiterates the need for mentoring and relevant professional experiences via internships and service learning, along with adjusting learning environments to reflect the needs of the audio industry.

Chapter 15 offers a thorough look at the issues affecting gender imbalance and representation in theatrical sound design and related fields. Clare Hibberd examines the lack of women participating in sound design in Scotland and frames the discussion in the broader context of audio education. Hibberd's review of the literature touches on brilliance bias, gender stereotyping, and current initiatives underway to expand participation for women and persons of color in music technology, sound design, music education, and similar disciplines. Hibberd's chapter details ways to address social justice issues and how her institution chose to incorporate sensitive topics into course modules. Hibberd's case study serves as a model for how audio educators might want to address social and political issues in and out of the classroom to advance a pluralistic and inclusive model of pedagogy and learning.

Chapter 16 closes the volume with a comprehensive review of the MSc in audio and music technology program at the University of York, UK. Jude Brereton and her colleagues trace three decades of audio education best practices while providing a fascinating look at how the *field* of audio education has changed (both at the University of York and elsewhere), and how institutions have to adapt to such changes to remain solvent and relevant in pedagogy and educational missions. Detailing York's rollout of a master's degree in audio and music technology, Brereton et al. outline the degree's learning outcomes, course design, admission requirements, and purposeful efforts to recruit, retain, and *graduate* a diverse and talented pool of students. Brereton and her colleagues advocate for a rigorous yet flexible learning pathway; pedagogy is designed to meet the particular needs of each student. For audio educators, Brereton et al. remind us that there is real value in topical coursework, supportive mentoring, peer learning, and projects that reinforce problem solving and creativity. Concluding the book, Brereton and colleagues offer innovative and practical approaches to assessment while reflecting honestly on their successes and areas for improvement as a department. Beyond employability, the synopsis here is both evidence based and reflective, two crucial resources that peer institutions can study and emulate.

It is our hope that *Audio Education: Theory, Culture, and Practice* provides valuable information for educators, students, researchers, creative practitioners, industry professionals, and academic administrators with a vested interest in audio-related topics in education. As with any major project like this, there are likely more research and pedagogical questions worth exploring in future editions and through related scholarly outlets. Certainly, one volume is but a glimpse of what audio education *might* be, and how the imaginative writings of the book's contributors can inspire dialogue, reflection, and action to further audio education's position as an emerging field in higher education.

Daniel Walzer, PhD, and Mariana Lopez, PhD, Co-Editors
May 2020

Reference

Boehm, C., Hepworth-Sawyer, R., Hughes, N., & Ziemba, D. (2018). The discipline that "became": Developments in Music Technology in British higher education between 2007 and 2018. *Journal of Music, Technology & Education, 11*(3), 251–267. doi:https://doi.org/10.1386/jmte.11.3.251_1

1

The Roots of Audio—From Craft to Established Field 1925–1945

George Brock-Nannestad

Introduction

Many are engaged in audio—some as professionals at one end or as customers, hobbyists, and enthusiasts at the other end. At times, it is difficult to differentiate between the two ends of the spectrum. At least to those who are not engaged in audio, it is somewhat enigmatic. One might try to say that audio is *applied acoustics*, but that is not all. Audio comes from Latin and means "I listen", but that says even less. What are the roots of audio as we know it today? One answer may be found by investigating how the field developed and the environments in which this took place. This chapter traces the historical development of the audio profession and field by means of contemporary English-language literature from the US and the UK, the hardware that was and became available, and the organization of the professionals as they developed the subject area.

Prehistory: A Sound Industry Without Amplification

When commercial sound recording and record production began, getting a useful signal recorded was a very complex operation. It was managed by using sound collecting funnels or horns that were connected to a diaphragm, which converted the pressure into the vibration of the stylus used for cutting "wiggles" into the surface of the blank, the *wax*. Those who did this were called "recording experts", and their result—the recorded wax blank—was the input to a very complicated manufacturing process that resulted in *phonograph records* (British English: *gramophone records*).

It was only when these had been manufactured that a test for quality could be performed. At times there was an exceedingly long delay in the feedback, which meant that the recording experts had to base their instant decisions as to loudness, placement of performers around the sound collectors, and similar on a vast personal experience (Batten 1956; Brock-Nannestad 2016). The reproduction took place by using mechanical equipment that converted the wiggles in the grooves in the manufactured record to sound, again employing a diaphragm and horn combination.

Significance of Amplification

Modern audio invariably seems to involve amplification in some way, which sets the protagonists apart from e.g. the acousticians who built the ancient Greek theaters (Hunt 1992), in which vases were built into the walls as passive resonators to improve distinctness of dialogue. Later theater acoustics were *passive* as well. The acoustical properties of performance venues have been described by several acousticians, such as in Beranek's (1962) book *Music, Acoustics & Architecture*. *Passive* means that any manipulation of a sound to obtain a desired quality can only make it weaker. *Amplification* means that a signal is obtained with a greater power than the un-amplified signal possesses itself. Embedding an amplifier makes *passive* circuits into *active* circuits. Amplification as we mostly use it today is electrical or electronic, but audio is no part of the field of *electronics* as such, only to the degree that the electrical signals are "low frequency" to distinguish them from radio signals, which are "high frequency". But what about the scientific discipline *electroacoustics*? This only refers to a transducer from an electrical signal to a sound signal or vice versa (Hunt 1954).

A transducer is inevitable for conversion of an electrical signal into an audible signal. The first practical transducer of this kind was the telephone receiver. It does not say anything about how the electrical signal came to be. For instance, Thaddeus Cahill's Telharmonium for wire distribution of music (ca. 1900 and for a few years) used electrical generators at specific frequencies (combined into chords by a keyboard) and at distribution strength directly to the transducer, the loudspeaker, or telephone in the home or venue (Williston 2000). This was in principle not very different from the radio frequency "alternator" devised by Ernst Alexanderson of the General Electric Company (https://en.wikipedia.org/wiki/Alexanderson_alternator) that without an oscillator provided a high-frequency voltage, which was modulated and used directly for long-distance signaling-radio. Common to these two approaches was the non-presence of amplifiers.

The electrical (not yet electronic) world waited for an amplifying element that could handle the small variable signal voltages in telephone systems and make more powerful signals, not the least because the telephone signal got fainter with distance. For faint telegraph signals, so-called *repeater relays* were already used to receive a weak current and control a strong current for retransmission. Such repeaters were placed along telegraph lines at suitable distances. The task of amplification was simple, because there were only two states that a telegraph line could sensibly be in: full voltage or zero, and the relay "on" or "off". It was simple to make a repeater. But all the small variable signals in speech sounds were in need of amplification.

Amplifiers for sound have existed since Chichester Bell described the water jet amplifier (Bell 1886), which was given publicity by lectures of C.V. Boys at the

London Institution in 1889 and 1890. Boys subsequently published a corresponding book, *Soap Bubbles and the Forces Which Mold Them*, which was in print in many editions from 1902 to 1960 (Boys 1960). At the time, amplifiers generated a large amount of extraneous noises. If the intention was to make a sound recording, the noise would be recorded along with the intended signal, and for this reason it was impossible to use fluid, pneumatic, or mechanical amplifiers (such as the Auxetophone) for recording. Instead, they could be used for amplifying the sound picked up from a record or in a loudhailer. The traditional loudhailer was merely a funnel without amplification where you shouted into a mouthpiece at the narrow end. The available power was radiated in one direction only, which was efficient if the listener was "in the beam", but not outside.

Amplifiers as we know them only came with the commercialization of the *vacuum tube* (British English: the *amplifying valve*) in the early to mid 1920s. The commercial availability of this component made possible the development of several related technologies: broadcasting of speech and music and the use of easily controlled components in sound recording. The basis of the recording industry started around 1895, originally without the use of amplification (Hoffmann 2005). When amplifiers first appeared, running them was a specialty, and two persons became responsible for a recording instead of one—an electrician and the recording expert (Columbia Graphophone Co. Ltd. 1930). Neither could really identify him- or herself as "an audio person". It was only when amplifiers became so reliable that switching them on and using a simple level control enabled one person to work the system; this might then signal the birth of the "audio person" in the mid to late 1920s. Working with an amplifier to control the level on the fly reduced the "guesswork" and reliance on experience and compensated to a degree the long delay in feedback as to the quality of a recording, and it reduced guesswork.

These were entertainment and also military uses; spin-offs were in medicine and physiology. Amplifiers could be made to perform better and more consistently after the introduction in 1927 of negative feedback (Kline 1993). Firm design rules became available in 1934. The improvements were so tangible that these technological developments were adopted very quickly.

Written Sources to Early Audio Technology

If we accept the thought that audio involves acoustics and electronic amplification, we must look for technical books dedicated to the subject written for those in the early periods who were seeking information that was otherwise not easily available. Those at the forefront, developing the field, wrote articles for journals (most frequently the text of papers they had read at meetings of the professional societies, such as the Institute of Radio Engineers, or the American Institute of Electrical Engineers). Unfortunately, to access the information, one had to be a member of such societies to read the articles. These persons also took out patents, which are very detailed descriptions and drawings of technical solutions. Textbooks as such, for study purposes, did not exist. The easiest access to valid information on central components of modern audio around 1930 were two British landmark books: *Modern Gramophones and Electrical Reproducers* by Percy Wilson and George W. Webb in 1929 and *Loud Speakers: Theory, Performance, Testing, and Design* by Norman W. McLachlan in 1934. These books are specifically mentioned because they were in English. The whole field was so fertile that there was a parallel development in for instance Germany and France, with a literature in German and French.

Percy Wilson was the permanent technical consultant to the magazine "the Gramophone", founded in 1923. He was educated at Oxford University and had a solid mathematical foundation. Wilson was capable of explaining fairly complex electro-acoustical phenomena to persons with an interest in disc phonographs and records, but without the in-depth mathematical knowledge required to solve differential equations. Wilson and Webb also covered the acoustic reproduction of records that were already out of fashion in the US in the 1930s but which were still used by 25% of the buyers of records from EMI—the customers—in Great Britain and in the (then) Dominions as late as 1947 (Mittell 1947).[1] The book had a long service life; it was only replaced in 1957. This special structure of the British market made EMI unable to adhere to the most recent recording standards, because such records sounded distorted on acoustic equipment.

Norman William McLachlan was industrially employed after having obtained a DSc degree (with a solid mathematical foundation) from the University of Liverpool, and while acting as the research director of TELCON, a British communications cable laying operation that was hit by the 1930s depression, he measured a huge number of loudspeakers. His book is full of experimental results, and he neatly explained the properties of the prevailing loudspeaker systems of the time. He was the person in the UK that did the most to promote the moving coil loudspeaker (the Rice-Kellogg construction from the General Electric Company), which became the surviving construction among many transducer constructions.

Mathematics as a Tool for Technical Development

All technical fields involving electricity and mechanics benefited from the application of mathematics to structure the understanding of the behavior of the various elements. This took a great upsurge in the second half of the 1800s. Using mathematics correctly was complex, yet its ability to predict performance was very desirable, and the new endeavor of electrical communication (telegraph, telephone) took full advantage of it. Oliver Heaviside in the mid-1880s worked on telegraph and telephone transmission and predicting by modeling and calculating the performance of cable networks (Heaviside 1894).

Heaviside determined that under certain given conditions, the mathematics used for describing the phenomena could be considerably simplified. He introduced the term *impedance* and created a so-called *operational calculus system* around the impedance for AC to make it function similar to a *resistor* for DC and could use a simple modification to Ohm's law, avoiding the use of the differentials and integrals inherent in traditional mathematical modeling of AC systems. Heaviside's use for electrical (soon to become electronic) circuits was noted by A.G. Webster in 1914 (Webster 1919), who used the impedance concept for acoustical components. The ease with which Oliver Heaviside had made use of impedance for solving problems of electrical circuits could be transferred to acoustical and electro-acoustical problems.

The physicists at the Western Electric laboratories (in the Bell System) adopted Webster's novel use of the impedance concept and applied it to telephone components that became much easier to design. A spin-off was the design of an electrical way of cutting "wiggles" in the grooves representing sounds, an *electrical cutterhead*. They published in the traditional way (papers, journals) for the professionals (Maxfield and Harrison 1926). For amateurs and other interested parties, Wilson and Webb provided a popular and easier understandable version in their book. The book has been considered so historically important that it was reprinted in a facsimile edition in 1980 (Wilson et al. 1980). We must conclude that without the mathematical foundation, technical audio development would have been difficult and slow.

Sound and the Movies

The general public had taken to watching movies and accepted that dialogue was likely written on intertitles (written dialogues shown on the screen between scenes of action). The availability of vacuum tube amplifiers also enabled the provision of "talkies" both on the recording side and on the reproduction side, like those found in film theaters. There, huge PA systems (amplifiers and loudspeakers) were hidden by the projection screen, and the illusion of people having a dialogue became nearly perfect. Film projection was a craft performed by film projectionists who had to learn new tricks when the sound signal was combined with the film (Cameron 1933).

Sound-on-film (where a soundtrack along the edge was a graphical representation of the sound waves) was developed in parallel to *sound-on-disc*, predominantly in the US. Here, the projectionist had to attend to a disc reproducer to run in synchronism with the moving images. *Sound-on-film* was the simpler and more efficient technology that survived, although there were fierce patent battles with European patent holders. The professionals in the 1930s would follow a practical handbook like *Motion Picture Projection and Sound Pictures* by James R. Cameron, written in 1933. Films were handled by professionals, and film technology was among the first areas in the entertainment industry to establish industry standards that ensured that films manufactured on equipment from one manufacturer would give good results on other equipment. For this reason, the *Journal of the Society of Motion Picture Engineers* featured a lot of audio information, even when they took television under their wings and changed the name to the *Journal of the Society of Motion Picture and Television Engineers* in 1950.

The publishing history of the standard handbook on film projection by James. R. Cameron, *Motion Picture Projection and Sound Pictures*, is recounted in the introduction of its fifth edition from 1933. The first edition from 1918 was merely called *Motion Picture Projection*, because that was its subject (i.e. *not* manufacturing, and obviously no thoughts about sound), and the second edition from 1921 was the same. The demand for this information was so great that it saw three printings. In 1923, the third edition appeared, also in three printings. In 1928, the fourth edition came out, and it saw four printings. The demand reflects the growth in the number of film theaters already in the silent film period. Each film theater was able to convert to sound by retrofitting suitable equipment. In 1930, an independent book, co-edited with John F. Rider, *Sound Pictures* was introduced, and the fifth edition of *Motion Picture Projection* from 1933 then incorporated the second edition of this book to give an expanded title and a handbook of 1,535 pages. It was still in print in 1947 and possibly later.

The film industry involved audio persons in the sound recording on the set and in the reproduction in film theaters. The sound had to be clear and so the microphones had to be close to the dialogue. Yet, the microphones were supposed to be invisible; so the system of booms was developed—cranes that could move a microphone close, but outside the frame and remove it when the frame grew or when panning. Post-processing evolved, permitting spoken dialogue recorded separately in a proper sound studio to be dubbed into the correct places, and the background music made to fit. This led to dubbing of dialogue into vernacular languages for export of movies—the tradition in early sound films had become to keep the original plot, but to use native performers for foreign versions (it had been much easier and cheaper with silent films, because one simply needed to change the intertitles into the desired language). Background noises were recorded in separate studios and dubbed-in where needed, very similar to the radiophonic broadcasting studios where sounds for plays were provided. These techniques have been refined and are presently used in day-to-day production of film and video.

To professionalize studio employees, the Academy of Motion Picture Arts and Sciences (founded 1927 and known for the *Oscar*) created the Academy School in Sound Fundamentals. The lectures were published to reach outside the school by the Academy in a Technical Digest series, and the book *Recording Sound for Motion Pictures* (Ed. Lester Cowan) was published in 1931. In 1938, it was determined that the technical development had advanced and that new literature was required, and the Research Council of the Academy of Motion Picture Arts and Sciences (established 1934) published a series of lectures in the form of the book *Motion Picture Sound Engineering*. This was a step on the way to professionalization of the field.

Driving the Development

The Second World War increased the requirements for and an interest in radio communication; several professional and trade magazines catered to both professionals and hobbyists. Radio communication of speech invariably had an audio channel, and for this reason radio magazines frequently had articles about the low-frequency side of communications. The only professional way to store sound (film was too bulky) between 1935 and 1945 was by means of recorded discs, and many articles about disc recording were published in the radio magazines. Broadcasting was one part of radio, and program exchange occurred by means of recorded disks transported by carrier from one station to the other, which required internal standardization to be truly useful. One magazine was more important professionally than the others. The magazine *Radio* had started publishing in 1921–1922, but from May 1947 it changed its focus completely, renaming itself *Audio Engineering* and concentrating on low-frequency technical matters.

Those professionals who recorded sound on disc (for broadcasting and record production) and manufactured components for this purpose had joined as a group in New York in 1942 called the "Sapphire Club", a reference to the cutting stylus for cutting disc records. In 1947, a similar group was established in the intense audio-visual environment of California (the film industry). Reports of their activities were given in *Audio Engineering*, and when the Audio Engineering Society was founded, based on these activities in 1948, a special section of *Audio Engineering* was set aside for society matters. In 1954, the focus of *Audio Engineering* shifted from essentially hardware and technical principles to including much more editorial matter on recorded content (commercial records), and the title of the magazine changed to *Audio*. However, the Audio Engineering Society did not have a problem with this loss of publication means; it had established its own professional journal for its members in 1953, the *Journal of the Audio Engineering Society*.

The Audio Engineering Society was a much younger association than the well-established Society for Motion Picture (added later: and Television) Engineers, yet the specialization was good for the development of the audio field. Audio had broken loose from the film, and sound was not to be subordinated to a video format until 30+ years later when digitization was introduced.[2] The AES dealt with all types of audio: broadcast transmission, recording and reproduction, film sound (to a lesser degree), and public address.

Conclusion

Audio developed from a nascent craft into a fully realized professional field; complete with theory and hands-on experience. Audio's rapid growth evolved over 20 years, or one full generation. The first area that developed a professional approach was the film industry, both by the professional association the Society for Motion Picture Engineers and the Academy of Motion Picture Arts and Sciences. This area quickly moved away from apprenticeship as the basis of education. As broadcasting came to be dependent on program exchange, this area, too, established professional standards. One might indeed say that the identity of an audio person is linked to an involvement with amplifiers and the inevitable transducers. Although it has not been mentioned before, the critical professional ear is obviously all-important, irrespective of the technology used.

Epilogue: Among Peers

We may continue the conclusion by reflecting that we have defined an identity of an audio person by means of the literature surrounding the subject. This is a 2019 view—how did the people who were in the midst of developing the field see it while it was happening? Very few engineers have written personal memoirs, like Joseph Batten, and the risk is that such texts can only really be understood by persons who have a similar level of knowledge.

Around 1960, a British loudspeaker manufacturer asked a number of his business acquaintances if they would write their experiences for a book he wanted to publish called *Audio Biographies*. This was Gilbert Briggs, who was in essence a polymath, who via his company, Wharfedale Wireless Works (Briggs 2012), had published books throughout the 1950s on high fidelity, loudspeakers, and the acoustics of the piano. His contacts all thought it was a good idea, and the book was printed in 1961 with more than 60 technical mini-biographies.

Briggs mostly brought entries from the UK, but it seems probable that if he had been closer to the US scene, he would have brought even more from the US. He did not intend the book to be a "Who's Who in Audio" (Briggs 1961), and it appears more as the presentation of a closely knit circle of professionals. And that is really what identity is about. These persons all had a broad outlook from a solid base of detailed knowledge in the audio field. This knowledge (apart from personal experience) can be obtained by studying the existing literature.

Notes

1. Electrical and Musical Industries, Ltd. (EMI) was the result of a merger between the British companies the *Gramophone Company* and the *Columbia Graphophone Company* in 1931. The company was a huge factor on the market. Its head of the International Artists' Department, Brenchley E.G. Mittell, reported on modern recording and reproduction in 1947.
2. The standards for sampling rate and bit depth in commercial audio were derived from the video field (except in broadcasting). Later, the compression standards for distribution of sound were similarly related to video requirements. Note that *mp3* is more correctly given as *MPEG-1 or MPEG-2 Audio Layer III*; the acronym meaning *Motion Picture Expert Group.*

References

Arthur, G. and Webster, A. G. (1919) "Acoustical Impedance, and the Theory of Horns and of the Phonograph", *Proceedings of the National Academy of Sciences*, 5, 275–282.

Batten, J. (1956) *Joe Batten's Book: The Story of Sound Recording*. London: Rockcliff Publishing Corporation.

Bell, C. A. (1886) "Sympathetic Vibration of Jets", *Philosophical Transactions of the Royal Society*, 177, 383–422.

Beranek, L. L. (1962) *Music, Acoustics & Architecture*. New York: John Wiley & Sons., Inc.

Boys, C. V. (1960) *Soap Bubbles, and the Forces Which Mold Them*. New York and London: Heinemann, 103–106.

Briggs, D. (2012) *A Pair of Wharfedales: The Story of Gilbert Briggs and His Loudspeaker*. Chichester: IM Publications.
Briggs, G. A. (1961) *Audio Biographies*. Idle, Bradford: Wharfedale Wireless Works Limited, 7.
Brock-Nannestad, G. (2016) "Against All Odds: Commercial Sound Recording and Reproduction in Analog Times", *Acoustics Today*, 12(3), 12–20.
Cameron, J. R. (1933) *Motion Picture Projection and Sound Pictures*, 5th Ed. Woodmont, CT: Cameron Publishing Company.
Columbia Graphophone Co. Ltd. (1930) *Recording Expedition Instructions*. London: Columbia Graphophone Co. Ltd.
Cowan, L. (Ed.) (1931) *Recording Sound for Motion Pictures*. New York and London: McGraw-Hill Book Company, Inc.
Heaviside, O. (1894) *Electrical Papers*, Vol. 2. New York and London: Macmillan and Co., 61–66.
Hoffmann, F. (Ed.) (2005) *Encyclopedia of Recorded Sound*, 2nd Ed. New York and London: Routledge.
Hunt, F. V. (1954) *Electroacoustics: The Analysis of Transduction and Its Historical Background*. Boston and New York: Harvard University Press.
Hunt, F. V. (1992) *Origins in Acoustics: The Science of Sound from Antiquity to the Age of Newton*. New York: Acoustical Society of America, 32–35.
Kline, R. (1993) "Harold Black and the Negative-Feedback Amplifier", *IEEE Control Systems Magazine*, 13(4), 82–85.
Levinson, N. (presumed Ed.) (1938) *Motion Picture Sound Engineering*. New York: D. van Nostrand Company, Inc.
Maxfield, J. P. and Harrison, H. C. (1926) "Methods of High Quality Recording and Reproduction of Music and Speech Based on Telephone Research", *Transactions of the American Institute of Electrical Engineers* 45, 334ff.
McLachlan, N. W. (1934) *Loud Speakers: Theory, Performance, Testing, and Design*. Oxford: Clarendon Press.
Mittell, B. E. G. (1947) *Commercial Disc Recording and Processing, Informal Lecture Delivered to the Radio Section of the Institution of Electrical Engineers on 9 December 1947*. London: EMI.
Williston, J. (2000) "Thaddeus Cahill's Teleharmonium". www.synthmuseum.com/magazine/0102jw.html, last visited 31 July 2019.
Wilson, P. and Webb, G. W. (1929) *Modern Gramophones and Electrical Reproducers*. London: Cassell.
Wilson, P. and Webb, G. W. (1980) *Modern Gramophones and Electrical Reproducers*. Ann Arbor, MI: University Microfilms International.

2

The Child as Musical *Affector*

Sound Perception and Manipulation as Artistic Process in Primary School Music Education

Elissa Johnson-Green

Introduction

> My notion would be, that anything which possesses any sort of power to affect another, or to be affected by another, if only for a single moment, however trifling the cause and however slight the effect, has real existence; and I hold that the definition of being is simply power.
>
> —Plato (360 BCE)

In the Socratic Dialogue, *Sophos*, Plato put forth an enduring definition of human consciousness. As aware beings, we have the power to affect change and to understand how change affects us. Daily interactions with objects and others teach us how to use this power (Damasio 1999, 2000). This learning process is deliberate and experimental. It relies on our ability to perceive and gather information, to synthesize that information, and to use that information in new ways so that we can continue to learn and grow.

Children are expert *affectors*.[1] Their natural inclination to absorb, process, and use information is paired with highly sensitive perception and superpowers of observation, which makes them ideally suited for creative thinking. As imaginative beings, children take cues from and then act on their environment to compel profound change. In Plato's terms, children are powerful beings.

People such as I, who study children's artistic processes, do so because we are interested in understanding how that process works—what happens in the transition from cognition to action. We examine their outward behaviors and the things they create for clues about how they seem to enact change so quickly and effortlessly. We notice how their views of the world take us by surprise with joyful unexpectedness and we endeavor to understand the world from their perspective.

In my work with children, I have noticed how seriously they take their musical creations; how their music seems to express directly what they are thinking and, by extension, *who they are*. The strength of children's relationship to and with music has always fascinated me. I have been studying the ways in which children perceive sound in their environment and then to express themselves musically using that sound as a kind of found object to create highly complex musical constructs.[2] When I began to learn about the audio arts and audio education, I realized that there is great potential for the youngest children to learn deeply about sound, its properties, and what it can do as an artistic medium.

Through my earlier research on children's musical lives (Custodero and Johnson-Green 2003, 2008), I came to understand that children are inherently musical beings (Campbell 1998; Malloch and Trevarthen 2009; Marsh 2011). This condition makes music a particularly good lens for studying the connection between children's perception of sound and their musical expression (see Figure 2.1).

Studying the processes and products of children's music making reveals the inner workings of children's musical thinking (Bamberger 1995; Barrett 1997, 2000; Carroll 2018; Hickey and Webster 2001; Kerchner 2000). The complex nature of children's musical participation has led researchers to study their active involvement in music from different perspectives (see Table 2.1).

All of these activities are integrated throughout children's daily lives such that examining any one of them will provide a view into how children think musically. However, I decided to focus mainly on children as composers because I wanted to study children's perception of sound and how they turn thought and experience into musical expression. Through my teaching, I discovered that composition allows children to act on musical material, to use music as a medium for creative development and as an agent for transformation.

Looking further into how children develop creatively, I also learned that they use every affordance available to them in their environment to further their own social, emotional, and intellectual growth (Shonkoff and Phillips 2000). My students

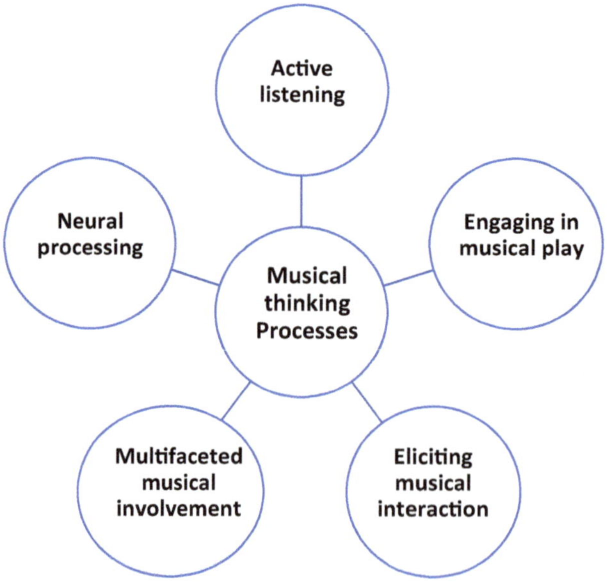

Figure 2.1 The components of musical thinking processes.

Table 2.1 Types of children's active musical involvement and the corresponding research

Children's musical involvement	*Research*
Singing	Custodero (2006), Dzansi (2002), Fox (1988), Ilari (2009), Niland (2015), Sims (1995), Rutkowski (1996)
Moving	Campbell (1991), Flohr and Brown, (1979), Metz (1989)
Playing instruments	Habibi et al. (2018), Alemán et al. (2017), Conway (2004), Rickard et al. (2010), Woody and Lehmann (2010), Young (2008a)
Listening	Pogonowski (1989), Pramling and Wallerstedt (2009), Randall et al. (2014), Reybrouck et al. (2009), Sims (2005), Sims and Nolker (2002)
Improvising	Custodero (2008), Guderian (2012), Wilson and MacDonald (2012, 2017), Brophy (2001), Burnard (2000), de Vries (2005), Giddings (2013), Russell and Austin (2010)
Composing	Barrett (1996, 2003), Bugos and Jacobs (2012), Burnard and Younker (2002, 2004), Deutsch (2016), Hickey (1999, 2003, 2012), Hogenes et al. (2016), Johnson-Green (2018, Forthcoming), Jorgensen (2016), Kaschub and Smith (2009), Marsh (1995), Shouldice (2014), Stauffer (2002), Strand (2016), Webster (2016)
Use of music technology	Acker et al. (2015), Addessi and Pachet (2005, 2005–2006), Burton and Pearsall (2016), Lum (2008), Merkow (2013), Young (2008b, 2009, 2012), Young and Gillen (2006)

generally were growing up in environments with access to many types of learning materials, including all kinds of technology. Tablets, computers, smartphones, and smart boards permeated their daily lives inside and outside of school. Using these devices had become second nature to many of my students, most of whom were growing up in technologically rich environments.

I became more aware of children's relationship to technology when I found myself heavily integrating it into my classroom life. Further, I began to consider the impact of my students' familiarity with technology on their music education. To compose music, my students used electronic keyboards, which gave them many options for sound production. I recorded performances of their pieces using my laptop computer, an iPad, and an iPhone. My students used these recordings to self-assess their work, which furthered their compositional development (Johnson-Green: In press). I taught them using an interactive smart board, which they used to work out compositional problems and to record themselves playing. The technology in my classroom became a compositional tool and strategy that was inseparable from the children's learning process.

The Connection to Audio Education

Children's complex, essential relationship with and to music makes them musical experts from an early age. Their finely tuned perception of sound, their innate ability to be musically creative, and their experience with technology make them ideally suited to begin audio education younger than music teachers might expect. To give context for children's developing sound/music cognition and perception, I will discuss their musical development, the connection to composition, and to audio education.

Musical Development

Human beings typically begin to hear and process ambient sounds at around 25 weeks gestational age (GA) (Draganova et al. 2018), recognize the mother's voice at 33–34 weeks GA (Jardri et al. 2012), and develop fully matured hearing between 35–38 weeks GA (Birnholz and Benacerraf 1983). At the same time, fetuses show remarkable sensitivity to musical sounds (Chorna et al. 2019; Kisilevsky et al. 2004; Lopez-Teijo et al. 2015; Woodward 1992) and there is evidence that they may retain musical memories after birth (Chorna et al. 2019; Granier-Deferre et al. 2011; Wilkin 1995). The prenatal sound environment generally includes the types of music inherent in the parents' culture (Campbell 2000; Byron 1995; Teie 2016) and that which parents value (Custodero and Johnson-Green 2003). As research on prenatal care has become more mainstream, parents may also play music specifically for their baby-to-be.[3] That infants are immersed in music before they are born has implications for their lifelong musical development (Parncutt 2016).

After they are born, babies are musically attuned to closest caregivers (Malloch and Trevarthen 2018), which supports attachment (Trehub et al. 1997), the foundation for healthy relationships. In infancy (typically 0–toddler) children's musical vocalizations and movement function as emotional expression (Adachi and Trehub 1998; Trehub and Nakata 2001–2002), in communication with (Malloch 1999; Malloch and Trevarthen 2009; Trevarthen 1999–2000; Trevarthen and Malloch 2009) and in social connection to caregivers (Ilari 2016; Lamont 2016; Volchegorskaya and Nogina 2014; Trehub 2001, 2016). By four months old, typically developing infants have developed similar hearing perception to that of adults (He et al. 2009). This finely tuned perception of musical sounds allows an infant to gather important information about the emotional state of key people (Lamont 2016; O'Neill et al. 2001) and to learn appropriate social responses (Cirelli et al. 2018; Ilari 2016; Malloch 1999; Trainor and Cirelle 2015).

As children grow, musical interactions remain essential to their development as people (Malloch and Trevarthen 2018). Music permeates children's lives; their relationship to and with it develops as they do, through constant exposure and use. Within the family, children may sing, dance, or listen to music daily (Custodero 2006; Johnson-Green 2012). With friends, interactive musical play becomes important as a social strategy (Alcock 2008; Barrett 2016; Marsh and Young 2006; Niland 2009; Young 2008a). Children's musical preferences become part of their musical identity starting in infancy (Trevarthen 2002) and ultimately form the foundation for ongoing development of their personal identity (Barrett 2011; Davidson and Borthwick 2002; Hargreaves et al. 2016; Niland 2015).

As children grow up in immersive musical environments, they become comfortable with using music as an expressive medium. Their musical perceptions and understanding are complex and profound. Throughout their development, children observe, learn, and communicate through musical sound as a material that helps them to transition from thought, reaction, and interaction to expression. Children's musical inventions are deliberate and original; they are improvisatory with some development of musical material directly connected to their play. In other words, children are natural composers. As composers they draw on all of their experience of daily life to create and communicate through music.

Musical Thinking

While it is impossible to know what another person is thinking and perceiving, it is possible to gather data from various perspectives to create a blueprint of what happens in the immediate transition from musical thought to musical expression. Webster and Richardson give a working definition of "thinking in sound":

> Musical thinking means making sense of the perceived sounds through such activities as grouping, comparing, predicting what is going to come next, and evaluating the effect of the music. Musical thinking in the performing or composing mode must also be marked by the ability to manipulate these sounds in new ways, to change the sounds in ways consistent with or in response to the task at hand.
>
> (1993: 9–10)

This process must also have a product for the musical thinking to become apparent. For children, that product may be what is expressed as an outward manifestation of musical thinking.

Studies of children's musical expression have focused on observations of both the processes and products of their musical creation (Barrett 2001; Burnard and Younker 2004; Hickey and Webster 2001; Johnson-Green 2018, In press; Kaschub and Smith 2009; Kuo and Chuang 2013; Lau and Grieshaber 2010; Pramling 2009; Reybrouck et al. 2009; Smith 2008; Webster 2009). With infants and very young children, the products of musical thinking are mainly sung improvisations (Sole 2017) and movement (Alcock 2008; Metz 1989) as musical exploration and communication. Children's improvisation skills grow as they do (de Vries 2005; Knudsen 2008), extending their range of musical ideas and incorporating learned musical material (Campbell 1998).

Improvisation is closely tied to music composition, where musical ideas are developed over time. Through composition, children may learn to develop awareness of musical mechanics and to express music through deliberate and structured creative decision making. Research on children's composition has focused on various perspectives of children's musical creativity, modes of artistic expression, and musical thinking.

The body of work on invented notation and creativity examines young children's musical expression from the perspective of music writing (Barrett 1996,

1997, 2001, 2004, 2005; Upitis 1987, 1990). Studies show that through use of invented notation young children's written musical expressions become more relevant to them and therefore more meaningful. In addition, the barriers to written musical expression as a symbolic or mnemonic device (Barrett 2002; Upitis 1987) are removed with the use of invented notation.

Walker (1981) discusses the activity of writing music notation as a cognitive process by which a composer translates visual images into sound. He states that, "in cognition of musical events some kind of mental imagery is first involved. Of particular interest is the nature of such imagery . . . which points to a relation between event and representation having a high degree of fidelity" (1981: 110). He goes on to demonstrate that young children attach symbolic images to auditory input, which are directly connected to the sounds that they hear. He concludes, "This assumes that such externalized symbols are reflections of internalized thought as claimed by Piaget and Arnheim" (1981: 111). Sadek (1987) supports these findings and goes on to say that while Western music notation may be the purview of musically trained people, people without formal musical training may express musical concepts through movement.

Bamberger's (1996) study of children's composition focuses on giving children access to a computer program that allows them to use images as musical building blocks to learn how to compose mathematically. She states:

> The kinds of elements and relations novices attend to in making sense of music as it unfolds in real time, are highly aggregated, structurally meaningful entities such as motives, figures, and phrases. These are the "units of perception"—the elements that novices have ready access to, their focus of attention. We don't listen to "notes" any more than we listen to letters printed on the page . . . I take this as further evidence that we intuitively begin by hearing structurally meaningful figures as musical entities, and only with further effort do we move in on the "notes".
>
> (1996: 13–14)

Here, Bamberger presents a highly useful framework for understanding how to listen to and analyze children's composition. If we consider children to be composers in their own right, we may look at their compositions from a theoretical perspective and understand them as the complex pieces of art that they are meant to be. Further, children show us through their composing how they perceive the world through a musical lens—we can hear their intentions in the performance of their work as much as we can see it in their musical drawings.

Further work looked at children's composition from a developmental perspective. Gromko and Poorman's (1998) research showed indications of a developmental link between aural perception and written expression of musical sound. In this study, children first were able to select a written version of a perceived pattern and then to draw their own representation of that pattern. Both activities showed that children directly connected what they heard to what they saw and to what they imagined the sound to look like.

In the aggregate, these studies of children's musical expression through composition show that children are highly sophisticated musical beings. Their musical perception and ability to use music effectively as an artistic medium make them ideally suited to early audio education. Children's innate understanding of what music is and does gives them a natural interest in learning about all aspects of music.

Music Composition-Focused Music Education

When children go to school, they may have an added dimension to their musical lives through formalized music training (Johnson-Green 2012; Littleton 2015). Generally speaking, school music focuses on musical literacy, meaning the types of skills that children will need to read music, to sing in a chorus, and to play an ensemble instrument (Education 2015). Recently, more attention has been paid to musical creativity, which may be loosely defined to include improvisation and composition (Education 2015). However, the types and quality of instruction around "creativity" may vary widely depending on the interest and ability of the teacher.

Hickey (2012) describes a "utopian" (p. 147) general music curriculum where improvisation and composition are at the center of everything the children do. Hickey's belief in children's innate propensity toward musical expression prompted her to design a curriculum based on the assumption of musical competence. She recognized the need for increasing students' creativity and agency to engage them fully in their own music education. Based on a similar philosophy, I set out to create and implement my own music education curriculum. My teaching incorporated developmentally appropriate and differentiated practice with a balance of structure (arrangement of the physical space, behavioral expectations, and scaffolded lessons) and creative freedom. Beginning with the premise that children are natural composers, I focused the music curriculum on improvisation and composition. Drawing on the children's natural ability to create music, I placed musical thinking at the center of every lesson.

The Study of Children's Musical Thinking Through Composition

This chapter focuses on a study that was implemented with students in grades K–5 (five to ten years old in the US) over four years. The full data set comprises the participants' (N=300) original scores, photos of those scores, video and audio recordings of their performances and of the children talking about their process, as well as my lesson plans and written observations. To address the ethical considerations of implementing a long-term study with children:

- I applied for and received permission to gather data from the school administration, my students' parents, and my students.
- I made clear to the adults and children that any data I collected would be used for conference presentations and research publications only.
- At the beginning of each year, parents were asked to sign a school-wide photo/video release form. However, I verbally indicated to my students during each class that I would only record them if they wanted to be recorded.

- I asked my students' permission to take photos of their scores and other work during each class.
- For transparency, the administrators and children had access to the recordings and photos, which they could see at any time.

This study highlights four compositions that were created using electronic keyboards. The children practiced and edited their pieces so that the final compositions had gone through several iterations before they consented to be video recorded for the final performance. Each student decided when a piece was finished, and then told me when they were ready to perform. Throughout the compositional process, my students asked for help when they either got stuck on an idea or when they were ready to move on and needed next instruction. Otherwise, I walked the room observing their progress and acted as a facilitator.

In keeping with differentiated teaching practice, I created an open scaffold whereby the instructions were simple, clear, direct, and general enough to engage my students individually. For example, a general instruction might be, "Work at the keyboard to find a short musical phrase. When you find something you like, write it down any way you can remember it. The music needs to be the same every time you play it" (see more specific examples of instructions in the "Projects" section of this chapter). This open structure allowed my students responsibility for all of their creative decisions and removed constraints on their musical thinking in terms of what they wrote within certain guidelines. Using this teaching method, I was able to remove myself from the actual creation of the music, which was necessary to give my students space to express themselves. This creative freedom produced compositions that represented each student's individual musical process.

The work included here is representative of the types of compositions that my students produced in general across grade levels. I chose these pieces because they are good examples of the work and also because they are complete sets of documentation (photos of scores, video recordings of the performances, and notes on or recordings of my students describing their compositional process).

Analysis of Musical Thinking

Analyzing children's musical creations to determine how they think and perceive musically may be an effective way to determine how children process ideas into music. As the literature suggests (Barrett 1997; Gromko and Poorman 1998; Upitis 1993), I focused on children's score writing as a direct expression of their musical ideas. I would argue, though, that what children write is not actually notation, which is a construct adults place on their means of expression, but is a semiotic device (Seeger 1951) used as a direct expression of musical thought. Each student wrote scores that they specifically designed to match the music they composed. Once my students finalized a score, it served the same function as a notated score would for an experienced music reader. Their scores were true representations of their music as evidenced through repetition, that is, their pieces sounded the same each time they played them.

Videos

I closely analyzed the videos for gesture (how the students moved while playing their pieces) and gaze (where the students were looking while they were playing). The students did not use a movement or eye-tracking device during data collection because it would have been too intrusive. The children's comfort with the documentation technology was critical to the study's success. The video analysis proved to be effective as it revealed that the act of performing did not rely solely on memorization. There are several instances throughout the video data set where the children clearly hesitated in their playing while they looked at their scores for confirmation. There are also instances where the students made mistakes in their playing according to the score and asked for their performance to be re-recorded with the correct playing. I did not teach them these skills and practices and we never discussed self-correction as part of the composition lessons.

Music

Each music/score pairing intertwine to form a complex relationship that would be impossible to show using notated transcriptions for the children's compositions. I determined that I needed to find an alternative way to analyze the moment of translation from thought to musical idea. To this end, I used spectrographic analysis to help me draw conclusions about how children perceive and express musical sound.[4] This type of analysis translates sound into a visual imprint, effectively "freezing" the sound to show how it acts through time. The resulting spectrogram graphs show sound as a product of frequency (x axis) in relation to time (y axis), which produces a patterned color spectrum of musical sound. The spectrographic analysis allowed me to look at my students' written musical expression as a form of visual movement, which brought me closer to understanding the musical representation. It also provided a way to determine how they perceive, process, and internalize musical sound to then express it as a direct reflection of their musicality.

Very few studies (Alm and Walker 2002; Hall-Craggs 1979; Liu et al. 2016; Thibeault 2011) have examined the use of spectrographic processing to illuminate the properties of musical composition. As in Seeger's (1957, 1951) formative use of signal processing to graph folk melodies, the present study used signal processing alongside children's drawings to show the immediacy of their musical thinking on paper.

All of the children's pieces presented here were analyzed using *Sonic Visualiser* software (Cannam et al. 2010), which uses fast Fourier transforms of windowed sections of original audio files to visualize frequencies. In other words, the *Sonic Visualiser* analyzes music in small chunks of sound and uses color to show where tone frequencies fall in time, creating an illustration of the music. After experimenting with the software, I determined a set of parameters that produced the clearest visual representations of the children's compositions (see Table 2.2).

Table 2.2 Analysis parameters and their definitions (Cannam et al. 2010)

Parameter	*Definition*
Melodic range spectrogram	Discerns individual musically meaningful qualities of the composition.
Color = Sunset	An intense blue and red color combination to show musical features more clearly.
Color scale = Linear	Maps the color to the frequency of each tone, where voltage levels and audio output levels are proportional.
Column normalization	Frequencies with the highest value in a given time frame receive the brightest available color. This will increase the overall brightness for quiet frames and may decrease it for loud frames.
Moderately small window size value	Gives a finer picture resolution in time rather than in frequency.
Window overlap = 93.7%	The overlap between consecutive windows, which focuses the spectrogram.
Frequency scale = Linear	The y axis is proportional to the frequency.

Note: For all of the figures shown here: x axis = the piece unfolding over time; y axis = each sound as it happens at each moment in time, measured in frequency.

The Projects

Overview

I began with the premise that the youngest children, in this case kindergarteners, would come to their school music class already having had years of immersive musical experience. While they might not classify music using certain symbols, they understood inherently what music is and what it does. My objective was to teach my students *how music works*. While the curriculum was structured and scaffolded, my students were free to decide the content of their musical creations. As they developed musical skills over time, the instructions became more general to allow them room to practice and develop these skills. Ultimately, my students were able to focus on developing one composition, one hour per week, over several weeks as they went through a compositional process beginning with improvisation.

The included pieces show ways in which my students described their experiences and environment through musical creation. Up through second grade, my students used only semiotic and graphic representations (Seeger 1951) because they needed to focus on developing compositional skills without any barriers. I was also not concerned with the general communication aspect of musical notation, in other words, they did not need to write music so that someone else would be able to read it. In this way, their drawn scores became a direct reflection of their musical thinking. In the examples I have given here, the spectrographic analysis is in one-to-one relationship with the children's performances of their pieces and of the scores they created.

I have chosen to represent three age groups: kindergarten (five years old), grade 1 (six years old), and grade 5 (ten years old). In formal schooling, moving from kindergarten to Grade 1 is significant for young children due to a stronger academic focus. The difference is also significant in both general and musical development. The wide time difference to Grade 5 will show the trajectory of musical thinking into later childhood.

"Bune" (Bunny): Anna (Age Five)

The kindergarteners studied animals as part of their general curriculum and so "animals" became a connected unit theme in the music class. Lessons began with the study of *Carnival of the Animals* (Saint-Saëns 1922), where students analyzed various pieces by drawing interpretively while listening to music, moving, and playing rhythm instruments. The students finished this unit with composition at the keyboards. For the example shown here, I had asked the children to choose an animal and then to write a piece of music that sounded like that animal. Anna wrote a piece about "a bunny hopping" (see Figure 2.2).

In her visual representation of the music, Anna's discreet circles are connected to show the bunny's movement as it hops across the page, labeled "hop". She has also given some instruction for playing the piece: We see a bunny, labeled "Bune",

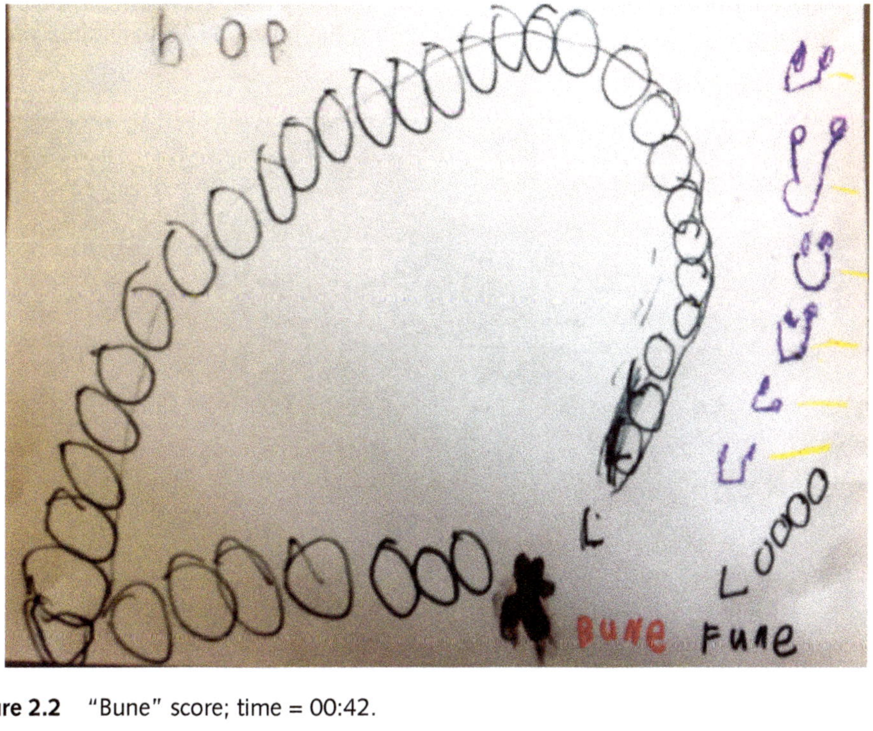

Figure 2.2 "Bune" score; time = 00:42.

at the bottom of the score where the music starts and an arrow to show which way the circles should be read. Next to the "Bune" we see another set of circles with an arrow: These seem to indicate that the circles *are* the "Bune". Anna has also thoughtfully provided her version of music notation to indicate that this is a musical score.

The playing itself is patterned yet erratic within that pattern, as if Anna captured how she imagined a bunny to move in the wild. Overall dynamic movement happens generally from low to high range frequencies and shows the motivic tone clusters interspersed with short silences. Anna plays with the motives as well using the following compositional devices: inversion, resting tones chosen from the tonal clusters, repetition, and augmentation.

When looking closely at the spectrogram analysis, we see the "Bune" hopping in a distinct pattern of music that follows the score (see Figures 2.3–2.5). For example, there are 36 circles drawn on the page and 33 statements of two clear motives and motivic development played in the performance: Motive 1 consists of single, short declarative note clusters that are typically played twice in high cluster-low cluster patterns; Motive 2 consists of quickly played note clusters, usually five in a row using either the same tones or a pattern of rising tone clusters.

In time, the exact center of the composition is at 0:21, which correlates with the eighteenth motivic statement—the exact center of the circle drawing on the score. The score shows that after the eighteenth circle, there is a general movement upwards and then a slope downwards to the end of the piece. In the performance, Anna plays the first motivic augmentation immediately after this point in the piece, which is situated at the highest frequency within the piece. The music then moves

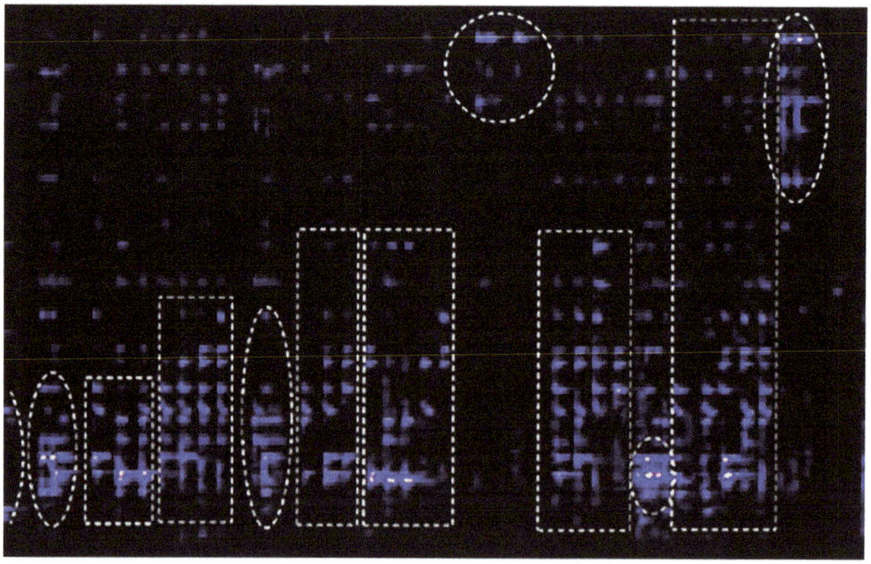

Figure 2.3 First page of "Bune" motivic analysis: Pattern of Motives 1 (ovals) and 2 (rectangles) alternating. Each shape contains clearly defined musical ideas that repeat.

Figure 2.4 Second page of "Bune" motivic analysis: The first oval contains a resting tone that is related to Motive 1; the other ovals contain the original Motive 1; the rectangles contain Motive 2; and the rounded rectangles contain augmentation of Motive 1 (as marked on the spectrogram).

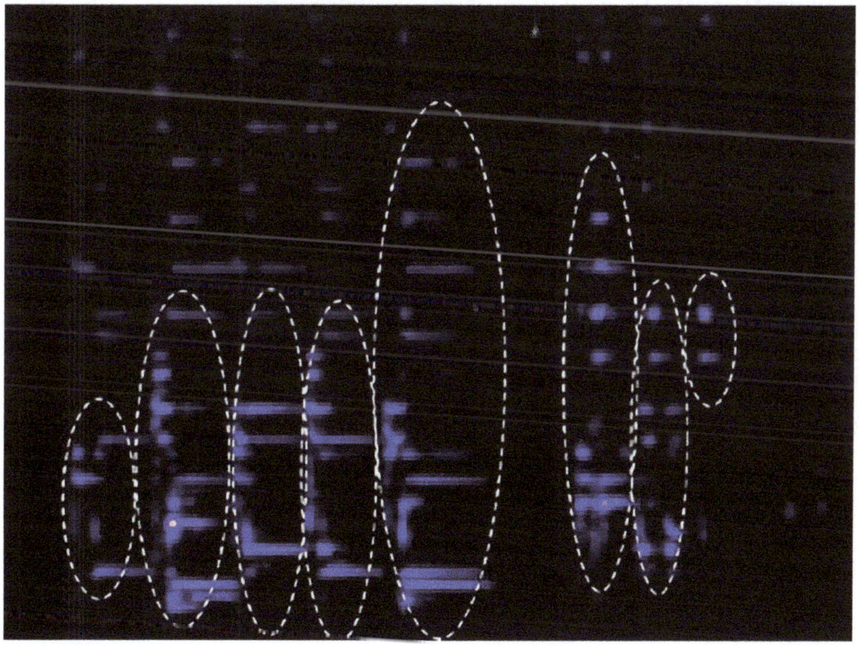

Figure 2.5 Final page of "Bune" motivic analysis: The ovals contain repetitions of Motive 1 with inversion and slight augmentation. The last oval contains diminution of Motive 1 to end the piece.

to a strong and energetic statement of Motive 1, more augmentation, in frequency parallel to the eighteenth motive and then moves in a downward slope in energy to the end of the piece. The last spectrogram image shows single, slightly augmented tone clusters played alternating with silence, which disperses the tension to the conclusion.

In this instance, complex music fills the open spaces in the drawing (where the empty circles indicate where to play), which indicates that Anna's internal "hearing" and then performing of her piece is more sophisticated than her drawing might show. Just as with adult composers, Anna's written score captures the least complex aspect of the musical creation and yet gives the listener a frame for understanding the intent behind that creation. Since the performance of the piece was essentially the same each time she played it in music class, the markings on the score, its connection to her playing, and its inherent compositional balance indicate that Anna deliberately constructed this piece to express her visual and aural images of "Bune".

"Butterfly": Nora (Age Six)

The first-grade general curriculum in literacy focused on reading and writing poetry and stories. I included both of these types of literature in my grade 1 music curriculum and taught music composition through the study of poems and stories. The poetry unit culminated in a project where the children wrote their own poems and then set them to music. Several of my students chose to draw pictures of their poetry subjects to use for a score. While these pictures do not necessarily include references to specific musical instruction, they do contain clues as to the composer's musical intentions. "Butterfly" is a good example of this technique (see Figure 2.6).

I have a little butterfly,
It flaps up and down,
Really, really fast.
It twirls around and around
until it gets dizzy.
It smells the flowers.
It drinks the pollen.
It comes to me,
Sits right beneath me.
It flies up and down again,
Does some tricks,
Puts on a show just for me.

Figure 2.6 "Butterfly" score and poem.

Source: Time = 02:39

Nora's poem tells us about *her* butterfly. She describes its movement from a personal standpoint where she is the anchor for activity. The poem is balanced:

- Statement of personal relationship (use of "I"): Line 1 (rest)
- Description of activity: Lines 2–6 (active)
- Statement of personal relationship: Lines 7–8 (rest)
- Description of activity: Lines 9–10 (active)
- Combined activity with personal relationship: Line 11 (active)

Nora personifies her butterfly's behaviors, which are specific to the kinds of movement that she imagines it to make deliberately. The butterfly is aware that she is present and acts accordingly.

The score shows Nora's illustration of her butterfly and the movement associated with it. Notice the active lines and spirals that surround the central drawing, which reflects the personification in the poem. The butterfly looks like a person: We see a triangle body with arms and legs with a face surrounded by wings and antennae.

The musical composition combines the ideas of the poem and the score. Each line of the poem becomes a musical phrase that depicts the butterfly's movement. Spectrographic analysis shows that the music is in a direct relationship with the lines and shapes of the score. The piece is modal, most prominently in Aeolian with movement into Ionian and Locrian. Structurally, Nora matched the music to the poem, so that both are through-composed. The descriptive elements within the poem and music create internal coherence without need for a superimposed form (e.g., ABA, etc.). In performance, Nora recites the poem while she is playing so that the music connects directly with the imagery. To create coherence, she repeats and develops the following three motives throughout the piece:

- Motive 1 = Left hand and right hand begin apart and play toward each other in a scalar pattern.
- Motive 2 = Quick, repeated notes played with one hand.
- Motive 3 = Alternating, single notes played in left then right hands. Combines elements of Motives 1 and 2.

Nora writes most of the piece in free rhythm to mimic the butterfly's movements.

For each phrase, I have given the combination of motives and tonalities that Nora used to illustrate her poem musically (see Table 2.3).

Nora's spectrograph shows a one-to-one relationship between the music and the score. The prominent melodic lines reflect the butterfly shape, the spirals, the sweeping lines, and the curves that make up Nora's illustration. She seems to have "drawn" a butterfly and its movements in sound. While the poem and the score provide concrete expressions of Nora's ideas, the spectrogram depicts the dynamic movement of a butterfly captured in music. The following panels show the spectrograms for each line of poetry/musical phrase. The dotted lines highlight

Table 2.3 Modal and motivic analysis of "Butterfly"

Introduction	Aeolian	1
I have a little butterfly	Aeolian	1, 1 (modified to one hand)
It flaps up and down	Aeolian, Locrian	2
Really, really fast	Aeolian (Locrian at the end)	2, 3, 1 & 3 combined, 2
It twirls around and around until it gets dizzy	Ionian	2, 2 (modified to both hands), 3, 1
It smells the flowers	Ionian	3
It drinks the pollen	Ionian	3 (repeat of previous phrase), 3 (repeat)
It comes to me	Aeolian	3
Sits right beneath me	Aeolian	1 (modified, hands play away from each other), 1 & 3 combined
It flies up and down again	Ionian	1 & 3 combined
Does some tricks	Aeolian	3
Puts on a show just for me	Aeolian, Ionian	3, 1, 3

Figure 2.7 The introduction with poetry lines/musical phrases 1–2. Long, sweeping and curving lines illustrate Motive 1 (hands moving toward each other in a scalar pattern). Notice the shape above, "I have a little butterfly". The spectrograph captured Motive 2 (quick, repeated notes) with straight, separated lines, which provide contrast.

the sound shapes as they unfold in time. These shapes reflect the written score and are in a one-to-one relationship to the text. The motives (see Table 2.3) have been marked at the top of the figures to provide a visual analysis in real time (see Figures 2.7–2.12).

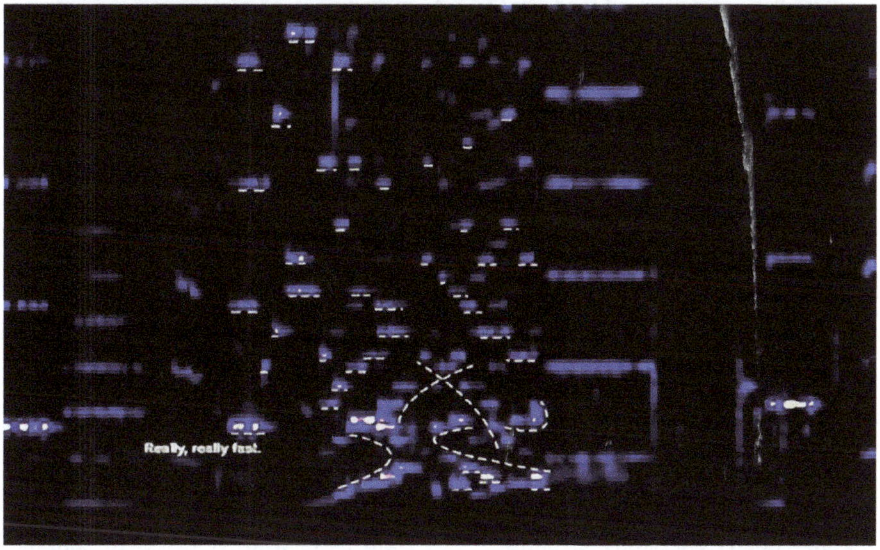

Figure 2.8 Poetry line/musical phrase 3. Nora introduces Motive 3 (alternating, single notes), which comprises elements of Motives 1 and 2. We also see Motive 1's long lines in the center of this panel as it is combined with Motive 3.

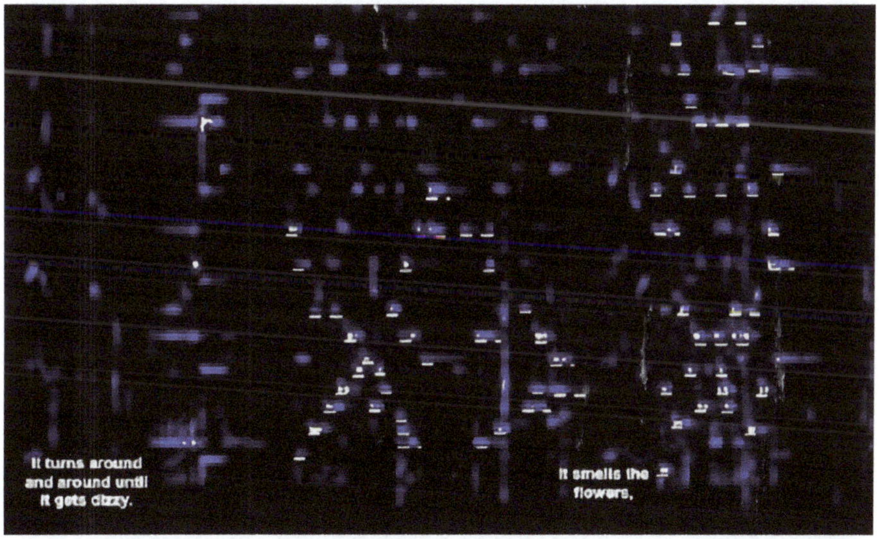

Figure 2.9 Poetry lines/musical phrases 4–5. This panel shows the last statement of Motive 2. After this point, the poem's perspective changes toward calmer action and description of Nora's relationship with her butterfly. We see this change in more open shapes with more space between each tone.

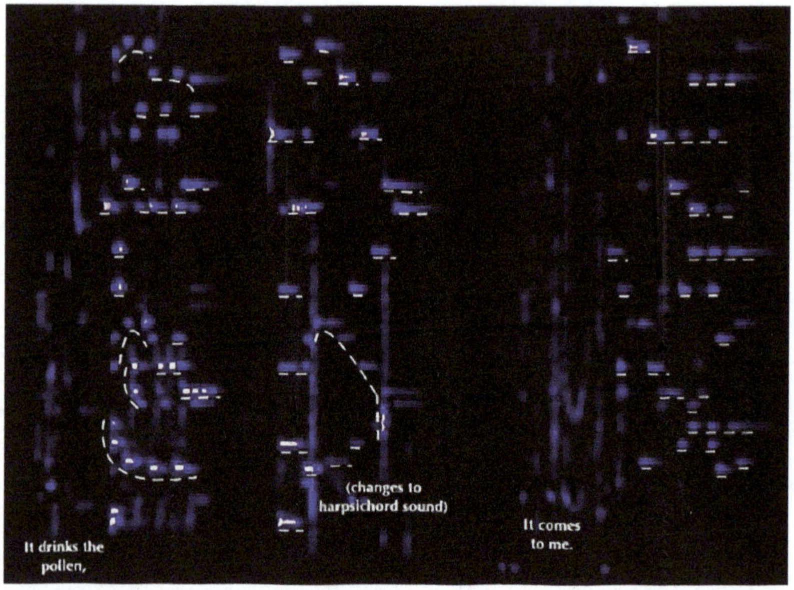

Figure 2.10 Poetry lines/musical phrases 6–7. Nora plays with Motive 3 to continue the sense of open, calm action. She creates coherent lines that mimic how she imagines a butterfly might move while collecting pollen. Notice how the musical shape changes when the butterfly's action changes with, "It comes to me". The panel illustrates a flatter affect as the butterfly moves toward Nora.

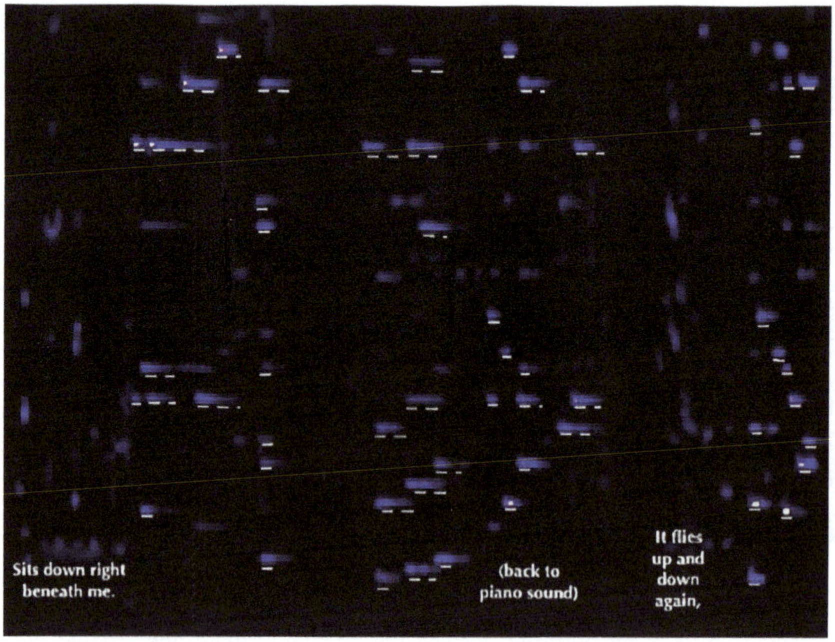

Figure 2.11 Poetry lines/musical phrases 8–9. This panel shows the most open illustration of the piece. The butterfly rests for most of the panel: the action at the end remains calm even though the butterfly moves again.

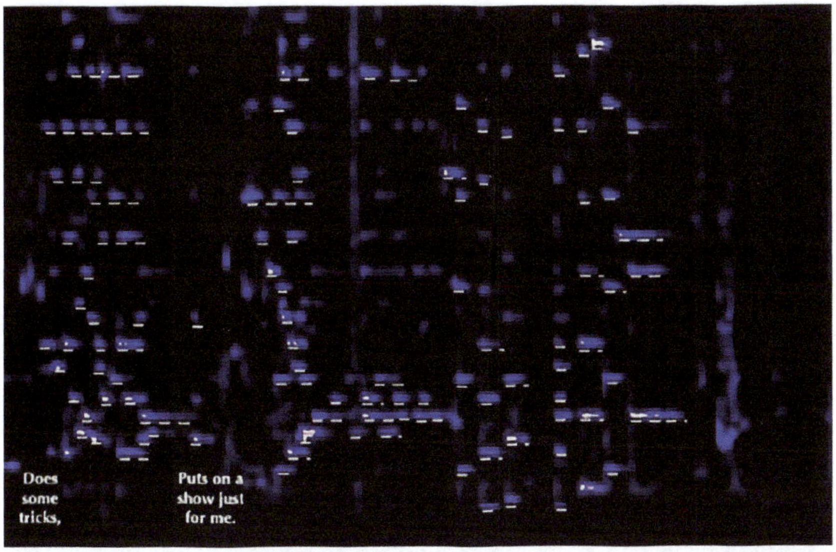

Figure 2.12 Poetry lines/musical phrases 10–11. In the final illustration, we see why the butterfly rested earlier. Apparently, it needed to conserve its energy for the tricks and show that it would put on for Nora. This panel shows dense clusters of sound that are faster paced and that happen in all frequencies. We even get an imprint of the butterfly at the end.

Emotions: Amy, Andrew, David, Mia (Age Ten)

The fifth-grade year focused on songwriting. At this point in their music education, my students had been through several years of learning compositional techniques. At the end of this year, they had developed their musical skills to a point of expertise appropriate to their developmental age. The song-writing curriculum focused on emotional expression in music: how to describe or elicit certain feelings through composition. We worked through several stages of writing, which included using imagery and evocative vocabulary as prompts for motivic development. Over the year, my students developed their ability to write music that could express abstract concepts such as "joy".

The set of compositions shown here were written at the end of the year. For this assignment, I gave my students only one class period to write their music and only one instruction: "Write music that describes an emotion. You may not use words". The pieces they wrote were immediate and visceral, which makes them relevant to this chapter. Their music flowed from direct expression of an emotional state perceived as sound. Unlike the previous two compositions, I will not provide a close musical analysis here as I believe, in this case, the pictures are worth thousands of words (see Figures 2.13–2.16).

The illustration indicates an open musical structure comprised of leaps and scale patterns played in the mid to high range of the keyboard. The combination of regular line, clear form, open shape, and leaps upward describe *Joyful*.

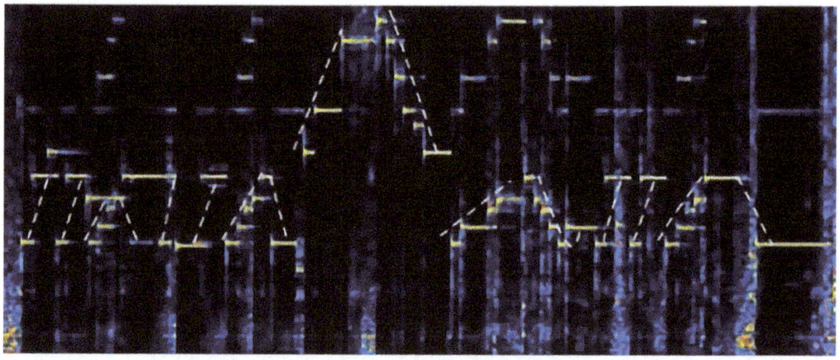

Figure 2.13 *Joyful* (Amy). The line analysis of this panel shows AABA form.

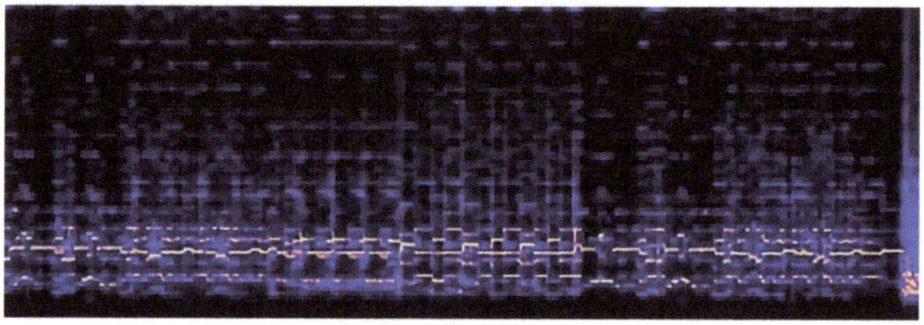

Figure 2.14 *Mad* (Andrew). In complete contrast to *Joyful*, this panel shows close, layered musical lines.

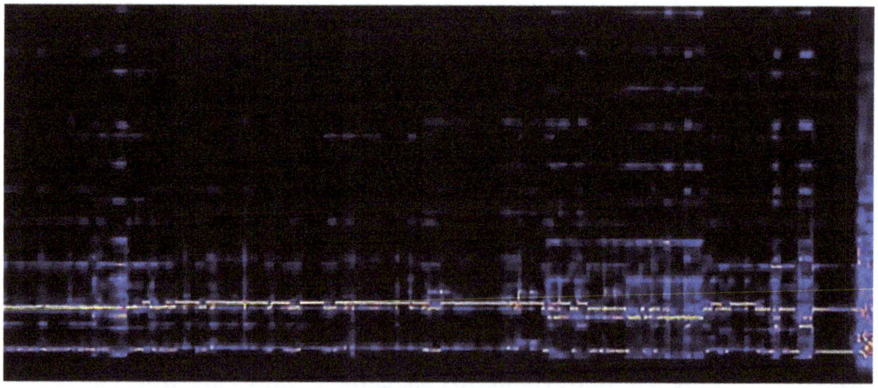

Figure 2.15 *Anxious* (Harry). This piece has two sections, divided here with a white dotted line. The panel shows a linear composition without much deviation from that line. Harry played the piece in the mid-lower range on the keyboard, also without deviation in frequency. The line also seems to "shake" as if Harry was anxious while playing the piece.

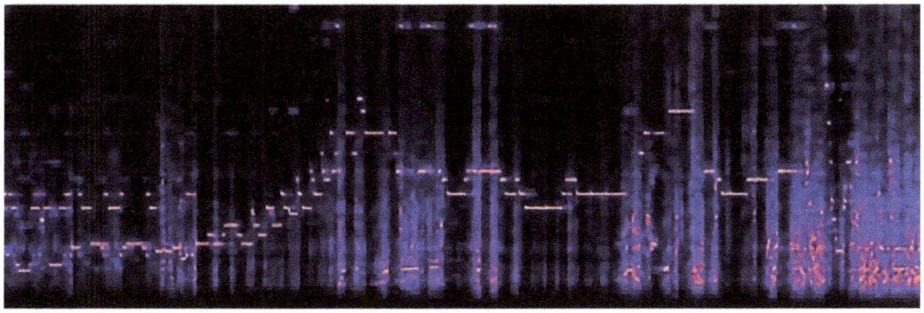

Figure 2.16 *Relaxed* (Mia). The open, sweeping lines and the play among ranges and frequencies illustrate Mia's musical description of *Relaxed*. This piece clearly shows three related yet contrasting sections, with general upwards movement. I have not provided a line analysis for this panel because I wanted to present it in an unbroken visual representation.

However, the structure is difficult to discern because the piece is composed in the lowest register on the keyboard without contrasting frequencies.

Conclusions: Children's Perception, Musical Expression, and Audio Education

The students' compositions show clear connection between the imaginative (i.e. visualization of an idea), the perceptive (i.e. experience of an idea), and the expressive (i.e. description of an idea). In the transfer of ideas into sound, my students deliberately and effectively chose musical material to illustrate specific ideas. The spectrograms illustrate the direct transition from thought to expression, which reveals the complexity of my young students' musical understanding.

Given young people's innate musical understanding and their fine-tuned ability to use music as a mode of artistic communication, I believe that there is great potential to begin audio education immediately in kindergarten. As I have been thinking about, researching, and writing this chapter, I have become convinced that young students would benefit from a curriculum that weaves audio education into aspects of their formalized music education.

Audio education might be regarded as integral to performances of original compositions, solo or ensemble instrumental or singing performances (both in class and on stage), and any other situation where music teachers might record their students' work. For example:

- Teach children how to record and edit their own or their peers' work (e.g. original compositions, instrumental or singing performances) using specialized audio editing applications for tablets or smartphones.
- Use video or sound recording and playback to teach children to listen for particular aspects of the sound quality; then teach them how to experiment with changing the sound.

- Bring in various types of sound recording equipment to use in the classroom; teach the students to use the equipment and then to analyze the differences among the recordings they make of an in-class performance.
- Use any type of recorder to do sound "field work". Have the students collect sounds inside or outside of the school environment and then teach them to edit their "artifacts" into a composition.
- Teach various sound editing techniques using a pre-recorded piece of music and computer software. Then teach students to download and edit their own compositions.

These examples may conform to developmentally appropriate practice starting in kindergarten. The youngest children may be encouraged to listen carefully to a performance of their own work, critique what they hear, and then revise sound and content.

Teachers may also use audio education techniques to give their students more agency over their own music education. Children's innate musicality, their ability to process complex musical ideas and then to articulate those ideas effectively make them ideally suited to sophisticated learning in a music classroom. Ultimately, introducing audio education early on in children's musical experience may provide a stronger foundation for their developing musicianship: and may allow them to become more powerful artistic affectors.

Notes

1. Here, I use Plato's language to create a descriptive term. I define *affector* as someone who has the power to affect others through artistic creation.
2. *Construct*: a hybrid of improvisation and composition, where children retain musical elements over time and use them for different expressive reasons in various contexts of daily life.
3. Most of the many websites for expectant parents include instructions about how and what to play during pregnancy (e.g. www.tinylove.com/us_en/articles/the-importance-of-music; www.babycenter.com/0_music-and-your-unborn-child_6547.bc; www.whattoexpect.com/pregnancy/ask-heidi/week-20/influences-on-the-womb.aspx, among others).
4. Thank you to William Moylan for helping me to develop this idea.

References

Acker, A., Nyland, B. & Niland, A. 2015. The relationship between children's learning through music and the use of technology. *Australian Journal of Music Education* (1).

Adachi, M. & Trehub, S. E. 1998. Children's expression of emotion in song. *Psychology of Music*, 26, 133–153.

Addessi, A. R. & Pachet, F. 2005. Experiments with a musical machine: Musical style replication in 3 to 5 year old children. *British Journal of Music Education*, 22(1), 21–46.

Addessi, A. R. & Pachet, F. 2005–2006. Young children confronting the Continuator, an interactive reflective musical system. *Musicæ Scientiæ: The Journal of the European Society for the Cognitive Sciences of Music* (Special Issue), 13–39.

Alcock, S. 2008. Young children being rhythmically playful: Creating musike together. *Contemporary Issues in Early Childhood*, 9, 328–338.

Alemán, X., Duryea, S., Guerra, N., Mcewan, P., Muñoz, R., Stampini, M. & Williamson, A. 2017. The effects of musical training on child development: A randomized trial of El Sistema in Venezuela. *Prevention Science*, 18, 865–878.

Alm, J. F. & Walker, J. S. 2002. Time-frequency analysis of musical instruments. *Society for Industrial and Applied Mathematics*, 44, 457–476.

Bamberger, J. 1995. *The mind behind the musical ear: How children develop musical intelligence*. Cambridge, MA: Harvard University Press.

Bamberger, J. 1996. Turning music theory on its ear: Do we hear what we see; Do we see what we say? *International Journal of Computers and Mathematics Education*, 1, 33–55.

Barrett, M. S. 1996. Children's aesthetic decision-making: An analysis of children's musical discourse as composers. *International Journal of Music Education*, 28, 37–62.

Barrett, M. S. 1997. Invented notations: A view of young children's musical thinking. *Research Studies in Music Education*, 8, 2–14.

Barrett, M. S. 2000. Windows, mirrors, reflections: A case study of adult constructions of children's musical thinking. *Bulletin of the Council for Research in Music Education*, 145, 43–61.

Barrett, M. S. 2001. Constructing a view of children's meaning-making as notators: A case-study of a five-year-old's descriptions and explanations of invented notations. *Research Studies in Music Education*, 33–45.

Barrett, M. S. 2002. Invented notations and mediated memory: A case-study of two children's use of invented notations. *Bulletin of the Council for Research in Music Education*, 153/154, 55–62.

Barrett, M. S. 2003. Freedoms and constraints: Constructing musical worlds through the dialogue of composition. Why and how to teach music composition: A new horizon for music education, Reston, Virginia. *The National Association for Music Education*, 3–27.

Barrett, M. S. 2004. Thinking about the representation of music: A case-study of invented notation. *Bulletin of the Council for Research in Music Education*, 161–162, 19–28.

Barrett, M. S. 2005. Representation, cognition, ad communication: Invented notation in children's musical communication. In: Miell, D., Macdonald, R. & Hargreaves, D. J. (eds.) *Musical communication*. Oxford: Oxford University Press.

Barrett, M. S. 2011. Musical narratives: A study of a young child's identity work in and through music-making. *Psychology of Music*, 39, 403–423.

Barrett, M. S. 2016. Attending to "culture in the small": A narrative analysis of the role of play, thought and music in young children's world-making. *Research Studies in Music Education*, 38, 41–54.

Birnholz, J. C. & Benacerraf, B. R. 1983. The development of human fetal hearing. *Science*, 222, 516–518.

Brophy, T. S. 2001. Developing improvisation in general music classes. *Music Educators Journal*, 88, 34–41.

Bugos, J. & Jacobs, E. 2012. Composition instruction and cognitive performance: Results of a pilot study. *RIME: Research and Issues in Music Education*, 10.

Burnard, P. 2000. Examining experiential differences between improvisation and composition in children's music-making. *British Journal of Music Education*, 17, 227–245.

Burnard, P. & Younker, B. A. 2002. Mapping pathways: Fostering creativity in composition. *Music Education Research*, 4, 245–261.

Burnard, P. & Younker, B. A. 2004. Problem-solving and creativity: Insights from students' individual composing pathways. *International Journal of Music Education*, 22, 59–76.

Burton, S. L. & Pearsall, A. 2016. Music-based iPad app preferences of young children. *Research Studies in Music Education*, 38(1), 75–91.

Byron, R. (ed.) 1995. *Music, culture, and experience: Selected papers of John Blacking*. London: University of Chicago Press.

Campbell, P. S. 1991. Rhythmic movement and public school music education: Conservative and progressive views of the formative years. *Journal of Research in Music Education*, 39, 12–22.

Campbell, P. S. 1998. *Songs in their heads: Music and its meaning in children's lives*. New York, NY: Oxford University Press.

Campbell, P. S. 2000. How musical we are: Blacking on music, education, and cultural understanding. *Journal of Research in Music Education*, 48, 336–359.
Cannam, C., Landone, C. & Sandler, M. 2010. Sonic visualiser: An open source application for viewing, analysing, and annotating music audio files. *ACM Multimedia 2010 International Conference*, Firenze, Italy.
Carroll, D. 2018. Children's invented notations: Extending knowledge of their intuitive musical understandings using a Vygotskian social constructivist view. *Psychology of Music*, 46, 521–539.
Chorna, O., Filippa, M., Almeida, J. S. D., Lordier, L., Monaci, M. G., Hüppi, P., Grandjean, D. & Guzzetta, A. 2019. Neuroprocessing mechanisms of music during fetal and neonatal development: A role in neuroplasticity and neurodevelopment. *Neural Plasticity*, 2019, 1–9.
Cirelli, L. K., Trehub, S. E. & Trainor, L. J. 2018. Rhythm and melody as social signals for infants. *Annals of The New York Academy of Sciences*, 1423, 66–72.
Conway, C. 2004. Parent, student, and teacher perceptions of assessment procedures in beginning instrumental music. *Bulletin of the Council for Research in Music Education*, 16–25.
Custodero, L. A. 2006. Singing practices in 10 families with young children. *Journal of Research in Music Education*, 54, 37–56.
Custodero, L. A. 2008. Living jazz, learning jazz: Thoughts on a responsive pedagogy of early childhood music. *General Music Today*, 22, 24–29.
Custodero, L. A. & Johnson-Green, E. A. 2003. Passing the cultural torch: Musical experience and musical parenting of infants. *Journal of Research in Music Education*, 51, 102–114.
Custodero, L. A. & Johnson-Green, E. A. 2008. Caregiving in counterpoint: Reciprocal influences in the musical parenting of younger and older infants. *Early Child Development and Care*, 178(1), 15–39.
Damasio, A. 1999. *The feeling of what happens*. New York: Harcourt Brace and Company.
Damasio, A. 2000. A neurobiology for consciousness. In: Metzinger, T. (ed.) *Neural correlates of consciousness*. Cambridge, MA: MIT Press.
Davidson, J. W. & Borthwick, S. J. 2002. Family dynamics and family scripts: A case study of musical development. *Psychology of Music*, 30, 121–136.
Deutsch, D. 2016. Authentic assessment in music composition: Feedback that facilitates creativity. *Music Educators Journal March*, 102, 53–59.
De Vries, P. 2005. Lessons from home: Scaffolding vocal improvisation and song acquisition with a 2-year-old. *Early Childhood Education Journal*, 32, 307–312.
Draganova, R., Schollbach, A., Schleger, F., Braendle, J., Brucker, S., Abele, H., Kagan, K. O., Wallwiener, D., Fritsche, A., Eswaran, H. & Preissl, H. 2018. Fetal auditory evoked responses to onset of amplitude modulated sounds: A fetal magnetoencephalography (fMEG) study. *Hearing Research*, 363, 70–77.
Dzansi, M. P. 2002. Some manifestations of Ghanaian indigenous culture in children's singing games. *International Journal of Education & the Arts*, 3. Retrieved from www.ijea.org/v3n7/index.html.
Flohr, J. W. & Brown, J. 1979. The influence of peer imitation on expressive movement to music. *Journal of Research in Music Education*, 27, 143–148.
Fox, D. B. 1988. Our swings are singing different songs: A child's view of music. *Psychomusicology: A Journal of Research in Music Cognition*, 7, 201–202.
Giddings, S. 2013. Inherent creativity and the road to happiness: Improvisation and composition in the music classroom. *Canadian Music Educator*, 55, 44–46.
Granier-Deferre, C., Bassereau, S., Ribeiro, A. L., Jacquet, A.-Y. & Decasper, A. J. 2011. A melodic contour repeatedly experienced by human near-term fetuses elicits a profound cardiac reaction one month after birth. *PLoS One*, 6, 1–10.
Gromko, J. E. & Poorman, A. S. 1998. Developmental trends and relationships in children's aural perception and symbol use. *Journal of Research in Music Education*, 46, 16–23.
Guderian, L. V. 2012. Music improvisation and composition in the general music curriculum. *General Music Today*, 25, 6–14.
Habibi, A., Damasio, A., Ilari, B., Sachs, M. E. & Damasio, H. 2018. Music training and child development: A review of recent findings from a longitudinal study. *Annals of the New York Academy of Sciences*, 1423, 73–81.

Hall-Craggs, J. 1979. Sound spectrographic analysis: Suggestions for facilitating auditory imagery. *Condor*, 81, 185–192.
Hargreaves, D. J., Macdonald, R. & Miell, D. 2016. Musical identities. In: Hallam, S., Cross, I. & Thaut, M. (eds.) *The Oxford handbook of music psychology*. Oxford: Oxford University Press.
He, C., Hotson, L. & Trainor, L. J. 2009. Development of infant mismatch responses to auditory pattern changes between 2 and 4 months old. *European Journal of Neuroscience*, 29, 861–867.
Hickey, M. 1999. Assessment rubrics for music composition. *Music Educators Journal*, 85, 26–33.
Hickey, M. (ed.) 2003. *Why and how to teach music composition: A new horizon for music education*. Reston, VA: MENC: The National Association for Music Education.
Hickey, M. 2012. *Music outside the lines: Ideas for composing in K-12 music classrooms*. Oxford and New York: Oxford University Press.
Hickey, M. & Webster, P. 2001. Creative thinking in music. *Music Educators Journal*, 88, 19–23.
Hogenes, M., Oers, B. V., Diekstra, R. F. W. & Sklad, M. 2016. The effects of music composition as a classroom activity on engagement in music education and academic and music achievement: A quasi-experimental study. *International Journal of Music Education*, 34, 32–48.
Ilari, B. 2009. Music listening preferences in early life: Infants' responses to accompanied versus unaccompanied singing. *Journal of Research in Music Education*, 56, 357–369.
Ilari, B. 2016. Music in the early years: Pathways into the social world. *Research Studies in Music Education*, 38, 23–39.
Jardri, R., Houfflin-Debargea, V., Deliona, P., Pruvoa, J.-P., Thomasa, P. & Pins, D. 2012. Assessing fetal response to maternal speech using a noninvasive functional brain imaging technique. *International Journal of Developmental Neuroscience*, 30, 159–161.
Johnson-Green, E. 2012. *Ordinary magic: The musical foundations of transition in family life*. Doctor of Education in Music Education Doctoral Dissertation, Teachers College, Columbia University.
Johnson-Green, E. 2018. Musical architects: Immersive learning through design thinking in a kindergarten music composition curriculum. *Visions of Research in Music Education*, 31. Retrieved from www.rider.edu/~vrme.
Johnson-Green, E. iAssess: Using the iPad for self-assessment of composition and performance in an elementary school music program. In: Greher, G. R. & Burton, S. L. (eds.) *Mobile technology: The promise and possibilities for creative music making*. New York: Oxford University Press.
Jorgensen, E. R. 2016. Another perspective: The joyous composer. *Music Educators Journal*, 102, 71–74.
Kaschub, M. & Smith, J. 2009. *Minds on music: Composition for creative and critical thinking*. Lanham, MD: Rowman & Littlefield Education.
Kerchner, J. L. 2000. Children's verbal, visual, and kinesthetic responses: Insight into their music listening experience. *Bulletin of the Council for Research in Music Education*, 31–50.
Kisilevsky, B. S., Hains, S. M. J., Jacquet, A.-Y., Granier-Deferre, C. & Lecanuet, J. P. 2004. Maturation of fetal responses to music. *Developmental Science*, 7, 550–559.
Knudsen, J. S. 2008. Children's improvised vocalisations: Learning, communication and technology of the self. *Contemporary Issues in Early Childhood*, 9, 287–296.
Kuo, Y.-T. & Chuang, M.-C. 2013. A proposal of a color music notation system on a single melody for music beginners. *International Journal for Mathematics Teaching & Learning*, 31, 394–412.
Lamont, A. 2016. Musical development from the early years onwards. In: Hallam, S., Cross, I. & Thaut, M. (eds.) *The Oxford handbook of music psychology*. Oxford: Oxford University Press.
Lau, W. C. M. & Grieshaber, S. 2010. Musical free play: A case for invented musical notation in a Hong Kong kindergarten. *British Journal of Music Education*, 27, 127–140.
Littleton, D. 2015. *When music goes to school: Perspectives on learning and teaching*. Lanham, MD: Rowman & Littlefield.
Liu, L.-L., Pang, Y. & Hu, Z.-L. 2016. Application of spectrogram analysis in traditional vocal music teaching and multimedia animation vocal music teaching. *iJET*, 11. doi:https://doi.org/10.3991/ijet.v11i11.6242.
Lopez-Teijo, M., Garcia-Faura, A. & Prats-Galino, A. 2015. Fetal facial expression in response to intravaginal music emission. *Ultrasound*, 23, 216–223.
Lum, C.-H. 2008. Home musical environment of children in singapore, on globalization, technology, and media. *Journal of Research in Music Education*, 56(2), 101–117.

Malloch, S. 1999. Mothers and infants and communicative musicality. *Musicae Scientiae*, (Special Issue), 29–57.
Malloch, S. & Trevarthen, C. 2009. Musicality: Communicating the vitality and interests of life. In: Malloch, S. & Trevarthen, C. (eds.) *Communicative musicality: Exploring the basis of human companionship*. Oxford: Oxford University Press.
Malloch, S. & Trevarthen, C. 2018. The human nature of music. *Frontiers in Psychology*, 9, 1–21.
Marsh, K. 1995. Children's singing games: Composition in the playground? *Research Studies in Music Education*, 4, 2–11.
Marsh, K. 2011. Meaning-making through musical play: Cultural psychology of the playground. In: Barrett, M. S. (ed.) *A cultural psychology of music education*. New York, NY: Oxford University Press.
Marsh, K. & Young, S. 2006. Musical play. In: McPherson, G. E. (ed.) *The child as musician: A handbook of musical development*. Oxford, United Kingdom: Oxford University Press.
Merkow, C. H. 2013. Measurement of infants' behaviors with electronic music toys. *Texas Music Education Research*, 14–26.
Metz, E. 1989. Movement as a musical response among preschool children. *Journal of Research in Music Education*, 37, 48–60.
National Association for Music Education (NAfME). 2015. *Core music standards: PK-8 general music* [Online]. Retrieved from www.nafme.org/wp-content/files/2014/11/2014-Music-Standards-PK-8-Strand.pdf.
Niland, A. 2009. The power of musical play: The value of play-based, child-centered curriculum in early childhood music education. *General Music Today*, 23, 17–21.
Niland, A. 2015. "Row, row, row your boat": Singing, identity and belonging in a nursery. *International Journal of Early Years Education*, 23, 4–16.
O'Neill, C. T., Trainor, L. J. & Trehub, S. E. 2001. Infants' responsiveness to fathers' singing. *Music Perception*, 18, 409–425.
Parncutt, R. 2016. Prenatal development. In: McPherson, G. E. (ed.) *The child as musician: A handbook of musical development*, 2nd ed. Oxford: Oxford University Press.
Plato 360 BCE. *The dialogues of Plato translated into English with analyses and introductions by B. Jowett, M.A. in five volumes*. Oxford: Oxford University Press.
Pogonowski, L. 1989. Critical thinking and music listening. *Music Educators Journal*, 76, 35–38.
Pramling, N. 2009. External representation and the architecture of music: Children inventing and speaking about notations. *British Journal of Music Education*, 26, 273–291.
Pramling, N. & Wallerstedt, C. 2009. Making musical sense: The multimodal nature of clarifying musical listening. *Music Education Research*, 11, 135.
Randall, W. M., Rickard, N. S. & Vella-Brodrick, D. A. 2014. Emotional outcomes of regulation strategies used during personal music listening: A mobile experience sampling study. *Musicae Scientiae*, 18, 275–291.
Reybrouck, M., Verschaffel, L. & Lauwerier, S. 2009. Children's graphical notations as representational tools for musical sense-making in a music-listening task. *British Journal of Music Education*, 26, 189–211.
Rickard, N. S., Vasquez, J. T. & Murphy, F. 2010. Benefits of a classroom based instrumental music program on verbal memory of primary school children: A longitudinal study. *Australian Journal of Music Education*, 36.
Russell, J. A. & Austin, J. R. 2010. Assessment practices of secondary music teachers. *Journal of Research in Music Education*, 58, 37–54.
Rutkowski, J. 1996. The effectiveness of individual/small-group singing activities on kindergartners' use of singing voice and developmental music aptitude. *Journal of Research in Music Education*, 44, 353–368.
Sadek, A. A. M. 1987. Visualization of musical concepts. *Bulletin of the Council for Research in Music Education*, 149–154.
Saint-Saëns, C. 1922. *Le carnaval des animaux*. Paris: Durand.
Seeger, C. 1951. An instantaneous music notator. *Journal of the International Folk Music Council*, 3, 103–106.

Seeger, C. 1957. Toward a universal music sound-writing for musicology. *Journal of the International Folk Music Council*, 9, 63–66.
Shonkoff, J. P. & Phillips, D. A. (eds.) 2000. *From neurons to neighborhoods: The science of early childhood development*, 1st ed. Washington, DC: National Academy Press.
Shouldice, H. N. 2014. Teachers' beliefs regarding composition in elementary general music: Definitions, values, and impediments. *Research Studies in Music Education*, 36, 215–230.
Sims, W. L. 1995. Children's ability to demonstrate music concept discriminations in listening and singing. *Journal of Research in Music Education*, 43, 204–221.
Sims, W. L. 2005. Effects of free versus directed listening on duration of individual music listening by prekindergarten children. *Journal of Research in Music Education*, 53, 78–86.
Sims, W. L. & Nolker, D. B. 2002. Individual differences in music listening responses of kindergarten children. *Journal of Research in Music Education*, 50, 292–300.
Smith, J. P. 2008. Compositions of elementary recorder students created under various conditions of task structure. *Research Studies in Music Education*, 30, 159–176.
Sole, M. 2017. Crib song: Insights into functions of toddlers' private spontaneous singing. *Psychology of Music*, 45, 172–192.
Stauffer, S. L. 2002. Connections between the musical and life experiences of young composers and their compositions. *Journal of Research in Music Education*, 50, 301–322.
Strand, K. 2016. Composition in an integrated arts program. *Music Educators Journal*, 102, 66–70.
Teie, D. 2016. A comparative analysis of the universal elements of music and the fetal environment. *Frontiers in Psychology*, 7, 1–8.
Thibeault, M. D. 2011. Learning from looking at sound: Using multimedia spectrograms to explore world music. *General Music Today*, 25, 50–55.
Trainor, L. J. & Cirelle, L. 2015. Rhythm and interpersonal synchrony in early social development. *Annals of The New York Academy of Sciences*, 1337, 45–52.
Trehub, S. E. 2001. Musical predispositions in infancy. In: Zatorre, R. J. & Peretz, I. (eds.) *The biological foundations of music: Annals of the New York Academy of Sciences*, vol. 903. New York: New York Academy of Sciences.
Trehub, S. E. 2016. Infant musicality. In: Hallam, S., Cross, I. & Thaut, M. (eds.) *The Oxford handbook of music psychology*. Oxford: Oxford University Press.
Trehub, S. E., Hill, D. S. & Kamenetsky, S. B. 1997. Parents' sung performances for infants. *Canadian Journal of Experimental Psychology*, 51, 385–396.
Trehub, S. E. & Nakata, T. 2001–2002. Emotion and music in infancy. *Musicae Scientiae*, 37–61.
Trevarthen, C. 1999–2000. Musicality and the intrinsic motive pulse: Evidence from human psychobiology and infant communication. *Musicae Scientiae*, 155–215.
Trevarthen, C. 2002. Origins of musical identity: Evidence from infancy for musical social awareness. In: MacDonald, R. A. R., Hargreaves, D. J. & Miell, D. (eds.) *Musical identities*. Oxford: Oxford University Press.
Trevarthen, C. & Malloch, S. (eds.) 2009. *Communicative musicality: Exploring the basis of human companionship*. Oxford: Oxford University Press.
Upitis, R. 1987. A child's development of music notation through composition: A case study. *Annual Meeting of the American Educational Research Association*, Washington, DC.
Upitis, R. 1990. Children's invented notations of familiar and unfamiliar melodies. *Psychomusicology*, 9, 89–106.
Upitis, R. 1993. Children's invented symbol systems: Exploring parallels between music and mathematics. *Psychomusicology*, 52–57.
Volchegorskaya, E. & Nogina, O. 2014. Musical development in early childhood. *Procedia–Social and Behavioral Sciences*, 364–368.
Walker, R. 1981. The presence of internalized images of musical sounds and their relevance to music education. *Bulletin of the Council for Research in Music Education*, 4, 107–111.
Webster, P. R. 2009. Pathways to the study of music composition by preschool to precollege students. In: Hallam, S., Cross, I. & Thaut, M. (eds.) *The Oxford handbook of music psychology*. Oxford: Oxford University Press.

Webster, P. R. 2016. Pathways to the study of music composition by preschool to precollege students. In: Hallam, S., Cross, I. & Thaut, M. (eds.) *The Oxford handbook of music psychology*, 2nd ed. Oxford: Oxford University Press.

Webster, P. & Richardson, C. 1993. Asking children to think about music. *Arts Education Policy Review*, 94, 7–17.

Wilkin, P. E. 1995. A comparison of fetal and newborn responses to music and sound stimuli with and without daily exposure to a specific piece of music. *Bulletin of the Council for Research in Music Education*, 163–169.

Wilson, G. B. & MacDonald, R. A. R. 2012. The sign of silence: Negotiating musical identities in an improvising ensemble. *Psychology of Music*, 40, 558–573.

Wilson, G. B. & MacDonald, R. A. R. 2017. The construction of meaning within free improvising groups: A qualitative psychological investigation. *Psychology of Aesthetics, Creativity, and the Arts*, 11, 136–146.

Woodward, S. C. 1992. *The transmission of music into the human uterus and the response to music of the human fetus and neonate*. Unpublished Doctoral Dissertation, University of Cape Town.

Woody, R. H. & Lehmann, A. C. 2010. Student musicians' ear-playing ability as a function of vernacular music experiences. *Journal of Research in Music Education*, 58, 101–115.

Young, S. 2008a. Collaboration between 3- and 4-year-olds in self-initiated play on instruments. *Music Education: A site for collaborative creativity, International Journal of Educational Research*, 47, 3–10.

Young, S. 2008b. Lullaby light shows: Everyday musical experience among under-two-year-olds. *International Journal of Music Education*, 26(1), 33–46.

Young, S. 2009. Towards constructions of musical childhoods: Diversity and digital technologies. *Early Child Development and Care*, 179(6), 695–705.

Young, S. 2012. Theorizing musical childhoods with illustrations from a study of girls' karaoke use at home. *Research Studies in Music Education*, 34(2), 113–127.

Young, S. & Gillen, J. 2006. *Technology assisted musical experiences in the everyday life of young children*. Paper presented at the Touched by Musical Discovery: Disciplinary and Cultural Perspectives, Sydney, Australia.

3

A Timeline is Not a Continuum

A More Inclusive Electronic Music History

Brian Heller

Introduction

The study of history in music production and technology tends to be limited to strictly electronic music from the Western canon. But strong historical relationships exist between what is called electronic music, and the broader discipline of sound technology. When studies of electronic music history are inclusive of both non-electronic and non-musical content, new levels of historical continuity are revealed. This helps students see connections between common electronic music history topics, and their predecessors. Opening course content to the full range of sound technology history makes the syllabus more inclusive of artistic and technical voices that had largely been excluded from the canon and texts of electronic music history.

This discussion is focused on the reinventing of an existing electronic music history survey course around the principle of inclusivity. The course[1] is part of a specialized program called Sound Arts at a two-year undergraduate college. The goal of the course redesign was to reframe electronic music history within the wider context of the development of sound technology, driven by opportunities in the field and student interests.

The scope of this writing is not intended to be a full curriculum or a complete list of content. Rather, it is to make a case for this inclusive approach, by providing examples of subjects and historical figures that become relevant once "the history of electronic music" shifts to "the history of sound technology." Seeing course content in this way is an opportunity to unearth electronic music as a phenomenon that is really about the creative uses of sound tools, more than strictly musical applications of electricity.

Background

The Sound Arts Program

The course examined here is called History and Literature of Sound Arts. It is a required part of the Sound Arts program at Minneapolis College, an urban community and technical college that is part of the Minnesota State system of public colleges and universities. The Sound Arts program is a sixty credit Associate of Science (A.S.) degree. Associate degrees are two-year undergraduate programs, generally serving two purposes: to acquire technical or vocational skills, and for liberal arts education. The A.S. degree in Sound Arts contains sixty credits, split evenly between program-specific courses, and general education courses.

The Sound Arts program is premised on teaching the fundamental principles of sound in each of the three domains (acoustic, analog, and digital), with a slant towards artistic ends. In its five production courses, emphasis is placed on understanding the basic nature of how sound and technology operate together, and how broad concepts can be adapted to particular needs. There are also additional program-specific courses, in related areas. Table 3.1 shows the sequence of required courses for the Sound Arts program. The History and Literature of Sound Arts course appears in the second semester of the first year, with Sound 1 and Fundamentals of Music as prerequisites.

The Sound Arts curriculum was designed in 1997 by Dr. Stephen Solum, who also authored its integral text, *The Craft of Controlling Sound* (2004). I worked

Table 3.1 Sequence of required courses for Sounds Arts program

First Semester	*Primary Topic Areas*
Sound 1 (Production Course)	Wave mechanics, analog audio principles, compositional form
Fundamentals of Music	Introduction to music theory, reading and notation
English and Elective	Research and writing, general education
Second Semester	*Primary Topic Areas*
Sound 2 (Production Course)	Digital audio principles, MIDI, signal paths, synthesis
History and Literature of Sound Arts	*History of sound and technology*
Program Elective	Elective course in another discipline (theater, cinema, computer science, music)
Small Group Communication and Electives	Interpersonal group communication, general education
Third Semester	*Primary Topic Areas*
Sound 3 (Production Course)	Microphone technique, critical listening
Sound 4 (Production Course)	Digital studio recording, DAW production, compositional form
Project Studio Design	Room design and acoustics
Electives	General education
Fourth Semester	*Primary Topic Areas*
Sound 5 (Production Course)	Final capstone project, project management
The Business of Sound Arts	Small business operation, copyright
Internship	Industry experience and mentorship
Electives	General education

closely with Dr. Solum from the inception of the program and developed the History and Literature course, as well as several others. I became the program director upon his retirement in 2013, and designed updates to the formal learning objectives and course descriptions over the 2017–2018 academic year.

The History and Literature of Sound Arts Course

The History and Literature of Sound Arts is the only dedicated history class in the program, and so functions as a broad survey course. The Music Department—which is independent of Sound Arts—offers history course electives, but none are strictly required of Sound Arts students. The purpose of the class has evolved to be equal parts historical grounding and listening experiences. Historical content is treated as a stand-alone discipline. That is to say, the historical topics and production topics are not intermingled in the same courses. However, historical material does often function alongside the production course material. As an example, the history course's discussion of Moog and Buchla's analog synthesizer designs happens simultaneously with hands-on subtractive synthesis work in a production course. That production course introduces synthesis to students through demonstration and a sound design activity, rather than through historical models and imitative exercises. In this way, students' exposure to a topic is not wedded to its history, as a requirement.

An important feature of the course is that it demands 'ear stretching.' Students are asked to engage with a great variety of sound work and contexts—some familiar, and some alien; some musical, and some not. The objective is to prepare them to approach the field's infinite diversity of creative work in an informed and sensitive way. Even if they are not artists, themselves, having this range of experience is essential to working *with* artists successfully. For example. familiarity with the sound-world of Pierre Schaeffer is good preparation for working with a producer who uses found sounds in their work. When a film director asks for "otherworldly" music, Louis and Bebe Barron's sound design score to *Forbidden Planet* (1956) can offer a touchstone for the discussion. The skill of digesting the entire world of sound is part of their mandate to be the "sound expert" when collaborating. This is one way of framing the learning outcomes of the course.

Redefining the Course

With a few exceptions History and Literature of Sound Arts has essentially been a straightforward electronic music history course, with material primarily from the Western music canon. The timeline began with Edison's sound recording in 1877, and Futurist composer Luigi Russolo's *The Art of Noises* from 1913 (neither of which are strictly electronic, it could be noted). Over years of refinement, it was observed that when content was added, it was often to the *beginning* of that timeline, rather than the end. There was a growing sense that a significant amount of material relevant to the trajectory of electronic music history was not actually electronic; and further, that some wasn't musical. Beginning in 2015, it became clear there was a need to reexamine how the entire course was framed, and how it related to students' lives.

Electronic vs. Technology

While formal definitions can be of limited value, a look into the various uses of the word "technology" proved to be a crystalizing agent for reassessment of the course's approach. The field of anthropology defines technology as, "the body of knowledge available to a society that is of use in fashioning implements, practicing manual arts and skills, and extracting or collecting materials." (The American Heritage Dictionary of the English Language 2016)

This is a broad definition, but it provided a "wrapper" around the syllabus, to integrate a wider range of material. By this guideline, technology is simply know-how put into practice, and not limited to any particular medium, discipline, discovery, or historical marker. This was a natural fit for the course because the Sound Arts program is not dogmatic about its approach to technology. In *The Craft of Controlling Sound*, Solum acknowledges early on that the use of technology is not a binary, but a question of degrees—a *continuum*, rather than a timeline: "Technology has grown over historical time, but time and technology need not be linked. . . . We can glide back and forth as we wish." (Solum 2004: 14–15) This notion was given weight in the redesign of the course, by addressing topics that utilize the body of acoustic and mechanical knowledge for the creative use of sound. In fact, these topics now comprise the first 30% of the syllabus (six class meetings). The intention was to build context and historical continuity for discussions of technology, wherever they were useful.

Music vs. Sound

The program's title offered an invitation to rethink the distinction between music and sound, as they related to the course. Obviously, music and sound are forever intimately linked, and the curriculum embraced that. But the program title uses the word "sound," as opposed to "music," and this is indicative of its philosophy. About half of the creative projects in the production courses are non-prescriptive about being grounded in "music" or "sound." It was important that the History and Literature course's weekly content reflected that balance and made use of the space between the musical and the non-musical.

The faculty is strongly committed to support for any approach to working creatively with sound, and takes care not to impose a particular aesthetic or style on student work. In fact, the design of the program makes a point of not requiring entering students to have a formal musical background. The students' extremely diverse backgrounds, interests, and goals reflected this. In surveying second-year students between 2013–2018, 31% expressed aspirations of doing work outside of, or in addition to, traditional music. These interests covered production for visual media, advertisements, video games, theater, computer science, electrical engineering, and more. Therefore, it was necessary for the course to be relevant to as many types of students as possible: If they were interested in sound effects design, for example, it was imperative to include that tradition. Those who tended towards audio engineering should have their interests represented, too.

To reaffirm the identity of Sound Arts as more than exclusively music, there needed to be a concerted effort to broaden the day-to-day course material. An analysis of the syllabus from before 2015 revealed that about 20% of class time was spent on topics that were not strictly musical. In its present form—after these changes—the time devoted to non-musical material has almost doubled, to about 40%.

As the conception of the course moved towards a break with the constraints of electronic music, it became, broadly speaking, more *inclusive*. The determination of which topics were considered relevant broadened, based on the students, and the fields they would be entering. More inclusive subject matter then had the resultant effect of a more diverse roster of artists and craftspeople, in terms of gender, race, cultural affiliation, professional role, and more.

Expanding the Technological Scope

The Hydraulis

By shifting the focus from "electronic music" to "sound technology," the syllabus timeline could now reach back far earlier than before. The hydraulis of Ancient Greece is an excellent introductory example. The inventor Ctesibius produced this small water organ around 3 BCE. It used the power of air pressure, created by water, to force air through pipes, and was controlled via a keyboard. Operation required a hand-operated water pump, typically worked by an assistant (European Cultural Centre of Delphi). In this way, the hydraulis very comfortably fit the course definition of technology: It applied the Greek's knowledge of pneumatics to the problem of how sound could be controlled beyond human limitations. It also used their knowledge of extracting the materials and fashioning the implements to fabricate the instrument.

The technology of the hydraulis had implications for the Greek's music creation. The constant supply of air pressure meant the physical effort required to create sound decreased significantly. In fact, refinements over time acoustically amplified the volume of the hydraulis enough to make it useful in amphitheaters (Landels 1999). At the same time, the balance of skills needed to create sound shifted from physical performance, towards understanding and making use of the apparatus.

Compare the hydraulis to a common electronic music history topic, such as the telharmonium, and the two start to have more in common than one might expect. The telharmonium's use of electromagnetics continued to decrease the physical effort needed to produce sound. Significant operational know-how was necessary to use the telharmonium's various controls; and it, too, required an assistant (*Magic Music from the Telharmonium* [1998] 2006). Perhaps most notably, both instruments suggested new paradigms of powers and limitations. A continuous supply of air allowed notes on the hydraulis to be held far beyond the limits of breath or string resonance. But the hydraulis also had a more limited range than other breath-driven instruments of its time, as keys allowed less flexibility of pitch than finger holes (Hagel 2018). For its part, the telharmonium streamed its live performances in real-time to subscribing hotels, restaurants and theaters in Manhattan by using the existing network of telephone lines. This meant that a live performance could overcome the limitations of time and space. But at the same time, the telharmonium's electromagnetic tone generators lacked the timbral complexity and polyphony of existing acoustic keyboard instruments ([1998] 2006).

Studying the hydraulis introduces to students the idea that the fundamental concepts behind the use of technology are not tied to a particular type of tool or point on a timeline. A demonstration of its operation shows the power of applied technology—regardless of the medium—much more concretely than the abstraction of later electronic music topics, such as multitrack recording and MIDI. Moreover, the example of the hydraulis lays the basis for a conceptual continuity between eras far removed from one another.

The Piano

There is great pedagogical opportunity in the piano. It is familiar and observable in operation; yet also highly complex and well-developed. Its design and construction represented a high-water mark for acoustic music technology. Solum's *The Craft of Controlling Sound* (2004) suggests to students that, as sound technology becomes more powerful, it also becomes more complex. While he is referencing modern digital technology in that passage, the piano provides an excellent analogy of this power and complexity differential. It is, after-a-fashion, a seven octave, velocity-sensitive synthesizer, with pedal control and unlimited polyphony. Its mechanical nature also allows straightforward modification, as will be discussed shortly.

The piano's complexity is both its strength, and its liability. Aside from its size and weight—a Steinway Model D measures 40 square feet and weighs almost 1000 pounds—its Rube Goldberg-esque system of moving parts create many points of potential failure. This complexity makes it relatively expensive to maintain, and also presents a need for experienced instruction. In short, the piano has a high cost of ownership, similar to the delicacies of an analog synthesizer or tape machine.

The piano's central place in the history of sound technology can sometimes be taken for granted. It is worth making students explicitly aware that its dominance in the nineteenth century shaped the technology used today, over 200 years later. Historically, electronic instruments that used a piano-style keyboard (such as the Moog synthesizer) have fared far better for popularity and influence than those that did not (the telharmonium). MIDI, itself, was keyboard-based by conception, and that has been a key to its thriving (Chasalow & Cassidy 1997; Holmes 2016). The sonic and cultural power of the piano presents a strong example of the interplay between technology and the music of the day. The evolution of its capabilities were at the heart of the Romantic lexicon. The music changed to meet the piano, and the piano changed because of the music (Burkholder, *et al.* 2014).

The American Experimentalists

The work of the American Experimentalists—modernist composers from roughly 1900 through the 1940s—brought challenging and strikingly original work to the history of Western music (Morgan 1994). None of these composers produced electronic music (at least during this time period), but they have important linkages with it. Together, they covered an enormous amount of conceptual ground, presaging later developments: new timbres, novel approaches to rhythm, and unique instruments, among others (Iverson 2019). With few exceptions, these radical musical voices toiled in obscurity, and with meager resources, save for their own inventiveness. In their own ways, they were quintessential outsiders: outspoken, idiosyncratic individualists; artistically and culturally marginalized; and working underground (Morgan 1994).

Henry Cowell

As the senior member of this group, Henry Cowell had a major role in the development of American music. His book *New Musical Resources* (1930) acted as a blueprint for at least the next 50 years of experimental music-making (Sachs 2012). This volume gave few examples from the existing literature, instead offering systematic plans for musical systems and tools that *could* be adopted. He posited compositional use of the technologies such as the phonograph and player piano, outlined a system to extend the use of the harmonic series to other musical parameters, and described a proto-drum machine, later built by Leon Theremin (Cowell [1930] 1996; Holmes 2016). Another of Cowell's ideas involved unconventional playing techniques inside the piano: strumming, sliding, striking, and all manner of manual sound production. Appropriately, he called this technique the "stringpiano," of which *Aeolian Harp* (1923) and *The Banshee* (1925) are representative examples. In Cowell's work, the piano was deconstructed, and rendered nearly unrecognizable (Nicholls 1997).

Cowell's middle career was tragically interrupted by four years at San Quentin penitentiary. In 1936, he was prosecuted by the State of California for engaging in oral sex, called a "morals charge" at that time. This happened to be with another male, which unsurprisingly became the focus of the affair—although the text of the law said nothing about homosexuality, specifically. After a number of bizarre twists and a public spectacle, Cowell confessed, so there was no actual conviction. Cowell's was quite prolific in prison, and he founded a successful music program for his fellow inmates. Thanks to the persistent efforts of friends and his wife, Sydney Robertson, he was eventually pardoned. He continued to compose after his release, but his music was markedly simpler and less adventurous (Sachs 2012).

Like many others in this discussion, Cowell produced no electronic music, but might just as well have. The timbral richness of his techniques were far beyond the capabilities of any electronic instrument of the day, and his sound imagination was highly-developed. To modern ears, his stringpiano could be called a sort of expertly-crafted additive synthesis.

Conlon Nancarrow

Conlon Nancarrow represents one of most compelling, but consistently neglected, touchstones for the use of technology in modern music. His primary preoccupation was with constantly changing tempo relationships between musical lines. These tempo ratios could be as straightforward as 4:5 (*Study 14*), and as bewildering complex as $\frac{1}{\sqrt{\pi}}:\frac{2}{3}$ (*Study 41a*) (Tenney 2008: 24, 20). To attain the precision his music required—and to enable it to be heard—he turned to another suggestion from Cowell's *New Musical Resources*: the player piano (Cowell [1931] 1996; Sachs 2012).

By measuring how fast the punched paper roll moved through the piano's reading mechanism, he determined the distance between the holes that triggered the keys, and could thus derive the tempos. What the player piano lacked in timbre (although he did modify the hammers), it more than made-up for as a fully programmable and polyphonic instrument. He had complete non-realtime control of tempo and sequence, which was exactly what his music demanded.

The realization of Nancarrow's musical ideas were directly related to the circumstances of his life as an American political dissident. Nancarrow joined the Communist party and then Abraham Lincoln Brigade—a volunteer militia that joined the resistance against fascism in the Spanish Civil War of 1937. The Brigade suffered heavy losses, and Nancarrow himself was wounded, captured, and escaped, returning to America in 1939 (Gearhart 2013).

Having fought for Communism was, unsurprisingly, not looked upon favorably in 1940s America. Nancarrow saw his surviving Brigade comrades harassed by the government, blacklisted, and denied passports. He took a modest inheritance, bought two player pianos, and had them shipped to Mexico City. There, in exile, he worked single-mindedly until his death fifty years later. Nancarrow lived and worked in isolation, almost unknown to Western music until the 1980s. The discovery of his completely unique body of work was a revelation to the contemporary music world (*Conlon Nancarrow: Virtuoso of the Player Piano* 2012).

Useful comparisons between MIDI and the player piano are not uncommon in electronic music. After all, we owe the MIDI sequencer's piano roll notation model to it. Like the MIDI language, the player piano has a specific data format, and can respond to that limited vocabulary. The holes in the paper more than closely resemble MIDI note messages, and, not unlike the hydraulis, use a pneumatic mechanism to trigger sound production. Some player pianos even had additional tracks on the paper roll that controlled hammer action (velocity), volume, and sustain pedal (control change messages 7 and 64, respectively) (The Pianola Institute 2019). Just as with a MIDI system, the resultant sound is dependent on the user's programming choices, as well as the sonic qualities of the instrument, itself. Player piano rolls are still plentiful and relatively inexpensive, so it is quite possible to unfurl one in class and allow students to use its tactility as a connection to the far more abstract MIDI messages.

Nancarrow is a perfect example of a composer devoting himself to a specialized instrument that was essential for his work. He leveraged many benefits of digital technology, but in a fully acoustic-mechanical domain. He harnessed a complete control system: step sequencing, independence of pitch and tempo, editing, and memory. Continuing that trajectory, the RCA MKII synthesizer echoed many of the player piano's characteristics. In fact, Milton Babbitt has stated that one of the reasons for his commitment to the RCA MKII was "purely for the temporal control." (Chasalow & Cassidy 1997) Like Babbitt after him, Nancarrow's mastery of his instrument provides a vivid illustration of the fusion of music and technology.

John Cage

Discussions of John Cage in electronic music typically focus on his work after 1950, which is, indeed, the bulk of his electronic output (with the notable exception of 1939's *Imaginary Landscape No. 1*). This is certainly appropriate, but the omission of his earlier work misses one of his key contributions: the prepared piano. Cage inserted various household objects between the strings of the piano (erasers, screws, washers, etc.), which opened an entirely different sound-world (Isacoff 2012).

Cage had studied with Cowell, and found opportunities to subvert tonal music, even before the prepared piano (Iverson 2019). Like Edgard Varese, he explored the relationship between music and sound using percussion music: "Percussion music is a transition from keyboard-influenced music, to the 'all-sound' music of the future. Any sound is acceptable to the composer of percussion music." (*John Cage: I Have Nothing To Say and I Am Saying It* 1990)

As with many of these composers, necessity was the mother of invention for Cage. He described the prepared piano as a practical solution to the problem of a performance space being too small to accommodate other instruments (1990). His prepared piano music was notated conventionally, which made it easily playable without the exotic techniques and unfamiliar notation of Henry Cowell's stringpiano, or Nancarrow's specialized technology.

It should not be lost that the work of the American Experimentalist composers discussed here relied heavily on the piano, as it was both easy to manipulate, and ubiquitous. They worked in relative obscurity, and with few resources. This offered them no access to the burgeoning field of electronic musical instruments, or the necessary technical expertise. Yet, their musical interests feel quite at-home next to the electronic music that followed. That they are largely absent from electronic music history texts is a missed opportunity to demonstrate that connection for students.[2]

The Technologist

While the American Experimentalist composers were largely self-made individualists with little outside support, of course some artists did have skilled collaborators. This was more common in European electronic music studios, where there were already well-organized support structures through state-sponsored broadcasting. The GRM in Paris, WDR in Cologne, and other studios throughout Europe employed technicians who were essential to the realization of composers' music, often with little more than scraps and hand-me-down radio equipment. Jacques Poullin, for example, created the Phonogéne—an eccentric proto-sampler—for Pierre Schaefer. He also guided a young Karlheinz Stockhausen through his first *musique concréte* etudes (Holmes 2016). WDR technician Heinz Schütz was actually pressed into service as a composer, to assemble the studio's first-ever piece for a demonstration (Iverson 2019).

Discussing early artist-engineer collaborations acknowledges that the technologies we use now were the results of inspired, if less visible, minds. It allows student to see that the body of knowledge used to fashion the implements of today had actually been in development for years by the researchers and institutions who provided the resources: Yamaha did not invent FM synthesis; it was John Chowning at Stanford. Digital theory came from Harry Nyquist, Claude Shannon, and Max Matthews at Bell Labs. The tools for convolution reverb were developed at IRCAM.[3] The inspiration for computer music itself might rest with Ada Lovelace, Charles Babbage's collaborator on a mechanical computer. She—not Babbage—explicitly predicted in 1842 that "the [Analytical] [E]ngine might compose elaborate and scientific pieces of music." (Collins & Rincón 2017: 59) This is not to downplay the important work of more forward-facing companies, but to encourage support of experimental research, which can be rendered invisible to the very users who rely on it.

To the artists they served, these technical minds were both teachers and assistants. They were technicians who had skills that needed vision; and enabled artists who had vision, but needed skills. Widening the course's parameters allows inclusion of them as organic parts of the history of sound and technology.

Expanding the Musical Scope

Student perceptions of coursework relevancy has been shown to directly affect motivation—and thus completion rates—in community colleges (Levesque 2018). As previously mentioned, over 30% of Sound Arts students have interests outside of traditional electronic music composition, so it was imperative that the course be adjusted to reflect that. However, application of technology to areas such as sound design for visual media, dance, and advertisements is largely absent from traditional history texts. Increasing the inclusivity of course content has integrated these artists' voices into the syllabus. This both acknowledged their role in the development of electronic music and aimed to avoid the risk of students with these diverse interests feeling creatively marginalized. Between 2015 and 2018, student evaluation data on the relevance of the course averaged in the 90th percentile, and the Sound Arts program's 71% completion rate is one of the highest at Minneapolis College.

Early Field Recording

Field recording is an essential part of the history of sound capture. Including an introduction to historical field recording presents to students the idea that sound recording has been both a creative, and a cultural tool.

Field recordists have evolved their own diverse inspirations and conceptual approaches. Recordist and sound artist Lawrence English wrote that these practices reflect “a range of interests that flow from the scientific, to the theatrical, to the narrative and beyond.” (2014)

The craft of field recording has also been formative for composition, by exposing artists to unfamiliar sound-worlds they then utilized in instrumental music, *musique concréte*, and sampling.

Even before using electricity, mechanical sound capture outside of the recording studio was quickly adopted by recordists. For composers Bela Bartok and Zoltan Kodaly, field recordings provided a platform for the development of their own musical languages. Beginning as early as 1906, the two spent several years traveling the remote countryside of Eastern Europe, recording regional folk songs, and then transcribing them. Capturing an unfamiliar musical language from an oral tradition not particularly suited to Western notation was a formidable challenge. This also had to be done with efficiency, since playback of fragile cylinders was destructive process. As Bartok and Kodaly were composers, their transcriptions then became inspirational musical materials, as well as a cultural archive. They did not simply reproduce the folksongs in their compositions, verbatim. Rather, they analyzed and absorbed the folk music’s musical materials, and used it to expand their own (Nelson 2012).

The work of Bartok and Kodaly evidences the intertwining of the technical and creative histories of sound technology. The technologies of sound recording and Western musical notation, even as intermediaries, provided a path for them to build their own musical language. Their activities were an early demonstration of how the then-new tools of sound recording could be a creative vehicle for composers.

Contemporaneous with Bartok and Kodaly in Europe, American Frances Densmore (1867–1957) was one of the first to capture Native American music in the field. Although her mentor, Alice Cunningham Fletcher, had already made some recordings, Densmore’s work for the Smithsonian was astoundingly prolific. She made around 2000 recordings in almost 50 years of work, remaining active well into her 80s. Her voyages, sometimes taken alone, took her to unfamiliar geography in deep cultural conflict; although she demurred by referring to her work as “exciting but not dangerous.” (Patterson 2015: 30) Densmore fashioned herself into a consummate professional, maintaining a cordial but still somewhat detached relationship with the Native Americans she visited, and sometimes even paid them for recording their songs. She saw herself squarely as a scientist, writing to her colleague that “the finished work must be neither theoretical nor sentimental, it must stand the test of scientific scrutiny.” (2015: 55) Even in the relatively open field of ethnomusicology, women and science were still considered to be a suspect combination, although she rarely addressed it directly in her writings (2015).

Needing complete self-sufficiency, Densmore chose her technology carefully. She purchased a high-quality cylinder phonograph before her first recording trip in 1908, and continued to use it until 1940, even after battery-powered recorders had become a possibility. She felt that the relatively light and highly-portable cylinder recorder was still the best suited for remote areas and allowed the equipment to keep a low profile during the recording process. Whenever possible, she worked to give her recordings of Native American music exposure to the public. She saw the archive as a potential source of inspiration for American composers to draw upon—remarkably similar to how Bartok and Kodaly were using their own (Gray 2015).

Introducing the magic of historical field recordings to students is best done by listening. Bartok and Kodaly's capture of Magyar folk tunes; Densmore's cylinders of Chippewa songs from Minnesota; and wildlife recordist Ludwig Koch's (1881–1974) early multimedia sound-books are all easily available, even at 120 years old. Alan Lomax (1915–2002) became the giant in the field, and for good reason—his astounding life's of work reaches from prison chain gangs in Mississippi, to Inuit cultures in Siberia. As field recording's greatest ambassador, Lomax made a case for his experience of sound technology as a cultural empowerment tool:

> The incredible thing is that when you would play this material back to people, it changed everything for them. They realized that they were just as good as anybody else. And I found out that what I was really doing . . . was giving an avenue for those people to express themselves and tell their side of the story.
>
> (*Lomax [1991] 1997*)

Hearing these artifacts can encourage students to connect the recording tools they already use with wider applications and geographies. To support that, the program recently made a point to increase their available field recording equipment and added related learning activities in the production classes. Field recording has a long history of connecting the medium of sound with other disciplines, which makes a good argument for its inclusion in a history course. That field recording is sometimes not directly musical makes it no less significant to the development of the technology, and culture of sound.

Raymond Scott and Bruce Haack

In developing new electronic instruments, and then applying them to a wide range of activities, two lesser-known figures in the field—American Raymond Scott (1908–1994) and First Nations Canadian Bruce Haack (1931–1988)—happen to have covered a great deal of similar ground, almost simultaneously. Raymond Scott's life and work entered a rediscovery period in the late 1990s (Collins & Rincón 2017; Winner 2008), but Bruce Haack's has received less acknowledgement in electronic music texts.

Though they worked independently, their biographies and interests have parallels. Both attended The Juilliard School, although Haack did so only for a short time. Both also had early commercial successes in the traditional music industry before turning to electronics—Haack as a pop songwriter, and Scott as a band leader.

Each of these artists spent significant time creating new electronic instruments. Including them in the history of electronic music completes important technological links to a number of developments soon after them. Haack and Scott were both self-made inventors, solitarily working out of their homes. Scott's Clavivox, Electronium, and Circle Machine were essentially complete analog workstations, with strikingly advanced approaches to multitrack recording, synthesis and sequencing

(Holmes 2016). Haack's instruments, while on a smaller scale, featured unusual live performance control sources, suggesting the work of Gordon Mumma and David Tudor. The Dermatron, like many of Haack's other inventions, was housed in simple household objects, and made use of light-sensing photocells and thermal skin response. His 1967 Musical Computer was even built into a suitcase, just as the Buchla 100 was (Brend 2012). All of these instruments were developed over a period from the late 1950s to the early 1970s, but the machines were constantly being modified, and parts repurposed. Neither seemed to have ambitions of selling their instruments commercially. Scott operated in almost paranoid secrecy, but often hired a young Robert Moog to build circuits for him. Moog considered Scott to be a significant influence on his synthesizer work (Winner 2008).

Scott and Haack each had success applying their instruments to television and radio advertisements. Between the two, they did work for several high-profile clients, such as Buffrin aspirin (with Jim Hensen), Cheer detergent, Goodyear tires, and Kraft cheese (Brend 2012; Winner 2008). Their early use of analog sequencers offered novel sounds that proved to be well-suited to advertisers' needs. As commercially successful entrepreneurs with businesses built around cutting-edge sound technology, Scott and Haack's legacies seem especially prescient today, where industry practitioners are expected to be creators, technicians, and small business owners—all on the tight deadlines of commercial clients (Collins & Rincón 2017).

A signature area of exploration both shared was the application of electronic music to children's activities. A significant amount of Haack's career was spent creating electronic music for children's dance, both live and recorded. With his long-time collaborator, dance pedagogue Esther Nelson, he regularly accompanied her children's dance classes, using the Musical Computer and a small collection of other homemade analog instruments. Nelson and Haack were featured on *Mr. Roger's Neighborhood*[4] in 1968, where again the sequencer was well-suited to the musical goals. Haack could trigger it by clapping, and then have his hands free for gestural control while singing, or playing the recorder. The fact that the sequence could be altered in real-time allowed the spontaneity necessary when working with children and dance.

For his part, Scott conceived the idea to compose a large work for infant development, called *Soothing Sounds for Baby*. Released in 1964, it was a series of three records, each for a different stage of a child's development between one and eighteen months. A collaboration with the Yale-affiliated Gesell Institute of Child Development, he was seen as "one of the first composers to merge the Brave New World of electronic sounds with a rhythmic pop sensibility." (Chusid 2000: 9–12) Although this was not a commercial or critical success at the time, it is now considered quite forward-thinking (Brend 2012).

Daphne Oram and Delia Derbyshire

Following Sound Arts students' range of eclectic interests, the historical models for sound effects and incidental music are also important components, and a platform for this can be found in the work of the BBC Radiophonic Workshop.

The Radiophonic Workshop was established by the BBC, somewhat reluctantly, as their radio and television sound production unit (Holmes 2016). With its official founding in 1958, the Workshop initially used the techniques of acoustic recording and tape manipulation. But, despite the fact that their work was contemporaneous with "classic" *musique concréte*, the Workshop was not considered by the BBC to be art, in itself. In fact the BBC's initial aim was to employ only engineers—and no musicians—which led to a resultant policy of not crediting any individual members for their work, or offering royalties (Holmes 2016; White 2001).

The Radiophonic Workshop was unusual in that it employed women very early in its existence—most prominently Daphne Oram and Delia Derbyshire. Oram was a founding member and became its first co-director. Despite this, she was frustrated with the BBC on several fronts. Since the BBC considered the Radiophonic Workshop to be a technical production unit, she found she was not given the creative latitude she sought, despite the imagination that the work required (Manning 2003). She also reported rampant institutional sexism in the industry, which was echoed by the BBC bureaucracy (Rodgers 2015). In one bizarre incident, she was told to take a leave of absence, because the male administrators thought the equipment would have a detrimental effect on her body, a demand not made of her male colleagues (Williams 2017). This culminated in her leaving after just one year.

Even before departing the Workshop, Oram had been working on the development of her own synthesizer: the Oramics Machine. The Oramics Machine used graphics that had been painted on 35mm film stock, decoded though a network of photocells. Shapes could be painted to control up to ten parameters, so the user could fully sequence and synthesize sound with these graphic "scores." Although not the first attempt at using optical technology to synthesize sound, it was certainly the most elaborate of its day (Manning 2003). Work produced with the machine dates from at least 1960, making it—along with Yevgeny Murzin's ANS Synthesizer—a groundbreaking image synthesis system (Crab 2016).

Studying the Oramics machine presents students with a radically different model for synthesis early in the development of the analog synthesizer. Exposing students to a range of interface paradigms is pedagogically valuable in a program with no prerequisite for formal music or technology training. Doing this acknowledges that highly creative work can come from artists who find the conventional model of subtractive synthesis to be an impediment, perceiving its modularity as disjointed, and its terminology arcane. While it is still a requirement that Sound Arts students have an initial grounding in classic synthesis, other avenues are then made available for their exploration. One of those tools is Metasynth[5], a powerful software image synthesizer that bears a strong relationship to Daphne Oram's invention.

Another early member of the Radiophonic Workshop was Delia Derbyshire, who's eleven year tenure began in 1962. By the time of her departure in 1973, she had been responsible for elements of nearly 200 BBC productions. Her most well-known work, the theme music of the *Doctor Who*[6] series, ran for 17 years, and was one of the first television themes produced entirely electronically (Computer

Music Journal 2001). Derbyshire did this with *musique concréte* techniques painstakingly applied to recordings of a plucked string, and the use of a voltage controlled synthesizer of sorts, created in-house (White 2001).

Derbyshire's output ranged far beyond *Doctor Who*—theater, radio drama, film, poetry, and music are all in her catalog. But *Doctor Who* represented unprecedented exposure of electronic music and sound design to a wider public. Ratings of *Doctor Who* between 1968–1969 averaged 7.8 million U.K. viewers per year (Dr. Who Guide 2019). By comparison, *Switched on Bach*[7] reached total U.S. sales of 500,000 units over that same time (RIAA, n.d.). But despite the exposure of their work, the contributions of Derbyshire and others doing similar work remained hidden in plain sight. In her discussion of sound design work at the WDR studio in Cologne, Jennifer Iverson put it succinctly: "Although it has been almost completely written out of the history, the sounds produced for radio plays in the 1940s and 1950s were the most direct precursors and inspirations for the electronic music of the early 1950s." (Iverson 2019: 36–37) Iverson cites the strict hierarchical laboratory model of postwar European studios as a contributing factor. Staff technicians (as sound designers were often referred to) were not to appear as equal collaborators with composers, regardless of the actual division of labor. When the "technicians" happened to be women, the problem of due credit was compounded. The irony is that radio drama sound design was some of the first work that many electronic music pioneers[8] did before being "promoted" to composers. At the GRM, WDR, and later at the BBC, the prevailing attitude was one of aesthetic classism: Programming from the electronic music studios was viewed as deserving of greater recognition than programming that appealed to wider audiences (2019).

In spite of the BBC's backwards approach to gender, several more women were employed at the Radiophonic Workshop over the years, including Maddalena Fagandini (who participated as early as 1959), Glynis Jones, and Elizabeth Parker, among others (Butler, n.d.). The lack of due-credit given to women over the history of sound technology—whether conscious or unconscious, individual or institutional—has been a liability for the history of sound technology. Women's contributions have been historically present across diverse areas of the field. When education is inclusive of those areas, it is, in turn, inclusive of those women (Iverson 2019).

The BBC Radiophonic Workshop is but one example of historical sound design that History and Literature of Sound Arts now covers. Sound in film, beyond just musical scores, is represented by Murray Spivak's groundbreaking work on *King Kong*, in which he manipulated recordings on optical film tracks, as early as 1933 (Platte 2017). Students also view the science fiction film *Forbidden Planet* (1956), with a score by Bebe and Louis Barron. Similar to much of the students' work, *Forbidden Planet* was created in the Barron's home, and not in the institutionalized studios so prevalent in electronic music history. A bridge to video game sound design is also made when these film landmarks are included in the class. Many of the demands and skill sets overlap between the two disciplines, and video game sound has become both a key area of student interest, and of industry growth (Schmidt 2017).

Public Enemy

Historical continuity with modern artists must also be integrated into the course. Despite being squarely electronic, music that makes heavy use of digital sampling is often not placed alongside earlier electronic music. Hip-hop, from its formative days in the 1980s, can be seen as both a continuation of recording and electronic music history, and a departure from it (Collins, *et al.* 2018) The pioneering group Public Enemy provides a prime example. Their work didn't merely use sampling; sampling was its very bedrock.

In an electronic music cannon rife with abstract music and privileged creators, *It Takes a Nation of Millions to Hold Us Back* represented a politically aware and marginalized voice. Thirty-two years after its release, its biting sociopolitical critique continues to be as urgent as ever. Layers of samples fueled the tension that drove their work: the voice of Malcolm X, the use of heavily-distorted rhythm tracks, and the constant urgency in the air raid-style sirens running through the record (Mohaiemen 2008). Students have pointed out that the very use of sampling was suppressed by the courts, as told through the song *Caught, Can We Get a Witness?* (Weingarten 2010). In an urban community college, with a population of 55% students of color that have a 32% retention rate, including this work is a crucial opportunity for relevancy. In its way, it makes incremental progress towards adding excluded voices back to the largely privileged trajectory of electronic music (Ali 2015).

Yet, at the same time, a pedagogical continuity can be drawn between Public Enemy's use of the technological tools, and that of the seemingly distant musical past. In its composition, their work was as much reliant on sound recording as Bartok's, and as much electronic music as *musique concréte* (Collins, *et al.* 2018). *Musique concréte* mined relevant acoustic sounds ("objets sonores"), conceptualized their organization, and structured them using the tools that were available. *Musique concréte* also "sampled" preexisting recordings, such as the spark generators and *Parsifal* in Vladimir Ussechevsky's *Wireless Fantasy* (1960). Both in principle and practice, this is also the work of Public Enemy. In fact, Clyde Stubblefield—the drummer in the famous sample from James Brown's *Funky Drummer* (1970)—has said he was influenced by the rumbling trains he heard as a child (Weingarten 2010), as if making a call-back to Schaeffer's *Étude aux Chemins de fer* (1948).

Of course, the content of Bartok, Schaeffer, and Public Enemy's music are as different as the circumstances of their creation. But in their own ways, each are compelling cases of how artists' accessibility to a particular technology affected the creative expression. They used what tools were available to them and made an instrument where there was none before—the acoustic recorder for Bartok, the analog recorder for Schaeffer, and the digital recorder for Public Enemy. This is another a case where decompartmentalizing course content reveals connections across genres and cultures.

Towards the end of the semester, students deliver presentations on a specific electronic music release after 1975, from a given list. The presentation list includes *It Takes a Nation of Millions to Hold Us Back*, as well as Afrika Bambaataa's *Planet Rock*, and De La Soul's *3 Feet High and Rising*. One of the assessment rubric criteria asks them to relate aspects of the album they are presenting to previous course material. In the past five years, a full 100% of the presentations on those three albums have scored at the highest rubric level, up from 71% the previous three years. These three albums are substantive electronic music, and students are asked to see them in that light.

Conclusion

Electronic music history is but one subset of the history of sound technology, and the two have influenced each other. Much of electronic music has its roots in work that is neither electronic nor music at all. The goal of overcoming human limitations is centuries old; and pneumatics, mechanics, and electricity are all merely different power sources to do so. In practice, electronic music is more fluid than labels or categories may lead students (and instructors) to believe. Addressing the full history of sound and technology transforms that fluidity into a powerful pedagogical tool to better illuminate the development of electronic music.

Each of the creators discussed here saw technology as a tool to be used in whatever way was necessary to realize their vision. Individual circumstances—whether gender, race, socioeconomic status, political leanings, or historical period—dictated what was available to them. Their non-dogmatic approach to creative work—unconcerned with how the result might be categorized—made them experts in exploiting the malleability of their technology. Equally important, they were undaunted by the cultural and institutional forces working against them. The study of electronic music history should not impose the very barriers that its subjects worked to overcome. Engaging the full spectrum of sound work demonstrates for students that the principles behind creative efforts need not be limited to a particular era or medium.

Making a place for more diverse content reveals the artists and craftspeople who are behind it. When entire areas of the field are marginalized, so are the people working in them. Increasing inclusivity in electronic music history courses gives students the opportunity to connect with *all* types of creative sound work, and to see themselves as a part of its rich history.

Notes

1. This writing will use terminology from the U.S. system of higher education. Therefore, a "program" is a degree with a specialized major, a "course" is a class within that program, and a "module" is a unit within a course.

2. A survey was made of five texts within the last six years, that were deemed suitable for an electronic music history survey course. The chart below indicates the number of pages each subject appeared in the index of the book:

Title	*Author(s)*	*Cowell/ Nancarrow/ Prep. Piano*	*Scott/ Haack*	*BBC Radio/ Oram/ Derbyshire*	*Bartok/ Densmore/ Lomax*	*Hip hop/ Public Enemy*
Audio Culture (2017)	Cox.Warner	2	–	–	–	–
Electronic and Computer Music (2013)	Manning	3	–	1	–	–
Electronic Music (Cambridge) (2013)	Collins, Schedel, Wilson	1/1	3	–	–	–
Live Wires (2013)	Warner	–	–	–	–	–
Electronic and Experimental Music (2016)	Holmes	1	5	15/4	–	1

3. Institut de Recherche et Coordination Acoustique/Musique, a computer music research center in Paris.
4. *Mr. Roger's Neighborhood* was a popular American children's educational television program, which ran on the Public Broadcasting Service (PBS) network from 1968–2001. In each episode, host Fred Rogers invited guests to introduce and demonstrate their professional activities and hobbies.
5. Metasynth is developed by U & I Software (http://uisoftware.com).
6. *Doctor Who* is a British science fiction television program, running from 1963–1989, and again from 2005 to the present. The show has remained highly popular in England and amassed an international cult following. The show has not only had commercial success, but also garnered British Television Academy (BAFTA) and National Television awards, and a Peabody award.
7. Released in 1968, *Switched on Bach* was a collection of J.S. Bach pieces performed on the Moog Synthesizer by American composer and performer Wendy Carlos. The album created a sensation, reaching number ten on the U.S. Billboard chart, and holding number one on the Billboard Classical chart for three years. Due to its commercial success, it is widely credited with exposing the synthesizer to the general public.
8. The list of composers who drew early experiences from sound effects collection for radio dramas at European electronic music studios includes, among others: Luciana Berio, Bruno Maderna, Mauricio Kagel, John Cage, and even Pierre Schaeffer, himself (Iverson 2019).

Works Cited

Ali, S. (2015). Integrating Hip-Hop and Cultural Relevant Lessons into the Public School Curriculum. *Journal of Research Initiatives*, 1(3), Article 4. [Online]. Available at: https://digitalcommons.uncfsu.edu/jri/vol1/iss3/4

American Heritage Dictionary of the English Language, The (2016). *technology*. [Online] Available at: https://ahdictionary.com/word/search.html?q=technology [Accessed 6 June 2019].

Brend, M. (2012). *The Sound of Tomorrow: How Electronic Music Was Smuggled into the Mainstream*. London: Bloomsbury Academic.

Burkholder, J., Grout, D. and Palisca, C. (2014). *A History of Western Music*. 9th ed. New York: W. W. Norton & Company.

Butler, D., n.d. *The Women of the Workshop*. [Online] Available at: www.bbc.co.uk/historyofthebbc/women-pioneers/radiophonic [Accessed 30 May 2019].

Chasalow, E. and Cassidy, B. (1997). *Dave Smith on the "invention" of MIDI.* [Online] Available at: https://econtact.ca/12_2/video/DaveSmith_MIDI.mp4 [Accessed 4 June 2019].
Chasalow, E. and Cassidy, B. (1997). *Milton Babbit on the EMC and the RCA Mark II.* [Online] Available at: https://econtact.ca/12_2/video/MiltonBabbitt_1997.mp4 [Accessed 4 June 2019].
Chusid, I. 2000. *Manhattan Research Inc [CD Book].* Aalsmeer: Basta Audio-Visuals.
Collins, N., Manning, P. and Tarsitani, S. (2018). A new curated corpus of historical electronic music: Collation, data and research findings. *Transactions of the International Society for Music Information Retrieval*, 1(1), 34–43.
Collins, N. and Rincón, J. D. (2017). *The Cambridge Companion to Electronic Music*. Cambridge: Cambridge University Press.
Computer Music Journal (2001). News. *MIT Press*, 25(4), 13–20.
Conlon Nancarrow: Virtuoso of the Player Piano (2012). [Film] Directed by James Greeson. USA: University of Arkansas.
Cowell, H., [1930] (1996). *New Musical Resources*. Cambridge: Cambridge University Press.
Cowell, H., [1931] (2012). Music of and for the Records. In: T. Taylor, M. Katz & T. Grajeda, eds. *Music, Sound, and Technology in America*. Durham: Duke University Press.
Crab, S. (2016). *120 years of Electronic Music*. [Online] Available at: http://120years.net/the-ans-synthesisereugeniy-murzinsoviet-union1958-2/ [Accessed 6 June 2019].
Dr. Who Guide (2019). *Ratings Guide*. [Online]. Available at: https://guide.doctorwhonews.net/info.php?detail=ratings&type=date [Accessed 5 June 2019].
English, L. (2014). *A Beginner's Guide to Field Recording*. [Online]. Available at: www.factmag.com/2014/11/18/a-beginners-guide-to-field-recording/ [Accessed 5 June 2019].
European Cultural Centre of Delphi (n.d.). *The Ancient Hydraulis*. [Online]. Available at: www.archaeologychannel.org/video-guide/video-guide-menu/video-guide-summary/109-the-ancient-hydraulis [Accessed 5 June 2019].
Gearhart, G. (2013). *Samuel Conlon Nancarrow*. [Online]. Available at: https://encyclopediaofarkansas.net/entries/samuel-conlon-nancarrow-1724/ [Accessed 23 May 2019].
Gray, J. (2015). An Archival Dilemma: The Densmore Cylinder Recording Speeds. In M. W. Patterson and J. M. Jensen, (Eds). *Travels with Densmore*. Lincoln: University of Nebraska.
Hagel, S. (2018). *Ancient Greek Music: A New Technical History*. Cambridge: Cambridge University Press.
Holmes, T. (2016). *Electronic and Experimental Music*. 5 ed. New York: Routledge.
Isacoff, S. (2012). *A Natural History of the Piano*. New York: Penguin Random House.
Iverson, J. (2019). *Electronic Inspirations: Technologies of the Cold War Musical Avant-Garde*. New York: Oxford University Press.
John Cage: I Have Nothing To Say and I Am Saying It. (1990). [Film]. Directed by Allan Miller. s.l.: PBS.
Landels, J. G. (1999). *Music in Ancient Greece and Rome*. London: Routledge.
Levesque, E. (2018). *Improving community college completion rates by addressing structural and motivational barriers*. [Online]. Available at: www.brookings.edu/research/community-college-completion-rates-structural-and-motivational-barriers/#section2 [Accessed 25 May 2019].
Lomax, A., [1991] (1997). Interview with Charles Kuralt, CBS Sunday Morning. On: *The Alan Lomax Collection Sampler* [CD]. Nashville: Rounder Records.
Magic Music from the Telharmonium. [1998] (2006). [Film]. Directed by Reynold Weidenaar. USA: CreateSpace.
Manning, P. (2003). The Influence of Recording Technologies on the Early Development of Electroacoustic Music. *Leonard Electronic Almanac*, Volume 13, 5–10.
Mohaiemen, N. (2008). Fear of a Muslim Planet: Hip-Hop's Hidden History. In P. Miller, (Ed). *Sound Unbound*. Cambridge: MIT Press.
Morgan, R. (1994). *Modern Times: From World War I to the Present*. Englewood Cliffs: Prentice Hall.
Mr. Roger's Neighborhood. (1968). Episode 68. [TV Program]. National Education Television, 22 May 1968.
Nelson, D. (2012). Béla Bartók: The Father of Ethnomusicology. *Musical Offerings*, 3(2), Article 2. Available at: https://digitalcommons.cedarville.edu/musicalofferings/vol3/iss2/2.
Nicholls, D. (1997). Living in the whole world of music. In D. Nicholls (Ed). *The Whole world of music: a Henry Cowell symposium*. New York: Harwood Academic Publishers.

Patterson, M. W. (2015). She Always Said, "I Heard an Indian Drum". In M. W. Patterson and J. M. Jensen (Eds). *Travels with Frances Densmore*. Lincoln: University of Nebraska.

Pianola Institute, The (2019). *History of the Pianola*. [Online]. Available at: www.pianola.org/history/history.cfm [Accessed 6 June 2019].

Platte, N. (2017). A Start is Born: Max Steiner in the Studios. In M. Mera, R. Sadoff and B. Winters (Eds). *The Routledge Companion to Screen Music and Sound*. New York: Routledge.

RIAA, n.d. *Gold and Platinum*. [Online]. Available at: www.riaa.com/gold-platinum/?tab_active=default-award&se=wendy+carlos&col=certification_date&ord=asc#search_section [Accessed 5 June 2019].

Rodgers, T. (2015). Tinkering with Cultural Memory: Gender and the Politics of Synthesizer Historiography. *Feminist Media Histories*, 1(4), 5–30.

Sachs, J. (2012). *Henry Cowell*. Oxford: Oxford Univresity Press.

Schmidt, B. (2017). *GameSoundCon Game Audio Industry Survey*. [Online]. Available at: www.gamesoundcon.com/single-post/2017/10/02/GameSoundCon-Game-Audio-Industry-Survey-2017 [Accessed 6 June 2019].

Solum, S. (2004). *The Craft of Controlling Sound*. Minneapolis: Sound Lab Learning.

Tenney, J. (2008). *Complete Studies for Player Piano [CD Book]*. Otherminds.

Weingarten, C. R. (2010). *Public Enemy's It Takes a Nation of Millions to Hold Us Back*. New York: The Continuum International Publishing Group.

White, R. (2001). *BBC Radiophonic Workshop: An Engineering Perspective*. [Online] Available at: https://whitefiles.org/rwx/rws2.pdf [Accessed 2 June 2019].

Williams, H. (2017). *The Woman Who Could 'Draw' Music*. [Online]. Available at: www.bbc.com/culture/story/20170522-daphne-oram-pioneered-electronic-music [Accessed 4 June 2019].

Winner, J. (2008). The World of Sound: A Division of Raymod Scott Enterprises. In Miller, Paul D. (Ed). *Sound Unbound*. Cambridge: MIT Press.

4

Creating Authentic Learning Experiences Using Max

Mark Bassett

Introduction

A solid understanding of foundational audio engineering concepts is the cornerstone of successfully operating almost all audio engineering technologies and a crucial pedagogical goal for students to attain (Keyes 2013). To aid in understanding the audio systems they work with, audio engineering students are encouraged to create "mental models" of the signal path and other foundational concepts, thereby "making inferences about a system or problem which the user builds as he or she interacts with and learns about the system" (Borgman 1986: 48). Students who fail to develop accurate mental models and subsequent understanding risk learning audio technology functions based on repetition, yielding nontransferable and fundamentally inaccurate conceptual models that serve to inhibit rather than foster transferability (Keyes 2013). This is supported by Davis and Clark (2015: 1) who suggest that "students in audio or music production programmes often have little or no knowledge of the intrinsic technical or mathematical theories" underpinning the most common types of signal processing. Students may learn how to operate the system (memorize procedures) but may not understand how the system functions and thus will struggle to transfer their knowledge to other audio systems and solve unfamiliar problems.

The pervasiveness of digital technologies within the audio engineering industry has seen educators begin to use digital audio workstations (DAWs) to teach foundational audio engineering concepts such as pre- vs post-fader sends, in addition to teaching students procedures within the software to achieve the desired outcome (e.g. recording and mixing). Employers and students put tremendous pressure on education institutions to teach using only industry-standard software that students will use after they graduate. Hill (2006) warns against this understandable but myopic approach, arguing that DAWs are not suitable as teaching tools as their functions are virtual, seldom transparent, and obscure the underlying signal processing. Hill contends that this lack of transparency leads to significant issues for students when they encounter a novel situation or unfamiliar system as they do not have the "background and experience professionals have developed over the years that allow [sic] them to visualize underlying structures and operations of recording systems" (2006: 1). Hill urges education institutions not to use Pro Tools in foundational classes, for example, arguing that such practices inhibit students' performance and adaptability and produce audio engineers who learn by rote. Engaging students in active learning activities using Max, on the other hand, requires students to apply fundamental skills used by professional audio engineers, including constant problem solving and self-directed learning (Fick 2015).

For more than a decade I have been using Max software as an alternative to DAWs when teaching students foundational audio engineering concepts, as it is interactive, adaptive, and facilitates multiple modes of learning and interaction (Bassett 2016). Teaching audio engineering concepts with Max harnesses the advantages of digital technologies without many of the drawbacks that are associated with teaching using a DAW. For example, unlike many DAWs, Max provides transparency of controller data and audio signal processing, and facilitates authentic and meaningful learning tasks through contextualized learning.

Max is also ideally suited to project-based learning as exemplified by Fick (2015) through his interactive approach to teaching synthesis design. Creating learning/teaching activities with Max allows educators to easily and quickly construct "intentionally ill-structured problems" (Webber 2018) with more complex Max-based problems eliciting multiple solutions, further facilitating project-based learning and helping students to develop accurate mental models. Fick suggests that interactive project-based teaching methods that rely on building patches in Max can also remove the "negative associations with scientific material in the early stages of [the students'] degree program" (2015: 2).

However, challenging the near-ubiquitous use of industry-relevant DAWs to teach foundational audio engineering concepts requires educators "to re-think traditional teaching methods, and shift focus from 'the gear' to 'the process'—targeting adaptability, creativity and transferable skills more explicitly" (Webber 2018: 2). Additionally, educators considering using visual programming environments to develop learning/teaching tools are encouraged to subscribe to the "just in time" approach to teaching (as demonstrated in the Max examples that follow) where information is provided to students when relevant and at a time when they can immediately apply it to the problem at hand—as opposed to providing students with a range of information "just in case" they might need it (Webber 2018: 1).

The difference in the breadth and depth of learning (based on the author's experience) that results from students developing their own mixing interface in Max, for example, then using the very interface they created to mix a track, versus simply providing access to a DAW, is significant. In the former, students create their own accurate, contextualized, and transferable understanding of the signal path as they literally built the mixing interface they are using. The predictable rebuttal that Max should not be used to teach audio concepts as it is not used by professional audio engineers misses the point entirely. The goal of teaching foundational audio engineering concepts and non-procedural skills using Max, as opposed to a DAW, is to produce "engineers who understand fundamental operating models on which all recording systems are based" (Hill 2006: 24).

I was required to learn Max early in my research career and have since found it to be an excellent choice for educators seeking to develop the type of learning/teaching activities outlined in this chapter. However, it must be noted that it is the learning concepts that Max facilitates that are of interest here, not necessarily the software itself and there are other software packages that can facilitate the development of similar learning/teaching activities—see Davis and Clark (2015) for a review and comparison of available options for teaching signal processing to music technology and production students.

This chapter aims to provide an overview of the use of Max software as a foundational learning/teaching tool in audio engineering education. However, before looking at examples of how Max can be used in this manner, it is prudent to discuss the contemporary learning environment in which Max is used.

Thinking About Teaching

Over the past 20 years or so, in an effort to improve teaching and learning, higher education institutions have begun to shift their focus from the teacher to the learner. Intrinsic to this shift is defining the learning outcomes that students are to attain (Biggs & Tang 2011). This practice, referred to in the field as Outcomes-Based Teaching and Learning (OBTL), states the general institutional, course, and program outcomes in these terms. It is, at its heart, a student-centered approach to teaching and learning that relies on criterion-referenced assessment as opposed to norm-referenced assessment. A program's list of intended learning outcomes (ILOs) serves to inform the teacher and the student how well (or if) the student has learned to do what was intended (Biggs & Tang 2011). The difference between OBTL and the old-fashioned notion of transmission of information or list of items to "cover", is highlighted by examining the three levels of thinking about teaching illustrated in Table 4.1: what the student is, what the teacher does, and what the student does (Biggs & Tang 2011).

The first level of thinking about teaching is entirely focused on what the student is and its advocates believe in the inherent differences in students' ability and commitment to learning. These teachers see themselves as the knowledge gatekeepers

Table 4.1 The three levels of thinking about teaching (Biggs & Tang 2011)

	Focus	*Teacher's role*
Level 1	What students are	Present information
Level 2	What teachers do	Explain concepts and principles, and present information
Level 3	What students do	1. Specify what they intend students to be able to do after they have taught a specific topic 2. Stipulate the level of understanding that students are to achieve 3. Adapt learning/teaching activities to help students achieve those levels of understanding

whose responsibility it is to first know their content well and, second, to transmit that content knowledge to students. Their version of teaching (transmission of knowledge) is held constant and any subsequent variation in student success is the result of each student's individual background, intelligence, motivation, and so on, for which the teacher cannot be held responsible. If each item in the curriculum (list of topics) has been "covered", their job is done. Biggs and Tang (2011) suggest that this belief of what teaching is, feeds an attitude that teaching is a selective activity, not an educative one—it supports the erroneous idea that the goal of teaching is not to ensure that each student learns, but to identify and rank students who are good versus bad learners.

This "blame the student" approach, which tries to explain the differences in student learning and achievement by pointing to the differences in students themselves, requires no reflection on the part of the teacher. When a student does not learn (when teaching has not been effective), the explanation lies in the student and what they personally did or did not bring to the classroom. Instead of reflecting on their own practice to examine how they could better teach for the diverse range of learners, these teachers instead reason away the performance of students who, if they only tried harder, concentrated more, attended every class, submitted on time, etc., would have done better; again, citing their checked content list as evidence that they have most definitely "covered it all" in class. Critiques of this level of thinking about learning may lead one to believe that the focus should instead be on what the teacher does, but this approach to teaching, while more effective, is also lacking.

Focusing on what the teacher does appears at face value, at least, to address the issues detailed above. This approach focuses the spotlight on the teacher and encourages them to examine and refine their own craft to improve what they do in the classroom (Biggs & Tang 2011). Teachers subscribing to this view see their role as not just transmitting information, but also transmitting ideas and comprehension (Biggs & Tang 2011). This approach to teaching, while a significant improvement over the previous level, still frames the primary duty of the teacher as that of a "transmitter". When presented with a student who has not learned effectively, teachers evidence their teaching skill with (well-prepared) lists of things that *they* did in the classroom. Teacher evaluation tools that look at how well a teacher uses technology in the classroom, their knowledge of content, and their approach to pedagogy, for example, perpetuate the idea that effective learning will take place if only the teacher gets better at what they do—the focus is on what the teacher did, not on whether students learned effectively.

The third level of thinking about teaching does not focus on what the student is, or what the teacher does, but on what the student does. In this sense, it is truly a student-centered approach to teaching and learning as it focuses on student learning outcomes (Biggs & Tang 2011). This approach is interested in what students are to learn, as opposed to what content is to be taught, or what the teacher needs to do, for students to meet the intended learning outcomes (Shuell 1986). It requires educators to think about how students will demonstrate learning, which has obvious and direct implications for assessment design.

Finally, and most importantly, educators must decide on learning/teaching activities, that is, what the students must *do* before, during, and after class to achieve the desired level of understanding. As this approach focuses on what the student does, it naturally draws us to discussing the Intended Learning Outcomes (ILOs), which must be designed so that they describe what it is that students should be able to do and the level of understanding that is required at the end of the teaching period. This approach is based on the constructivist theory of knowing; the theory that learners actively construct their own contextualized knowledge (rather than acquiring it) based on their own activity (Biggs & Tang 2011).

Overview of Max

Cycling '74's Max software (previously titled Max/MSP) is a high-level, dynamic, real-time visual programming environment. The dual-platform software, which features an interactive, adaptive visual interface, has proven popular at education institutions as a media technology teaching tool, as well as an advanced research tool; in 2016, Max was being used by over 1,500 institutions and companies worldwide (Cycling '74 2016). Users create patches (programs) in Max by connecting high-level objects[1]—the basic logic elements in Max—small programs that perform tasks within the larger program—together using patch cords in the visual interface.

While it is possible to delve into the traditional programming (code) side of Max, as it is a high-level programming environment, users can create complex, professional programs such as the author's Parametric Equaliser Training application (Bassett 2019; Bassett & Martens 2016), without writing a single line of code. In addition to the numerous high-level objects that perform a variety of signal processing functions, Max also includes many user interface objects that process data, audio, or both, including faders, equalizers, sonograms, spectrograms, and a wide variety of meters.

Several objects in Max are specifically designed for audio engineering applications, including [atodb] and its partner [dbtoa], which convert linear amplitude values to decibel values and vice versa. As expected, Max contains mathematical operators such as [+], [*], and [/], however, the software also includes signal processing versions of these objects, such as [+~] and [*~] which can be used to sum and multiply two audio signals together respectively. These objects enable users to rapidly prototype synthesizer designs, for example, and hear the results of their design decisions in real time.

Max in Audio Engineering Education

Advantages of Max

There are several reasons why educators may favor Max as a learning/teaching tool within audio engineering education. The ability to create complex programs without the requirement to learn a programming language is a standout of the software, and patching (programming) in Max is fast and intuitive, enabling users to test ideas quickly by rapidly prototyping complex programs (Davis & Clark 2015). Patching in Max can be done in real time as there is no requirement to pause audio signal processing while patching—connections can be made, interrupted, and broken while audio processing is running with relatively minimal disruption to audio outputs. Perhaps the software's greatest advantages are its visualization of the signal flow and logical, electrical circuit-like design philosophy that makes it approachable for inexperienced users (Davis & Clark 2015).

It is common for audio education institutions to limit students' options to reconfigure audio systems to avoid adversely impacting the next student that uses the system (Eidsheim 2009). Max provides students with the freedom to not only build their own audio system, but also to reconfigure, rebuild, and refine it as they progress. Visual programming environments such as Max provide opportunities for educators to move beyond didactic to more experiential and situational modes of learning through the development of virtual technology simulators, for example, allowing instructors to place greater emphasis on being designers of the learning environment (Eidsheim 2009).

In the following examples, Max is not deployed to aid in the coverage of topics, as "covering topics" is hopefully not an intended learning outcome (ILO) of the class, but used to help educators design interactive learning/teaching activities that enable students to meet the ILOs. Max facilitates the third approach to teaching that focuses on what the student does, encouraging the teacher to design appropriate activities that will allow the student to achieve the desired level of understanding. This also serves as a reminder to educators that when writing ILOs, they must carefully consider what it is that the student should be able to do after teaching has taken place (Biggs & Tang 2011).

Teaching with Max I: A Basic Fader

This section presents a basic example of using Max to create learning/teaching activities through the construction of a series of small programs, each of which builds in complexity as the student progresses. For the example provided, it is assumed that mono and stereo audio signals and the concepts that traditionally precede these have been appropriately taught (learned by students) prior to this class.

Figure 4.1 shows the initial Max patch that is provided to students in which a mono white noise signal is split and sent to a left and a right meter, and to the DAC/Main Outs. The first challenge for students is to add level control to this patch using two objects, a signal multiplier and a floating-point number object, then add post-gain metering. Students are encouraged to consult the comprehensive interactive help files included in Max if they do not understand how to use an object.

Having solved the first problem, students soon discover that the number object is cumbersome to interact with, and when presented with the question "How can we improve the usability of this program?" the addition of a dial or slider to control

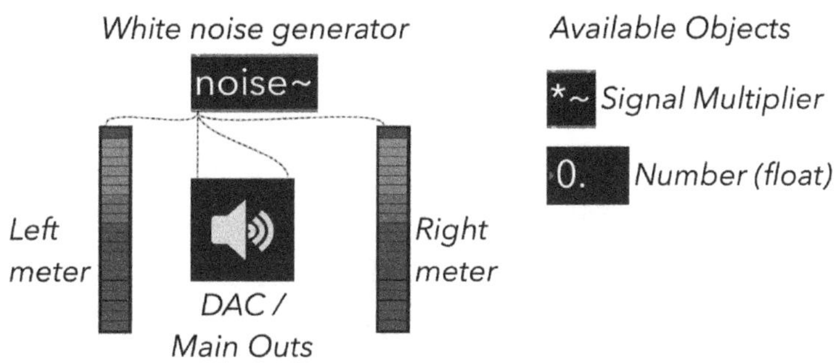

Figure 4.1 The first Max patch.

the gain value is a common reply. Upon implementing a slider, students quickly discover that the slider is practically unusable as its values range from 0–127 by default and this causes the signal to distort as multiplying the amplitude by a factor of 80, for example, is equivalent to increasing the level by approximately 38dB.

Students are asked to think of a suitable alternative range of values to multiply the signal by, keeping in mind that when the fader is at the bottom position, there should be no output. A question to help students arrive at the appropriate minimum value might be "What value must we multiply the amplitude of a signal by so that it is zero?" The answer is in the question of course: multiply a number by zero and you will get zero. As for the maximum value, do we want to double the amplitude at the top fader position? Triple it? As students can always go back and modify any object's parameters it is not vital at this stage to make a final decision on this, I chose a doubling of amplitude—a maximum value of 2.

Having solved the scaling problem (Figure 4.2), students are reminded of audio engineering conventions, one of which is to display the value of the fader position in decibels, not linear amplitude values. Students are introduced to the [atodb] object which converts linear amplitude values to decibel values and the next challenge is presented.

Appropriately implemented, the [atodb] object introduces practical usability to the program. Students can now adjust the level of the signal by a specific decibel amount, and are instructed to locate the 0dB (unity gain) point, which is halfway up the slider due to the linear scale. Students are again reminded of professional audio conventions and the position on the 0dB point on professional audio equipment. This introduces the next problem: to scale the slider's values exponentially by adding a fifth argument to the [scale] object so that the unity gain point is placed appropriately.

With the 0dB point on the slider now appropriately placed due to the exponential scaling, it is time to add an additional fader to provide independent control over the left and right channels. Students are encouraged to duplicate the objects available on screen as needed to achieve this task. The final step prior to creating a more user-friendly interface is to replace the white noise generator with a [playlist~] object, which will allow students to playback stereo audio files (Figure 4.3).

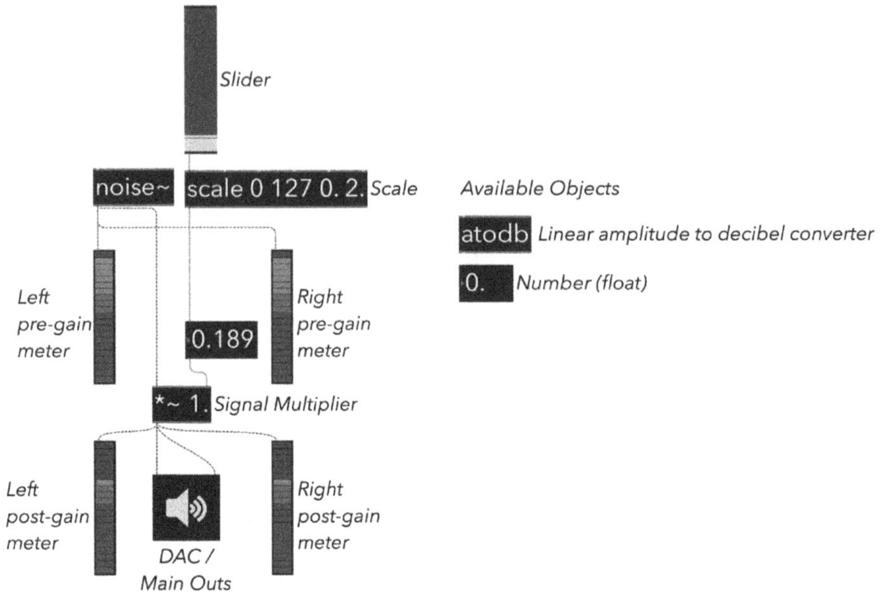

Figure 4.2 A slider has been implemented to control the gain value.

The patch, while functional, is not particularly user-friendly and the internal workings are unnecessarily visible to the user. Max features a "Presentation Mode" that hides all patch cords and only shows the objects that the user selects to be visible. The other primary advantage of Presentation Mode is that objects can be positioned and sized differently to how they appear in the "patching mode" shown in the figures above. Students are encouraged to think about which objects are necessary for the user interface and then add these objects to the presentation.

Teaching Audio with Max II—Mid-Side Processing

Students can use the patch shown in Figure 4.3 to change the level of the left and right channels of a stereo audio file; however, this does not have any real practical use. The following example demonstrates how Max can be used to teach students about mid-side processing through the construction of a series of small programs, each of which builds in complexity as the student progresses. The practical result is a stereo/mid/side switcher interface with metering and gain offset. The pedagogical result is that students will gain a meaningful, contextualized understanding of how mid-side processing works.

Prior to engaging with Max, students are provided with a theoretical overview of mid-side processing in class. The concept of *Mid=L+R, Side=L-R* is introduced; however, examples of the sum (mid) and difference (side) signals are deliberately *not* played to students in class. Students are left to discover for themselves what these signals sound like as they progress through the construction of their program and

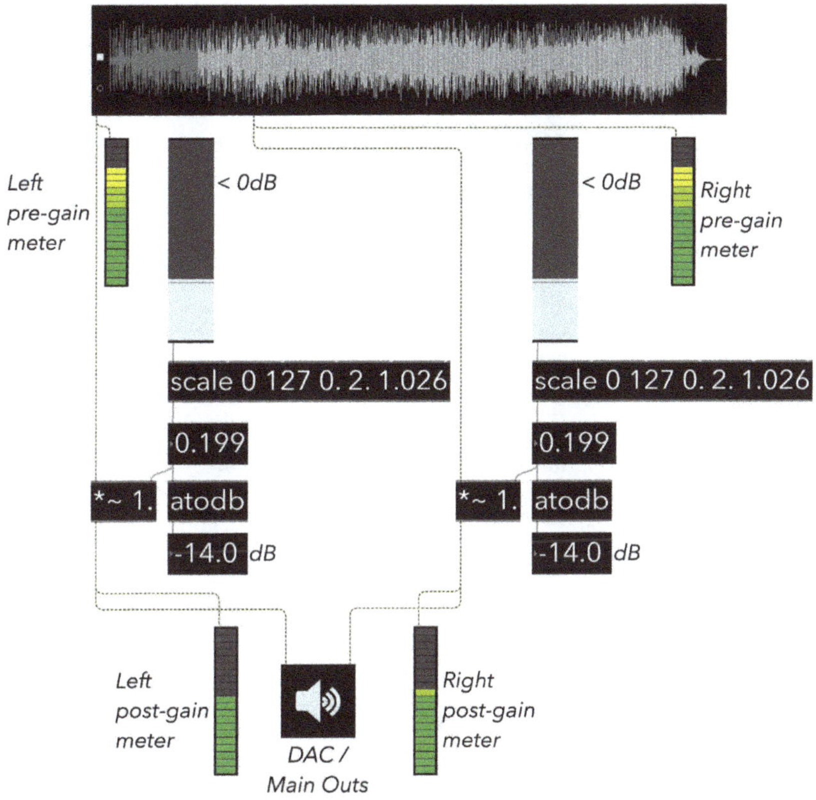

Figure 4.3 Once the patching is complete, the design of the interface can take place.

ponder what the benefits might be of processing these components of a stereo signal. Likewise, the information that can be learned about the processing that has been applied to a stereo file by listening to the mid and side signals is also not provided to students. Following the short in-class instruction on the theory of mid-side processing, students then engage with Max. A simple patch is provided that demonstrates how to meter and monitor the left channel of a stereo audio file (Figure 4.4). Signal sends and returns ([send~] and [receive~]) are also introduced at this early stage.

Students are instructed to modify the patch by duplicating and modifying objects as required so there is metering and monitoring of both the left and right channels. Having arrived at the solution, students are instructed to sum the left and right signals together using [+~] and to monitor only the output of this summing.

Authentic Discovery and Meaning

One of the advantages of students constructing their own signal processors in Max is that they will naturally encounter problems in an organic manner along

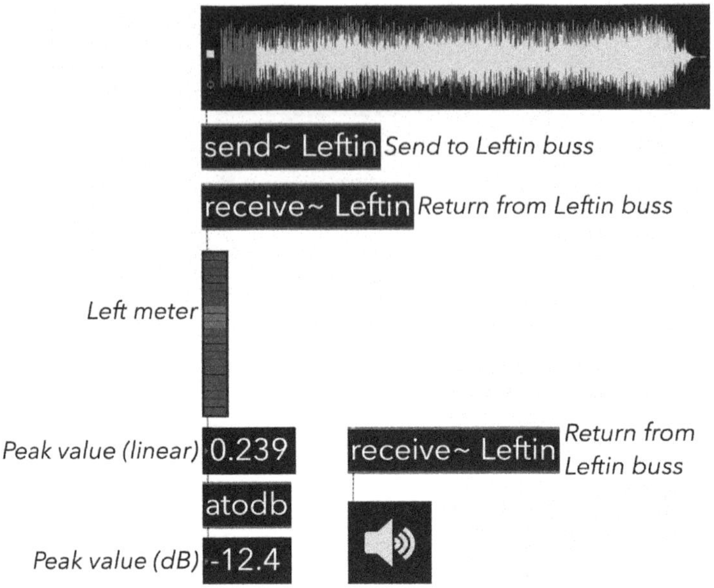

Figure 4.4 The starting mid-side patch.

the way. In this example, some students will have noticed that the summed signal has a higher level than the left and right signals, so high in fact, that it may clip. A question can then be put to the class: "What would happen if we switched from monitoring the stereo signal to monitoring the mid signal?" The obvious answer is that we may well hear an increase in level, but the more interesting questions are "Why is this the case?" and "How should we correct for this?" Giving students the chance to discover for themselves why summing the left and right signals can lead to an increase in level is an example of an authentic learning experience.

The features of DAWs are refined and intentionally limited to produce an efficient recording/mixing environment that allows users to do exactly what the software was designed to do, but little else. Max, on the other hand, is inherently purposeless; it is a program "that does nothing" (Zicarelli 2002: 1). A new patcher in Max presents users with a blank screen that forces them to ask "What do I want to create?" DAWs are not designed to teach students about audio engineering and neither is Max. However, as Max is not designed to do *anything* per se, but rather to be completely open-ended and allow *the user* to do anything, it affords students the opportunity to naturally create and solve audio-based problems in a near-endless number of ways, encouraging creative problem solving, while fostering a deep, contextualized understanding of concepts and principles.

In this way, Max affords students agency in their education and facilitates self-initiated learning that is driven by their interests. It allows students to make "bad" decisions and supports learning through intentionally ill-structured problems. The basic alternative is to simply tell students that the mid signal will often be higher in level than the individual left and right signals—a more advanced version is to demonstrate this in a DAW or on a console—but these teaching approaches can lack meaning and links to content (Hill 2006).

Students are then instructed to subtract the right signal from the left using [-~] and monitor only the output of this difference signal. At this point, some students may notice that the level of the side signal is different from the level of the left, right, and mid signals. Having successfully metered and monitored the left, right, mid, and side signals, the [selector~] and [radiogroup] objects are introduced with a new patch shown in Figure 4.5.

This patch allows students to view the sonogram of either pink or white noise, by clicking on the respective button of the [radiogroup] object to choose between the two signal inputs of the [selector~]. Students are then instructed to duplicate and modify the objects within this patch (excluding the sonogram) to create a program that allows the user to select between off, stereo, mid, or side signals and monitor the output of this selection. A solution to the problem is shown in Figure 4.6.

The program is functional, but the mid channel is clipping at times during playback. The class is instructed that the decision has been made to allow the user to modify the level of the mid and side signals using the [*~] signal multiplier object. Students can either place the [*~] object before or after the [+~] and [-~] objects, but only one of these solutions is correct. As clipping takes place within the [+~] object, the signal must be lowered in level *before* it gets to this processor. Turning down the output of the [+~] object will remove clipping from the meter, however, this has simply reduced the level of a signal that has already clipped. As such, the [*~] object must be inserted before the [+~] object, lowering the levels of the left and right signals before they are summed. One way for students to discover if they have placed the [*~] objects in the correct place in the signal path is to encourage

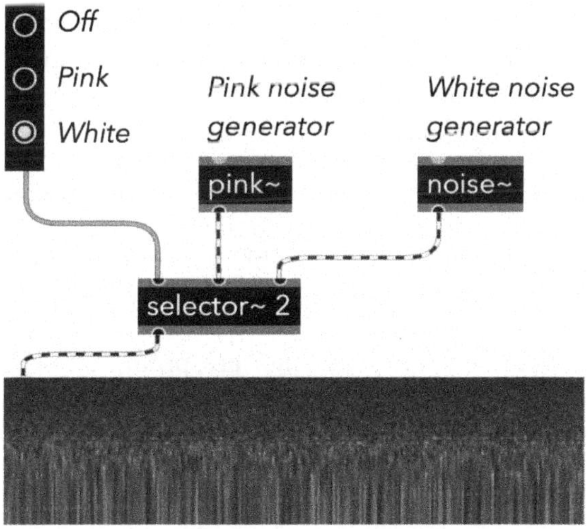

Figure 4.5 The [selector~] object in combination with the [radiogroup] object.

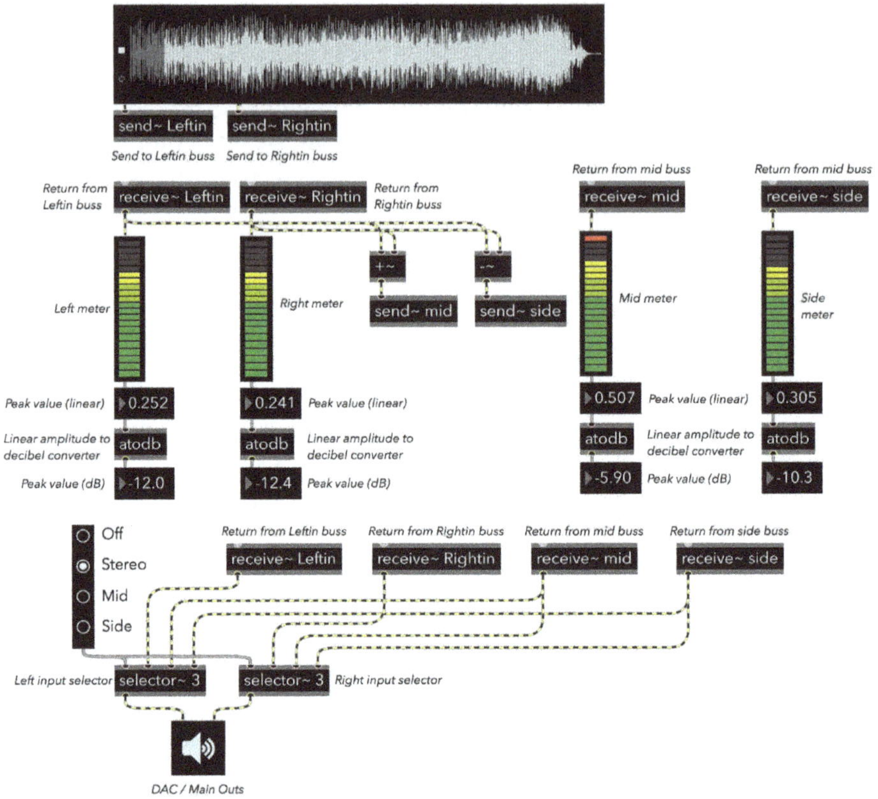

Figure 4.6 The [selector~] object facilitates switching between audio signals.

them to probe the level of the signals passing through the patch cords using Max's inbuilt Signal Probe, or by placing a meter before and after each signal processing object.

At this stage, the patch functions as intended as a stereo/mid/side switcher, with level control for the mid and side signals. However, the interface is not very visually appealing, some would say it is confusing, and the inner workings of the program are clearly and unnecessarily visible to the user. The program is simply not intuitive to use from a design perspective. This leads us to the interface design part of the class.

As with the previous example, students are prompted to think about which objects are necessary for the user interface and to add these objects to the presentation. Students are encouraged to explore the attributes of the objects in their interface, including background color and transparency, and to also add colored background panels using the [panel] object. Additional VU-style meters (with accurate ballistics) are also available in Max, as shown in the user interface in Figure 4.7.

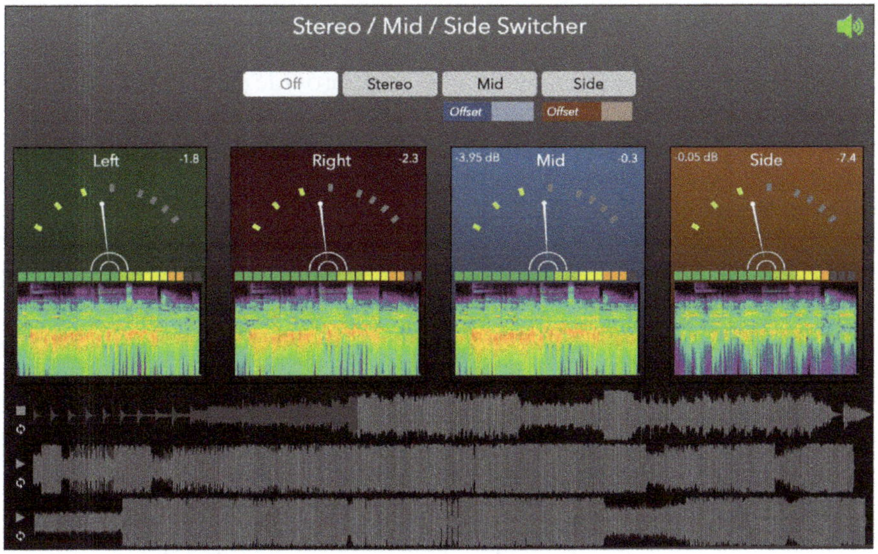

Figure 4.7 A simple stereo/mid/side switcher user interface.

The patch is a useful, fully functional stereo-mid-side switcher with VU and dBFS meters, spectrograms for each channel, and gain offset sliders for the mid and side channels. This program can be exported as a stand-alone OSX or Windows application that the end user can run without the need to own a copy of Max.

In the preceding example, students literally created the clipping phenomenon/problem in Max and because of this, their learning has *meaning*, and, as such, they develop "a much higher cognitive level of awareness and application" (Hill 2006: 7). In my experience, this approach serves to engage and motivate students more as they are not being simply told about a concept followed by an example, but thinking about how the phenomenon came about and whether it is a problem that needs to be addressed, and if so, how to solve it. This type of discovery learning, whereby students discover relationships between signal processors by experimenting and manipulating objects "results in deeper rooting of knowledge, enhanced transfer, the acquisition of regulatory skills, and thus better motivation" (Graf & Kinshuk in Eidsheim 2009: 13). Students are motivated as they are engaged in meaningful and worthwhile learning activities (Biggs & Tang 2011).

Conclusion

In classes where the goal is for students to learn foundational audio concepts and understand how an audio system works, there is a risk that by learning with a DAW, students will memorize procedures based on repetition as opposed to developing accurate mental models of underlying processes, as it is possible for students to memorize what controls to modify in what order to achieve an outcome, without understanding exactly what those controls do and how their functions interrelate.

Visual programming environments like Max provide an alternative to DAWs when used as learning/teaching tools in audio engineering education and offer distinct advantages, including facilitating a student-centered approach to teaching and fostering authentic and meaningful learning experiences. The transparency of controller data and audio signal processing in Max allows students to "get under the hood" of a mixer interface, for example, where they can explore, break apart, and reconfigure the signal flow. The almost endless freedom to modify signal processing within a Max patch encourages students to pursue their curiosities and allows educators to respond and adapt to students' learning needs by modifying the learning environment.

The very nature of the "build it yourself" ethos that Max inspires leads to students encountering problems in an authentic manner and forces them to visualize the signal path and experiment to discover solutions. This aligns with the third level of thinking about teaching as students are actively constructing contextualized knowledge based on their own activity.

Practical lab time is often limited and it can be argued that the more time students spend using Max, the less time they have to learn technologies that they use in the industry. However, Max is not proposed as a substitute to DAWs for recording and mixing, for example. Rather, Max is proposed as an alternative to DAWs for teaching foundational audio engineering concepts, which DAWs, as just argued, are inherently ill-suited to.

Note

1. Not to be confused with Object-Orientated Programming (OOP).

References

Bassett, M. (2016) *Max as an Interactive, Multi-Modal Learning and Teaching Tool for Audio Engineering*. Paper presented at the Audio Engineering Society 141st Convention, Los Angeles, CA.

Bassett, M. (2019) *SAE Parametric Equaliser Training* [Online]. Available from: https://apps.apple.com/au/app/sae-parametric-equaliser-training/id1132922597 [Accessed: 1 August 2019].

Bassett, M. & Martens, W.L. (2016) *SAE Parametric Equaliser Training: Development of a Technical Ear Training Program Using Max*. Paper presented at the Audio Engineering Society 141st Convention, Los Angeles, CA.

Biggs, J. & Tang, C. (2011) *Teaching for Quality Learning at University: What the Student Does* (4th ed.). Berkshire, England: Open University Press.

Borgman, C.L. (1986) The User's Mental Model of an Information Retrieval System: An Experiment on a Prototype Online Catalog. *International Journal of Man-Machine Studies*, *24*, 47–64.

Cycling '74 (2016) *Max in Education* [Online]. Available from: https://cycling74.com/ [Accessed: 1 February 2019].

Davis, R.E. & Clark, D.F. (2015) *A Comparison of Available Options for Teaching Signal Processing to Music Technology and Production Students*. Paper presented at the 26th UK Audio Engineering Society Conference on Audio Education, Glasgow, Scotland.

Eidsheim, N.S. (2009) Desktop Simulation: Towards a New Strategy for Arts Technology Education. *Journal for Learning through the Arts*, *5*(1).

Fick, J. (2015) *From Creativity to Science and Back Again: Supporting Audio Students Through Active Teaching Approaches*. Paper presented at the Audio Engineering Society 139th Convention, New York, NY.

Hill, B.R. (2006) Teaching with Pro Tools? Proceed with Caution! The Development of Mental Models for Recording Engineering Instruction. *The Journal of the Music & Entertainment Industry Educators Association*, *6*(1).

Keyes, C.J. (2013) *Alternate Software and Pedagogical Strategies for Teaching Audio Technology to Classically Trained Music Students*. Paper presented at the Audio Engineering Society 134th Convention, Rome, Italy.

Shuell, T.J. (1986) Cognitive Conceptions of Learning. *Review of Educational Research*, *56*(4).

Webber, C. (2018) *Master of All Trades, Jack of None* [Online]. *Navitas Insights*. Available from: https://insights.navitas.com/master-of-all-trades-jack-of-none/ [Accessed: 12 January 2019].

Zicarelli, D. (2002) *How I Learned to Love a Program That Does Nothing. Computer Music Journal*, *26*(4) [Online]. Available from: www.mitpressjournals.org/doi/10.1162/014892602320991365 [Accessed: 4 August 2019].

5

Gamifying the Learning Experience

Evaluating Teaching Methodologies in an Introductory Module to Sound for Film and Television

Mariana Lopez

Introduction

The Science of Sound and Cinematography is a first-year module within the three-year BSc in Film and Television Production at the Department of Theatre, Film, Television and Interactive Media at the University of York, UK. The module runs for a total of 13 weeks and its content is divided into "sound" and "cinematography". As part of their instruction on the topic of sound, students have a weekly one-hour lecture for 13 weeks and have five two-hour practical sessions. The module has been running as part of the degree program since 2010. Historically, students have reported through feedback forms, supervision meetings, and informal discussions finding the module challenging and feeling disengaged.

Furthermore, the module is the only one in the whole degree that is assessed through an exam, causing anxiety among students, who often express concerns even before the module has started. Such *test anxiety* can be defined as the perception of the assessment (an exam, for instance) as threatening (Putwain 2008). Research in the field has linked the level of anxiety experienced to a number of factors, including fear of failure, competence beliefs in relation to the subject of the test as well as test-taking skills, performance goals in relation to peers, mastery goals, and the messages communicated by family members and teachers (Putwain, Woods and Symes 2010).

Upon joining the academic department as a lecturer, I was keen to challenge this conception and anxiety regarding the module. I wished to emphasize the practical applications of its theoretical content as well as creating a unique experience during both lectures and practical sessions through gamification methods. In the UK, lectures are seen traditionally as less interactive than workshops and seminars, with lectures often consisting in the tutor presenting material to the students, and the workshops/seminars being occasions for teamwork and greater levels of interaction with the tutor and among peers.[1]

My goal was to reduce this distinction between session types and consider instead both lectures and workshops holistically, as opportunities for learning and engagement. Furthermore, the change was also prompted by the shifting student perception regarding lectures due to the introduction of lecture capturing at the university. Lecture capturing in this context refers to the availability of voice recordings of the lecture accompanied by the visual output of the projector screen immediately after every session. Such availability resulted in students informally commenting on no longer needing to go to the sessions, they could just experience them at home. Changes in technology and attitude presented an opportunity to provide students with an experience that required their active participation, motivating them to attend and, as a result, build a more cooperative learning environment. Making an introductory module on sound more engaging within the context of a film and television production degree could open up the possibility of more students appreciating its importance regardless of whether they wish to pursue sound as a specialization later.

This chapter explores the use of gamification strategies on the sound-focused part of the module in the academic year 2017–2018 with the aim of increasing student engagement and grounding the revision in real-world examples, following Jacobs' (2005) suggestion on making technical and scientific concepts accessible to students. The UK term "revision" is used to refer to activities that are carried out to solidify the understanding of content and prepare for an assessment instance, such as an exam. The first section explores the concept of gamification and its connection to motivation in teaching and learning. I then discuss the specific strategies applied in *The Science of Sound and Cinematography*, before exploring the results of a survey conducted amongst students to assess the effectiveness of the methodologies applied.

The Concept of Gamification in an Educational Context

In this chapter the term "gamification" refers to the incorporation of game elements to activities seldom associated with "play", in this case, a university setting in general and, more specifically, the lecture theater and practical sessions (Deterding et al. 2011). Although gamification is often associated with the use of technology, and literature seems to focus mostly on its incorporation to e-learning, as well as its combination with distance learning (Elias 2011; Jacobs 2005), this does not always need to be the case (Kapp 2012), and gamification strategies can be applied to a wide variety of contexts. Furthermore, the idea of using video games for teaching has been regarded as a way of joining educational content/context together with the motivation provided by games, making students feel interested in the subject matter and actively seek answers to questions (Prensky 2003).

As detailed later, gamification and "serious gaming" are connected but yet distinct. Landers (2014) explains that serious gaming is focused on communicating learning through its content. It involves the creation of a fully fledged game and often features simulations. An example of a serious game is the virtual environment *vHealthCare* developed by BreakAway Games,[2] which allows professionals in the healthcare sector to hone their skills and learn new techniques in relation to diagnosis and treatment recommendation. On the other hand, the process of gamification is focused instead on changing students' attitudes toward learning content by encouraging them to feel more engaged with the topic (Landers 2014).

Gamified experiences are not necessarily built like games but instead modify preexisting activities by borrowing elements in games. For example, a class focused on revising module content, involving questions and answers between lecturer and students, could be modified by dividing students into teams and awarding them points for every correct answer. When referring to gamified learning, Landers (2014) mentions that these activities tend to include elements that can be classified in relation to the attributes that are gamified; such classification can be seen in Table 5.1.

Successful gamification practices in education are also associated with four main elements of game design: (1) *freedom to fail*, (2) *rapid feedback*, (3) *progression*, and (4) *storytelling* (Kapp 2012; Stott and Neustaedter 2013). *Freedom to fail* refers to a design strategy by which a mistake does not result in a conclusively negative outcome, there are opportunities for trial and error, and several attempts are needed to master a technique. For example, in a game, players might have several lives or go back to a previous checkpoint in the storyline. This is also linked to the use in universities of formative assessments in which students are required to submit a piece of work and are given feedback and often a score, but the result does not count toward the mark given in that module or their final degree classification (Stott and Neustaedter 2013).[3] Formative assignments are there to help students understand their strengths and weaknesses, and so help develop their work further. Games and educational settings share this technique, but it is evident that asking students to write a technical report and not giving a mark does not turn the experience into a game.

Although *feedback* is likely to be a key component in all formal educational settings, game design techniques such as progress bars and leader boards can help

Table 5.1 Gamified attributes according to Landers (2014)

Action language	Refers to the interface used by the participant, e.g. the use of a controller in class
Assessment	The use of point systems
Conflict/challenge	Includes tasks such as issuing questions that need to be solved in competition with other teams so as to earn points
Control	It is the possibility of the students/players steering the game with their actions, such as by choosing the category of the next question in a quiz
Environment	It refers to, for example, turning the classroom into a simulation of a concert hall by projecting images on the walls
Game fiction	It indicates when, for instance, a revision question is phrased as a quest and its correct answer as the defeat of a mythical monster
Human interaction	This is interaction with other players, whether in person or online
Immersion	Modifying the space in which the activity takes place so as to increase the sense of being part of the experience, similar to *Environment*
Rules and goals	These can help propel learners/players forward and establish what they are aiming for while defining what success looks like

provide this more rapidly and often instantaneously. *Progression* is also key in gaming through, for example, the existence of levels, and can be said to have a parallel in scaffolded learning (Stott and Neustaedter 2013). Scaffolded learning refers to a pedagogical strategy by which the tutor provides students with the support structures (scaffolding) to access new material. As students become more confident, those support structures are gradually removed and students gain more and more control over the learning process (Hogan and Pressley 1997).

Storytelling can also help students remember concepts more effectively, and this can be done through the use of case studies and scenarios that allow them to apply what they have learned. In addition to these four elements, Kapp (2012) also discusses what he refers to as "the interest curve", which is the need to provide events, challenges, and opportunities that keep hold of the player's/learner's attention so that the benefits educators seek from gamified activities are maintained throughout the module.

In addition to these factors, researchers have acknowledged that participants' motivation to participate in games is related to their engagement in three main areas: *cognitive*, *emotional*, and *social* (Dominguez et al. 2013; Lee and Hammer 2011) (see Figure 5.1). The *cognitive* area refers to the area of knowledge the activity is focusing on, what is being learned. The *emotional* aspects relate to what is felt during the activity. Such emotions can include curiosity (wanting to learn more so as to master a level in a game), frustration (clicking on the wrong answer by mistake), joy (being awarded points), optimism ("Next time I'll get this"), pride (receiving a round of applause from classmates and teachers or being at the top of the leader board), and even failure. In games, failure can be seen as an opportunity, participants are encouraged to keep trying and experimenting until they have achieved their aim (freedom to fail). Furthermore, rapid feedback aids engagement

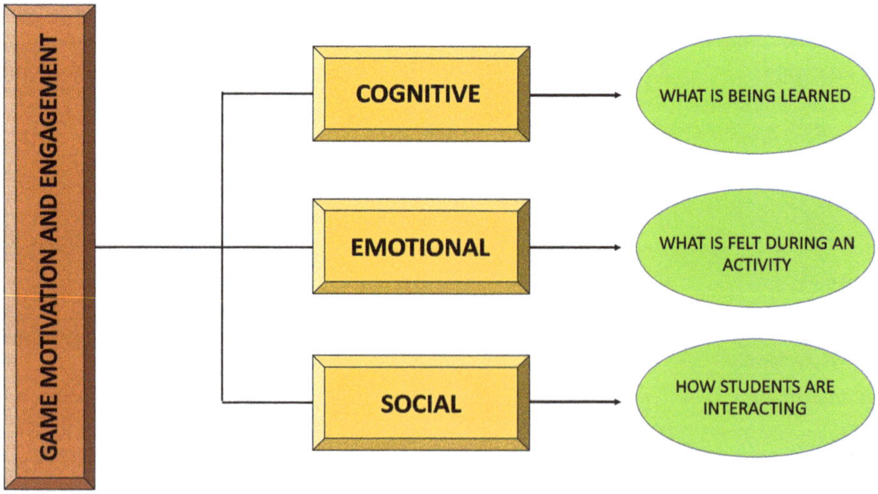

Figure 5.1 Different factors involved in game motivation and engagement.

in activities on an emotional level, as the answer is given at the moment of engagement. The *social* aspect is linked to modes of interaction, which can be *cooperative*, *competitive*, and/or *social*, as well as the fact that by introducing gamified school experiences students are given "permission" to assume a different identity (maybe in a role-playing game) and even see school in a different light (Dominguez et al. 2013; Lee and Hammer 2011). Researchers have shown that the use of gamification can increase interactivity and engagement within the classroom setting (Hitchens and Tulloch 2017; Iosup and Epema 2014) while also encouraging collaboration both in person and through virtual environments (Denis and Jouvelot 2004, 2005; Figueroa-Flores 2016; Knutas et al. 2014), which can be key in creative settings such as those linked to audio engineering and music practices.

The context in which gamified activities take place is crucial to their success (Deterding 2014). Because gamification in education is utilizing game elements in a context that is not often associated with play and playfulness, it is important that students do not feel embarrassed to pursue these activities. Creating a permissive environment is crucial, as is participants' sense of autonomy, to the success of gamified activities (Deterding 2014). As it will be seen in the following section, autonomy is linked to intrinsic motivation (Ryan and Deci 2000), and gamification can be linked to the concepts of both intrinsic and extrinsic motivation.

I was particularly drawn to gamification strategies due to my own experience as an audio student, who struggled to engage with the content and found that the explanations and activities provided by my tutors did not make the content accessible. Furthermore, having taught general study and academic skills at university level (building arguments, giving presentations and referencing, among others), I had used gamification techniques and witnessed how seemingly dry topics could be turned into interactive instances of learning. Hence, I was eager to transfer such experiences to the field of audio.

Types of Motivation and Their Role in Gamification

Most educators, regardless of the teaching methodologies they apply, will share a desire to motivate students to learn about the subject they are teaching. *Self-determination theory* differentiates between intrinsic and extrinsic motivation (Gillet, Vallerand and Lafrenière 2012; Riley 2016; Ryan and Deci 2000; Ryan, Rigby and Przybylski 2006; Vansteenkiste, Lens and Deci 2006). Intrinsic motivation refers to doing something because the activity itself is interesting and enjoyable. For example, a student might read about psychoacoustics because they are curious and want to find out more. Extrinsic motivation refers to doing an activity for the sake of a separate outcome. For instance, the student motivated to read the book on psychoacoustics does so because they want to achieve a high mark in their class. The student's motivation is not linked to a personal connection to or a passion for psychoacoustics. Rather, the student is motivated by the outcome (a high mark) resulting from completing the task. Motivation can be at the same level for both students, but the source of that motivation is different.

Cognitive Evaluation Theory, a sub-theory of self-determination theory, argues that actions such as rewards and feedback can generate feelings of competence, which, when accompanied by a sense of autonomy, can increase intrinsic motivation (Deci and Ryan 2008; Ryan and Deci 2000). Fostering relatedness, that is, a sense of belonging and connectedness by feeling respected by fellow students and teachers, for example, can help the acceptance and internalization of classroom values, and this means that an extrinsic goal can be thought of as one's own (Deci and Ryan 2008; Ryan and Deci 2000). The development of a sense of competence is also crucial as students need to feel that they have the skills to complete the task at hand (Deci and Ryan 2008; Ryan and Deci 2000). This often means setting up tasks that are at an appropriate level and giving relevant and effective feedback.

These concepts are crucial to gamification, because researchers have argued that if gamification is used to increase levels of competence, autonomy, and relatedness then it can indeed help increase intrinsic motivation in the classroom (Barata et al. 2013; de-Marcos et al. 2014). Still, the fact that points systems, badges, and leader boards are extrinsic motivators has also been discussed and thought of as potentially undermining intrinsic motivation (Barata et al. 2013; O'Donovan, Gain and Marais 2013). Yet, this is not necessarily a disadvantage of gamification in education, as intrinsic and extrinsic motivation are in a constant state of flux and can co-exist (Kapp, Blair and Mesch 2014). Students might have thus started reading the book on psychoacoustics with the aim of getting more points in a quiz, but then realized that they enjoyed reading that book and wished to continue to do so. The quiz gave them the initial extrinsic motivation and that was then transformed into intrinsic.

Gamifying *The Science of Sound*

As part of *The Science of Sound and Cinematography* module, I designed and delivered a series of gamified experiences which aimed to increase student engagement, generate a sense of progress, and help students revise the taught content. A game that does not have a clear educational aim is likely to be seen by students as a waste of time, which detracts from the playing and learning outcomes (Landers and Callan 2011). As a result, all gamification instances had the learning outcomes of the module at the center of their design strategy. Moreover, they were designed to encourage different types of interaction among students—*cooperative*, *competitive*, and *social*—(Dominguez et al. 2013; Lee and Hammer 2011; Pirker, Rifnaller-Schiefer and Gütl 2014), due to the desire of appealing to different students'/players' preferences. Furthermore, all of them fell under the attributes of *Assessment* and *Conflict/Challenge* identified by Landers (2014) and discussed in Table 5.1.

The module in question is one in which regular revision of concepts is crucial to students' progress. Every week lectures and workshops build on concepts presented in previous weeks. It is crucial that in every session, students are given the opportunity to go through topics again, ask questions, and reinforce their understanding, before moving forward. As a result, all lesson plans included sections in which I revised previously introduced concepts before presenting new content. The fact that the module is assessed through a final exam that accounts for 100% of the overall grade, means that students have in the past appreciated revision throughout the academic year as well as during the weeks immediately preceding the exam. Most of the gamified activities discussed here were used as a platform to facilitate such instances of revision on theoretical content.

Online Quizzes

The first type of activity I worked with was the use of online interactive quizzes through the free educational platform *Kahoot!*, an activity focused on a *competitive* type of interaction.[4] *Kahoot!* allows the tutor to set a number of multiple-choice questions, whose options are then coded with symbols that appear on the students' mobile devices upon entering the game pin provided by the instructor. The aim is for the students to choose the right answer by making sure they are selecting the symbol that corresponds to the correct answer (see Figure 5.2). In addition to this, the tutor can set the time students have to answer the questions. Students answer the questions individually and the top scorers are featured in a leader board at the end of each question (an example of rapid feedback).

The top scores are then calculated automatically at the end of the quiz and featured in a virtual podium. The presence of the leader board and the virtual podium are of the utmost importance as similar strategies have been found to promote a sense of achievement, foster competition, and create a sense of belonging (O'Donovan, Gain and Marais 2013). It is worth noting that even though I focused on

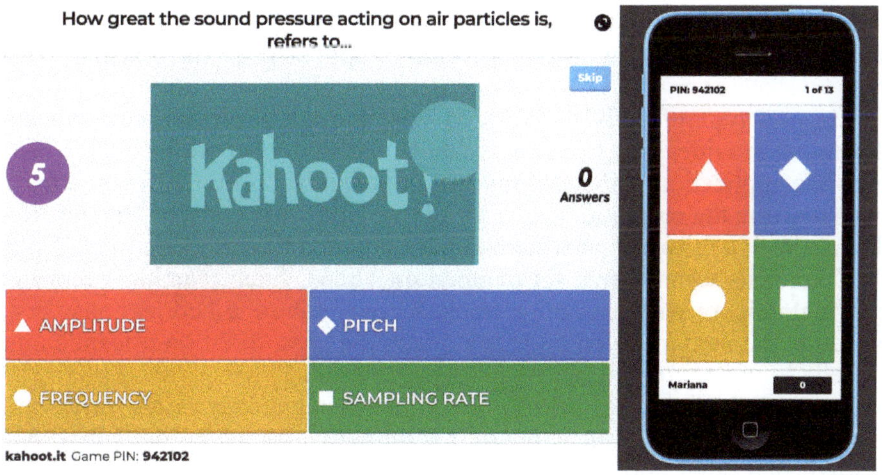

Figure 5.2 A screenshot of a *Kahoot!* quiz indicating the question at the top, the options at the bottom, and on the left-hand side we can see the number of seconds left. The right-hand side of the figure shows how the options are visualized in students' mobile devices.

Kahoot! as a platform to deliver interactive quizzes, educators might choose a different platform or even program their own (Cheong, Cheong and Filippou 2013). My particular choice was determined by the fact that it was freely available, I had prior experience utilizing it in class, and it had all the components needed to deliver my choice of activity.

In 2017–2018, I deployed the *Kahoot!* quizzes four times. The goal of each game, which was introduced as an end-of-class activity, was to gauge how well students had understood each lecture. Similar to the research by Cheong, Cheong and Filippou (2013) on the tool *Quick Quiz*, the focus was to keep the questions reasonably straightforward, have short time limits to answer, and participation being voluntary as the task was not being assessed. On this point, students could choose a fake nickname when entering the game and could choose simply not to enter the game at all. Most questions and answers were followed by brief explanations to clarify content and encourage students to look further into the question in their own time. The emphasis here fell particularly on questions in which the majority of the class had been mistaken, or those in which the answers were split, and thus showed the students did not fully grasp the content.

Kahoot! quizzes are designed so that everyone participating follows the same rhythm so that questions and answers are all available to everyone at the same time. The participants, as a group, progress to the results once either the time set for responding has passed (ten seconds, for example) or everyone has answered the question (whichever happens first). The tutor then triggers the next question, and this process is repeated throughout the quiz until the group reaches the final question and the virtual podium. Students were observed to enjoy the activity, cheering when I announced it as the next task. Furthermore, in meetings with individual students they would often comment on how much they enjoyed the quizzes and looked forward to them. In order to encourage a supportive environment during the activity, I would pause to congratulate the top scorer and at the end of the game I would ask the class to join me in a round of applause for the students with top scores.

The second type of activity explored was through *Socrative*,[5] an educational platform that provides free tools for teachers and students both through a browser or a mobile application. I focused my use of *Socrative* in two different ways. The first consisted in the preparation of a set of questions that allowed students to start the lecture with a refresher of what had been covered the previous week. Students worked in teams to answer a set of open questions. Each team was represented by a virtual unicorn within a racing track. As they submitted correct answers they advanced to the next question and, as a result, the position of their team's unicorn advanced.

Although the activity had a level of competitiveness to it, *cooperative* interactions were at the center of the game as it is through such interactions that students supported each other and managed or not to win the race. Unfortunately, due to temporary loss of internet connection, the virtual unicorns and the tracks were not visible to students, so the competitive aspect was lost. This type of technological mishap is one of the reasons why tutors might sometimes be reluctant to organize interactive sessions based on online platforms, as their failure often results in the class being slowed down and the students becoming disengaged. Another issue with this particular way of using *Socrative* is that in order for students to progress through the questions their answers had to be *verbatim*, and so even if they did know the answers they could not always progress, and this was observed to reduce engagement and cause frustration.

Socrative *as a Student Feedback Tool*

Socrative was also used as a platform to collect anonymous feedback through an Exit Cards system. Exit Cards are the collection of written feedback forms from students just before they leave a taught session. This method of collecting feedback is crucial to keeping the students' voice at the center of teaching and learning as it allows educators to introduce changes during the academic year in response to feedback, instead of waiting to the very end of the module to ask students what they thought about it. Here, students were asked two questions predefined by the system and a question determined by myself as the tutor.

The first question, set by the system, was "How well did you understand today's material?" and the answer options were "Totally got it", "Pretty well", "Not very well", "Not at all". The second predetermined question was "What did you learn in today's class?" and for this question students were presented with an open text box. For the question set by myself they were asked, "What topics from the module would you like us to revise in future sessions?" The Exit Cards method proved particularly useful to gauge student understanding of the topic and it allowed me to prepare the content of the revision lectures, that is, a set of lectures that do not incorporate new content but instead focus on revisiting topics.

Giving students a chance to provide anonymous feedback was meant to increase their confidence in requesting revision topics. Previously, students were just asked to email the lecturer, but this meant shy students might not necessarily request topics. Another advantage of the Exit Cards is their presence showed students that their opinions were valued. Furthermore, the Exit Cards encouraged a sense of autonomy among the students, as they were asked to assess their own learning (instead of this being assessed by the tutor) and provide guidance to teachers by requesting topics for revision. The results of the activity were shared with the students in a subsequent session in which I listed the topics that would be revised in future lectures based on the students' choices.

Buzzers

The third type of interactive activity involved a greater level of *cooperation* as it consisted of short quizzes using physical light and sound buzzers distributed throughout the lecture theater, with students working in teams in a question/answer setting. Students were provided with rapid feedback, orally by the tutor and immediately after they had answered the question. The buzzers were also used as sound sources themselves to introduce the mechanisms involved in sound localization in relation to human hearing. With the aid of volunteers, the buzzers were triggered from different parts of the room while the rest of the class remained seated with their eyes shut and had to point toward the direction of the source.

The buzzers were used when analyzing the soundtracks of films, to encourage students to request I stopped playback to explain what different elements they had recognized (Foley effects, voice over, ambience), with the objective of introducing them to the notions behind soundtrack analysis by asking them to take on an active role in the deconstruction of the variety of sound elements present within the sonic worlds that were presented. Students were observed to enjoy the use of the buzzers and laughed at the opportunity of triggering different types of sounds in response to challenges. Furthermore, students often came to me after class wondering if they could borrow them for an activity of their own as part of group work in other modules and other related activities, suggesting this was an experience that caught their attention and they wished to emulate in other contexts.

Treasure Hunt

Social interaction and *cooperation* was the focus of the fourth type of activity. It was called a Treasure Hunt to allude to the Easter period and Egg Hunt activities, but in this case focused on questions that were similar to the ones presented in previous exams. Students were divided into groups and challenged to answer the questions accurately and as fast as possible. They had to bring the answers in written form to the tutors at the front of the class and once they were given the green light they were handed out another question (the questions used are listed in Table 5.2). If more explanations were required, they were encouraged to go back to their groups and expand on their answers, incentivizing them to think about the variety of possible answers and how they might want to approach their own exam experience.

The activity was designed in this manner to encourage freedom to fail and a sense of progression. Students were observed participating enthusiastically and

Table 5.2 Questions used for the Treasure Hunt

1. You are given an audio file that has a broadband noise problem and you are asked to fix it. What would you do? Tip: Think about tools and different options on how to tackle the problem.
2. You have a dialogue recording you did on set using a radio mic and you want to process it so that it matches the acoustic characteristics associated with the visuals. What would you do?
3. Discuss three different microphone polar patterns, include clear examples on situations in which each of them would be useful. You may include diagrams.
4. You are hired to work on the sound for a film and you need to record sound to accompany an establishing shot of the seaside. What recording technique will you use? Why?
5. Explain the Inverse Square Law in connection to audio. What is it? How is it related to the critical distance? How would you calculate the effects of the Inverse Square Law in a room in a practical way (that is, without equations)?

Bonus Round: What do the terms low-pass and high-pass mean in relation to audio? What does it mean for sound to be "colored"? What is the relationship between frequency and pitch? What is the difference between mixing in and out of the box?

even raced to the front of the room to get their answers checked. They also discussed answers among themselves and went through notes while playing, and so helping different students take on different roles within their groups. The three top scoring teams were awarded Easter chocolates in respect of their successful "hunt for knowledge" and, therefore, the Easter Egg hunt.

Humorous Videos

Although not games per se, humorous videos were also used throughout the module to introduce topics such as microphone use and room acoustics. These were often used either as introductions to the topic or as breaks from content that I deemed to be quite dense, and so was used as a method to try to regain student attention. Previous research has found that the combination of humor and a relaxed environment can help the retention of module content (Garner 2006). Moreover, it can help reduce academic-related anxiety and as a result turn the classroom experience into one that is perceived as positive and fulfilling (Korobkin 1988), aiding well-being in general.

Survey Results

An online survey was distributed via email among the students that took the module and was designed to be completely anonymous so that participants felt free to express their opinions. Demographics were not collected as part of the survey, as this data had the potential of making some students recognizable. The survey was approved by the departmental ethics committee and completed by 21% of the cohort (17 students). The survey questions were aimed at finding out how the students perceived the gamified elements of the module and allow for reflection on which activities could be taken forward, which ones should be modified, and what could be expanded in the future.

When asked which of the activities (*Kahoot!* quizzes, *Socrative*, quiz buzzers, Treasure Hunt, videos with humorous content, other) had been their favorite, 65% chose *Kahoot!* quizzes, followed by videos with humorous content (24%). Furthermore, in support of this response, a large number of participants indicated having enjoyed the quizzes, finding them useful to monitor their progress and confidence throughout the module, and helping them revise for the exam (see Table 5.3). Such results are similar to the ones reported by Cheong, Cheong and Filippou (2013) who found that their newly designed tool, *Quick Quiz*, improved student learning as well as encouraging engagement and increasing enjoyment.

Table 5.3 Students' responses on the *Kahoot!* quizzes

Enjoyed the quizzes.	88%
Quizzes helped them monitor their progress in the module.	82%
Quizzes allowed them to gain awareness on the topics they felt more confident in.	88%
They aided more efficient revision for the exam.	71%

Table 5.4 Students' feedback on Exit Cards

Exit Cards made them feel their opinions were valued.	59%
Exit Cards helped them monitor their progress.	47%

Regarding the quizzes, one student reported:

> It sometimes helped to remember key facts about certain topics, such as definitions, measuring units and specific values to remember, such as with bit/sampling rates. However, since most of the . . . assessment was long answer questions, it was difficult to keep track of my progress of the whole module content.

The downside spotted by the student regarding the quiz format is one that I looked to address through the Treasure Hunt. Regarding the latter, 65% of the students thought it was fun and that it made the revision for the exam more accessible. However, only 53% of the students said they were able to use the questions and feedback to help them revise in their own time. This is likely to have been dependent on the dynamics adopted by each team of players. Some teams might have been overreliant on one or two team players, meaning that not all students benefited from the activity. A possible modification, which might help overcome this hurdle, is the creation of specific roles for each student within the team, which could be framed as "characters" within the "quest" for answers, incorporating more aspects of social interaction. The use of characters, in the form of avatars in e-learning has shown to have positive effects (Clark and Mayer 2016; Kapp, Blair and Mesch 2014). Outside e-learning, this is a method that is at the center of role-playing activities within the classroom, such as those explored through the *Reacting to the Past* activities (Barnard College 2019; Burke 2014).

Feedback on the *Socrative* platform was divided, with no significant differences between students that enjoyed the experience, those that did not and those that did not remember using the platform.[6] These responses are not surprising as the platform was only used twice and on one of the occasions technical difficulties meant that it did not fulfill its aim. However, it is worth noting that even though students did not express a particular enjoyment in using the platform, responses did indicate that over half of the students surveyed felt that the Exit Cards indicated that their opinions were valued and some felt it helped monitor their confidence in different topics as well as gauge their progress throughout the module (see Table 5.4).

Gamification as a Teaching Methodology

Survey participants indicated that they wished more lectures included interactive content, and the majority felt that the inclusion of games and other interactive content within the sessions was an indication that the lecturer was making an effort to engage students with the module (see Table 5.5). A similar finding was reported by Stott and Neustaedter (2013), who pointed out that students are likely to acknowledge the efforts of the teacher in trying to provide a better learning experience, and

Table 5.5 Feedback from students on gamification as a teaching method

They wished more lectures included interactive content, such as games.	82%
Felt that the inclusion of games and other interactive content within the sessions was an indication that the lecturer was making an effort to engage students with the module.	88%

that this in itself aids student engagement. It seems that even the initial steps of attempting to provide a gamified experience are seen positively by students.

When asked what they liked the most about the interactive nature of the classes, students commented on the engaging nature of the sessions and the ability to hold their attention:

> It made me feel more involved with and engaged with the subject, so I wanted to learn more and felt more able to do so. It was less intimidating to approach our lecturer with questions and made the module much more enjoyable.

Other positive factors mentioned by survey participants were that they had fun learning, there was a competitive nature to the tasks, the activities helped information sink in and aided exam preparation by encouraging weekly revision. The survey results also indicated that students would like to participate in a greater number of group activities, which is in accordance to findings by Pirker, Rifnaller-Schiefer and Gütl (2014), who reported that students tended to prefer team-based work. Students in the survey also reported that they wanted more activities of the same type, rather than new ones.

When asked what they liked the least about the interactive nature of the classes, the top concern was regarding those activities that required students to answer questions out loud in front of the class, as they required participants to have the confidence to do so, and those that did not have that confidence were being left out, this was the case of some of the activities that involved buzzers. There were also two comments on the games taking out time from revision, but this may be due to a lack of clarity on the side of the lecturer in explaining that these activities were indeed central to the revision process.

> It felt like time could have been better spent teaching more content rather than devoting large chunks of time to gamified sections that didn't really teach anything new. Practicals and seminars feel like more appropriate settings for interactivity.

Questions on timing and when they should be placed were also of interest:

> I think they worked well at the end of the lectures, but when part way through, it was weird transitioning from interactive segment back to proper lecturing.

The use of the term "proper lecturing" denotes a possible preconception by students of what is useful and what is a filler. It is of the utmost importance to consider the need for a clearer communication channel between lecturer and students, as the success of a gamified classroom lies mostly in students embracing the methodology.

There was also a comment that was specifically on the use of humorous videos:

> I'm not sure the brain is designed to be able to find something humorous when it is trying to process information in an environment as stressful as a lecture.

This comment, as well as others from the same respondent, which suggested that the games were not actually instances of learning and might have been overused, allow for reflection on the fact that gamification, as any teaching method, will not be suitable for every student. Even different modes of gamification might not be suitable to everyone. For example, although students in this survey, as well as in previous studies, indicated that competitive aspects were motivating, others have been found to be put off by competitiveness (Dominguez et al. 2013), so a variety of activities that cater to different preferences, including preference for a more traditional form of lecturing, are key to the success of the module. Moreover, these responses highlight a known issue in the use of games in educational settings, which is linked to the challenge of ensuring close links between learning objectives and gamified activities, so that not only high motivation levels are reached but that the content is also understood by the learners (de Freitas 2007).

Furthermore, I believe it would have been useful for me to explain at the start of the module to all students that this was a method that would be used so that they knew what to expect and that it was indeed grounded in pedagogical research. This is something that I did in the subsequent academic year and was seen to increase the engagement of the cohort with the activities. Reflection on these comments also allowed me to be transparent with subsequent student cohorts on the relaxed environment I was trying to create, encourage openness and transparency on what worried them and what they liked/disliked and as a result reduce the sense of teaching spaces as stressful. This allowed me to challenge possible student preconceptions of orderly, quiet classrooms equating real, academic work, and games needing to be separated from that (Willis 2006). By reconsidering what a classroom "should" be like, we are questioning current modes of practice and hierarchies and inviting teachers to embrace new fluid forms of teaching and learning and, as a result, challenge themselves (Allsup 2016).

Limitations and Future Considerations

It is worth noting that a limitation of the study is that only 21% of the cohort participated in the survey, making it challenging to determine whether the data gathered is representative of the full cohort. More data needs to be collected for further validation. However, this initial data allows me to reflect on what worked, what did not, and ways of moving forward. It is also worth noting that even though previous studies have demonstrated that gamification practices can help increase pass rates (Iosup and Epema 2014), there was not enough data available to conduct an analysis of this kind. However, this could provide a potential avenue for future research.

In terms of moving forward, a new type of activity was introduced in the subsequent academic year (in progress at the time of this writing). Such activity was introduced in response to the survey feedback analyzed in this chapter as well as module feedback, in which students commented on wanting more practice on long form questions and not just multiple-choice quizzes as this would help align their practice experience to the exam setup. Although this aim had been partially covered by the Treasure Hunt activity, it was decided that an additional gamified task would be added to address this concern.

Table 5.6 Questions asked for the activity "The Secret of Heslington Hall"

The first day of filming is fast approaching and as a sound recordist for the production you need to make sure you choose the appropriate microphones. The first on your list of things to do is to decide what microphones to use for the following recordings. Explain what you would use and why you've made the choices you made, refer to the specific theories, concepts that allowed you to make the decision:
Dialogue recording of a scene using a variety of shots (from close-ups to wide shots)
The recording of a piano playing the theme tune for the film
Very loud percussive sounds including starter pistols
Things have taken an unexpected turn in this gripping drama.
Scene 10 features a scene in which, in a *Narnia* style,[1] the main character opens a door meant to lead to a Heslington Hall office but instead finds herself in a rainforest.
You need to record the atmos track for the rainforest scene.[2] What recording technique will you use? Why?
You are having a busy week with the production and due to the fact that this is a very low-budget production you aren't just doing location sound, you are also the dialogue editor.
When going through the takes you notice there's a very loud hum in some of the dialogue recordings.
How will you fix this? Please note that re-recording isn't an option.
Following up from last week, the scene that presents the confrontation between Sarah and Lord Armstrong has now been shot and you need to work on the sound in postproduction. The scene takes place in a corridor and your job is to make the production sound match the acoustics of the corridor that we see. How would you go about achieving this? Don't just mention the technique, also think about the different parameters you will need to adjust and what workflow might be most appropriate considering you will need to make sure both Sarah's and Lord Armstrong's voices match the acoustics of the space.
Due to unforeseen circumstances you are told that you need to do an emergency voice over recording but there's no studio available. You are given an office to work in. You need to make sure the acoustic conditions of that office are as suitable as they can possibly be for your recording. What sort of things will you be checking? What microphone technique will you use? Why?

[1] This refers to the series of novels *The Chronicles of Narnia* by C.S. Lewis in which the protagonists open a wardrobe to find themselves in a magical land.

[2] By "atmos" I am referring to ambience tracks.

The newly introduced activity was called "The Secret of Heslington Hall" and used *storytelling* as the main aspect of gamification. "The Secret of Heslington Hall" (referring eponymously to a building on the university campus) presented an imaginary low-budget film production. The students were presented with a very simple introductory statement informing them that the production was a gothic mystery drama set in Victorian times and filmed in the North of England. Once the premise was introduced, students were issued every week with a set of questions linked to the production.

Questions alternated between those regarding the picture side of the module and those related to the audio aspects. The questions asked were linked directly to the topics that had been covered in the previous weeks and had the aim of helping students make direct connections between the theoretical aspects of the module ("the science of sound") and the practical aspects of sound work in film and television. Table 5.6 features the list of questions presented to the students. A deadline was set by which the students needed to send their answers via email, and they were encouraged to participate for its own sake. Those students that sent the questions on time got personalized feedback from the lecturers, those that did not could still benefit from a sample answer given in class by the tutors.

The level of participation in the tasks set in "The Secret of Heslington Hall" was disappointing, only five or six students submitted questions each week out of a cohort of 72, making engagement very low. The reason for low engagement might have been due to the lack of clear gamification procedures. As a new activity, and a slightly experimental one in the context of the module, there was no clear guidance provided to students as to what were the "fun" benefits of completing the task, in addition to the personalized feedback. Having not made those rules of engagement clear to the students made it difficult to engage their attention. Although the use of leader boards, trophies, and other scoring systems were considered, this was not presented to the students when the activity was first introduced, meaning that this was not used as a motivational tool. In the future, the activity will be restructured so that it has clearer goals, both educationally and playfully.

Conclusions

The introduction of gamified elements to the module *The Science of Sound and Cinematography* was received positively by 88% of students that completed the evaluation form. Students commented on how the interactive nature of the classes helped them retain the content and encouraged them to revise on a weekly basis. Furthermore, it allowed challenging topics to appear accessible and made lectures something they looked forward to, as well as improving levels of intrinsic motivation. However, it is worth acknowledging that the biggest challenge in the implementation of gamification is fighting student preconceptions on what an educational setting "should" be like, and thus showing it is possible to have fun in the process of learning without these aspects representing a waste of time or resources.

At the center of the process is a clear alignment between learning outcomes and gamification strategies that sees them work together to achieve higher levels of engagement, while also appealing to different types of student interaction and motivation. In the specific context of a degree in film and television production, this trains the creatives of the future, where a core understanding of the importance of sound is key. Consequently, the true engagement of students with the topic has the potential to have a long-lasting impression regardless of whether they choose to specialize in sound production or postproduction, or whether they follow a different path altogether. Gamification in audio education can therefore play a central role in the training of film and television professionals that understand the importance of sound, at the intersection between theory and practice.

Notes

1. In the context of this chapter, the words *tutor* and *lecturer* are used interchangeably to refer to any faculty member with teaching responsibilities.
2. *vHealthCare* by BreakAway Games, www.breakawaygames.com/vhealthcare/
3. At the University of York, as well as many other universities in the UK, assessments are classified as formative or summative. A formative assessment is often given a mark, but that mark does not count toward the final degree. They are instances offered to students as an opportunity to familiarize themselves with a format of assessment such as an essay and give them the opportunity to learn from mistakes. Summative assessments are those that are given a mark that counts toward the final degree.
4. https://kahoot.com/
5. https://socrative.com/
6. Significance was calculated through binomial tests and set at p=0.05.

References

Allsup, R.E. (2016) *Remixing the Classroom: Toward an Open Philosophy of Music Education*, Bloomington and Indianapolis: Indiana University Press.

Barata, G., Gama, S., Jorge, J. and Goncalves, D. (2013) "Engaging Engineering Students with Gamification", *Proceedings of 5th International Conference on Games and Virtual Worlds for Serious Applications (VS-GAMES)*, Dorset, September.

Barnard College (2019) *Reacting to the Past*, https://reacting.barnard.edu/ [Last accessed on 8 May 2018].

Burke, K.R. (2014) "Roleplaying Music History: Honing General Education Skills via 'Reacting to the Past'", *Journal of Music History Pedagogy*, 5 (1), 1–21.

Cheong, C., Cheong, F. and Filippou, J. (2013) "Quick Quiz: A Gamified Approach for Enhancing Learning", *PACIS 2013 Proceedings*, 206.

Clark, R.C. and Mayer, R. (2016) *E-learning and the Science of Instruction* (4th edition), Hoboken, NJ: Wiley.

Deci, E.L. and Ryan, R.M. (2008) "Self-determination Theory: A Macrotheory of Human Motivation, Development, and Health", *Canadian Psychology/Psychologie canadienne*, 49 (3), 182–185, doi:10.1037/a0012801.

de Freitas, S. (2007) "Using Games and Simulations for Supporting Learning", *Learning, Media and Technology*, 31 (4), 343–358, doi:10.1080/17439880601021967.

de-Marcos, L., Domínguez, A., Saenz-de-Navarrete, J. and Pagés, C. (2014) "An Empirical Study Comparing Gamification and Social Networking on E-learning", *Computers & Education*, 75, 82–91, doi:10.1016/j.compedu.2014.01.012.

Denis, G. and Jouvelot, P. (2004) "Building the Case for Video Games in Music Education", *Proceedings of Sound and Music Computing Conference*, Paris, France.
Denis, G. and Jouvelot, P. (2005) "Motivation-driven Educational Game Design: Applying Best Practices to Music Education", *Proceedings of the 2005 ACM SIGCHI International Conference on Advances in Computer Entertainment Technology*, Valencia, Spain, June, 462–465, doi:10.1145/1178477.1178581.
Deterding, S. (2014) "Eudaimonic Design, or: Six Invitations to Rethink Gamification", in Fuchs, M., Fizek, S., Ruffino, P. and Schrape, N. (eds.) *Rethinking Gamification*, 305–323. Leuphana University of Lüneburg: Meson Press.
Deterding, S., Dixon, D., Khaled, R. and Nacke, L. (2011) "From Game Design Elements to Gamefulness: Defining 'Gamification'", *Proceedings of the 15th International Academic MindTrek Conference*, 9–15, doi:10.1145/2181037.2181040.
Dominguez, A., Saenz-de-Navarrete, J., de-Marcos, L., Fernandez-Sans, L., Pages, C. and Martinez-Herraiz, J.J. (2013) "Gamifying Learning Experiences: Practical Implications and Outcomes", *Computers & Education*, 63, 380–392.
Elias, T. (2011) "Universal Instructional Design Principles for Mobile Learning", *International Review of Research in Open and Distance Learning*, 12 (2), 144–156.
Figueroa-Flores, J. (2016) "Gamification and Game-based Learning: Two Strategies for the 21st Century Learner", *World Journal of Educational Research*, 3 (2), doi:10.22158/wjer.v3n2p507.
Garner, R.L. (2006) "Humor in Pedagogy: How Ha-Ha can Lead to Aha!" *College Teaching*, 54 (1), 177–180, doi:10.3200/CTCH.54.1.177-180.
Gillet, N., Vallerand, R.J. and Lafrenière, M.A.K. (2012) "Intrinsic and Extrinsic School Motivation as a Function of Age: The Mediating Role of Autonomy Support", *Social Psychology of Education*, 15 (1), 77–95, doi:10.1007/s11218-011-9170-2.
Hitchens, M. and Tulloch, R. (2018) "A Gamification Design for the Classroom", *Interactive Technology and Smart Education*, 15 (1), 28–45, doi:10.1108/ITSE-05-2017-0028.
Hogan, K. and Pressley, M. (1997) *Scaffolding Student Learning: Instructional Approaches and Issues*, Cambridge, MA: Brookline Books.
Iosup, A. and Epema, D. (2014) "An Experience Report on Using Gamification in Technical Higher Education", *Proceedings of the 45th ACM Technical Symposium on Computer Science Education*, New York, 27–32, doi:10.1145/2538862.2538899.
Jacobs, K.L. (2005) "Investigation of Interactive Online Visual Tools for the Learning of Mathematics", *International Journal of Mathematical Education in Science and Technology*, 36 (7), 761–768, doi:10.1080/00207390500271149.
Kahoot! https://kahoot.com/ [Last accessed on 8 May 2019].
Kapp, K.M. (2012) "Games, Gamification, and the Quest for Learner Engagement", *T+D Magazine*, www.td.org/magazines/td-magazine/games-gamification-and-the-quest-for-learner-engagement [Last accessed on 22 July 2019].
Kapp, K.M., Blair, L. and Mesch, R. (2014) *The Gamification of Learning and Instruction: Fieldbook, Ideas into Practice*, San Francisco: Wiley.
Knutas, A., Ikonen, J., Nikula, U. and Porras, J. (2014) "Increasing Collaborative Communications in a Programming Course with Gamification: A Case Study", *Proceedings of the 15th International Conference on Computer Systems and Technologies*, 370–377.
Korobkin, D. (1988) "Humor in the Classroom: Considerations and Strategies", *College Teaching*, 36 (4), 154–158.
Landers, R. (2014) "Developing a Theory of Gamified Learning: Linking Serious Games and Gamification of Learning", *Simulation & Gaming*, 45 (6), 752–768.
Landers, R. and Callan, R. (2011) "Casual Social Games as Serious Games: The Psychology of Gamification in Undergraduate Education and Employee Training", in Ma, M., Oikonomou, A. and Jain, L.C. (eds.) *Serious Games and Edutainment Applications*, London: Springer-Verlag, doi:10.1007/978-1-4471-2161-9_20.
Lee, J. and Hammer, J. (2011) "Gamification in Education: What, How, Why Bother?" *Academic Exchange Quarterly*, 15, 1–5.

O'Donovan, S., Gain, J.E. and Marais, P. (2013) "A Case Study in the Gamification of a University-level Games Development Course", *Proceedings of the South African Institute for Computer Scientists and Information Technologists Conference*, 242–251.

Pirker, J., Rifnaller-Schiefer, M. and Gütl, C. (2014) "Motivational Active Learning–Engaging University Students in Computer Science Education", *ITICSE'14*, Uppsala, Sweden, June 21–25, doi:10.1145/2591708.2591750.

Prensky, M. (2003) "Digital Game-based Learning", *Computers in Entertainment: Theoretical and Practical Computer Applications in Entertainment*, 1 (1).

Putwain, D.W. (2008) "Examination Stress and Test Anxiety", *Psychologist*, 21 (12), 1026–1029.

Putwain, D.W., Woods, K. and Symes, K.A. (2010) "Personal and Situational Predictors of Test Anxiety of Students in Post-compulsory Education", *British Journal of Educational Psychology*, 80 (1), 137–160, doi:10.1348/000709909X466082.

Riley, G. (2016) "The Role of Self-determination Theory and Cognitive Evaluation Theory in Home Education", *Cogent Education*, 3 (1), doi:10.1080/2331186X.2016.1163651.

Ryan, R.M. and Deci, E. (2000) "Intrinsic and Extrinsic Motivations: Classic Definitions and New Directions", *Contemporary Educational Psychology*, 25, 54–67, doi:10.1006/ceps.1999.1020.

Ryan, R.M., Rigby, C.S. and Przybylski, A. (2006) "The Motivational Pull of Video Games: A Self-determination Theory Approach", *Motivation and Emotion*, 30 (4), 344–360, doi:10.1007/s11031-006-9051-8.

Socrative, https://socrative.com/ [Last accessed on 12 July 2018].

Stott, A. and Neustaedter, C. (2013) "Analysis of Gamification in Education", Technical Report 2013-0422-01, Connections Lab, Simon Fraser University, Surrey, BC, Canada, April, http://clab.iat.sfu.ca/pubs/Stott-Gamification.pdf [Last accessed on 8 May 2019].

Vansteenkiste, M., Lens, W. and Deci, E. (2006) "Intrinsic Versus Extrinsic Goal Contents in Self-determination Theory: Another Look at the Quality of Academic Motivation", *Educational Psychologist*, 41 (1), 19–31, doi:10.1207/s15326985ep4101_4.

vHealthCare by BreakAway Games, www.breakawaygames.com/vhealthcare/ [Last accessed on 12 July 2018].

Willis, J. (2006) *Research-based Strategies to Ignite Student Learning*, Alexandria: Association of Supervision and Curriculum Development.

6

Learning About Critical Listening Through Popular Music

Adam Scott Neal

Introduction

Popular music—by its very name, the *lingua franca* of our culture—is a strong entry point for learning how music works. Along with harmony, rhythm, and formal structures, pop music also places high importance on timbre, texture, and text-painting. This chapter reflects my practical experience teaching critical listening in elementary schools (K–5 in the United States) and universities. In both cases, I tailored my instruction to a general audience. In this chapter, I discuss the concept of critical listening, suggest categories for elements of music, and present strategies for listening lessons using popular music recordings. These listening lessons are adaptable for students from elementary school through college.

Background

While studying at Georgia State University, I was a teaching artist for an elementary school (K–5 in the United States) music integration program called Sound Learning (Beauchesne 2007). During the residency, my team taught third-grade students to compose original pop songs. The first part of the process involved listening and analyzing songs, such as James Brown's "I Got You (I Feel Good)". The activity included singing melodies and clapping rhythms. However, the main focus was identifying elements of the songs' arrangements, which included "The Beat, Bass Line, The Melody, and Other Fun Stuff." "I Got You" was a useful model because of its relatively sparse arrangement. The drums and bass guitar comprise the Beat and Bass Line categories, respectively. The vocal and solo saxophone serve as Melody. The rhythm guitar and horn section comprise the "fun stuff." After learning about these example songs, the students crafted their own songs by playing instrumental parts on keyboards and creating original lyrics related to school subjects, including science.

More recently, I have taught Introduction to Music Technology classes at the University of Florida (UF) and East Carolina University (ECU). The class was required of all music majors at UF and required of all jazz majors at ECU (non-majors also enrolled, space permitting). In these courses, I included units on notation software, audio editing, live sound engineering, and entrepreneurship. I also included recurring lessons (about four per semester) of critical listening practice with popular music selections. By listening to four or five songs in a single class, the students made connections between pieces; gained listening experience; and gained knowledge about audio techniques, synthesis, and sampling.

Toward a Definition of Critical Listening

The Foundation for Critical Thinking (2019, n.p.) defines *critical listening* as: "A mode of monitoring how we are listening so as to maximize our accurate understanding of what another person is saying". *Hearing* concerns the perception of sound and the faculties of the ear, while *listening* connotes attention and "silent obedience", according to Worthington and Bodie (2018, p. 3). This sense of unequal agency between speakers and listeners is reinforced by early studies of listening, which focused on retention and recall of academic lectures (p. 8). When a person speaks, the listener attempts to understand the message based on the spoken words, inflection, context, and other factors. However, as Ceraso (2014, pp. 102–103) writes:

> [W]hen listening is taught, it is usually treated as the practice of attending to audible words or sounds in order to make meaning of them. That is, teaching listening often involves teaching students to approach sound as another form of text; sound is simply more content to be interpreted. In addition to teaching students what sound *means*, I argue that it is critical to teach them how sound *works and affects*.

Ceraso proposes "multimodal listening", which "[attends] to not only the sensory, embodied experience of sound, but to the material and environmental aspects that comprise and shape one's embodied experience of sound" (p. 105). Ceraso contrasts listening to a band on her laptop speakers, then listening to the same band at a concert. Not only do the songs sound "flat and tinny" on the laptop, but her listening mode is concerned mainly with content. At the concert, she experiences a "heightened vitality" due to the amplification and full frequency range, but also to the participatory environment which encourages bodily engagement through clapping and singing (pp. 105–106).

Listening to music is a multimodal process, although not always taught that way. According to Kratus (2017), music teachers introduce listening concepts in isolation, such as melodic dictation. "Rather than focus on a single dimension of music", like a *sol-mi* pattern, "listeners in real-world settings create their own experience by shifting attention among a variety of simultaneous aspects of music as directed by their knowledge, mood, and context" (p. 47). At a concert, one may attend to the feeling of the amplified kick drum vibrating in the chest, to the changes of the guitar's timbre through various effects pedals, or to the charismatic commands ("make some noise!") from the lead vocalist. One may attend to the melody and rhythm, but these are only parts of the listening experience.

The emphasis of pitch and rhythm in listening is what Wishart (1996, pp. 22–34) calls the "lattice model". Pitch occupies the y axis and time occupies the x axis, while timbre is represented on the z axis, almost as an afterthought (and discounting the fact that timbre itself is three-dimensional!). One example I have used to illustrate the three-dimensionality of timbre is Rage Against the Machine's "Bulls on Parade". In the guitar break (at 0:22 and 2:52), the guitarist uses a wah pedal to quickly filter the guitar's tone. The reveal and attenuation of higher harmonics (y axis) creates the distinctive "wah" effect—which is revealed over time (x axis). Timbre is simply too complex to be described as hanging on a lattice of pitch and rhythm, especially when other elements of sound, like volume and distance, can also be represented on the z axis.

To propose a listening model beyond the time-pitch x-y lattice, Wishart (1996) draws the analogy between sound and landscape. When we listen, we notice the nature of perceived space, the disposition of sound-objects in space, and the recognition of individual sound-objects (p. 140). In other words, we recognize different timbres, their direction from our listening position, and cues like reverberation that tell us the size of the space we are in. Music recordings create sonic landscapes, either by replicating the space in which the recording was made (a choir in a church) or by creating an imaginary landscape. To describe such imaginary landscapes, Moore (2012) proposes a model called the "soundbox". A soundbox is an illustration of the three-dimensional space suggested by a recording—Moore draws icons of instruments meant to depict their lateral position (i.e. panning), their approximate frequency space, and the depicted distance from the listener (pp. 34–35).

Moore (2012) agrees that timbre and texture are often considered secondary to melody, harmony, and rhythm. However, in "recorded popular music, secondary domains are able to do much more than simply 'shape' content: indeed, they frequently *constitute* content" (29). While the most memorable element of some songs may be the melody, in others, the most memorable aspect is the sonic landscape. Consider "Strawberry Fields Forever" by the Beatles. Several aspects of the sonic landscape make this song memorable. One is the dramatic *shift* of landscape during the song, made by splicing two performances together. The first chorus and verse feature a standard rock instrumentation—voice, guitar, bass, drums, keyboard—while the remainder of the song features trumpets and cellos. Another memorable aspect of the song is its false ending: after the last chorus, the song fades out, only to fade back in. The creative production techniques, which include splicing tape and fading in and out, are as vital to the character of the recording as the melody, harmony, and lyrics.

While a song can be described in terms of pitch and rhythm, that does not reflect the multimodal experience of hearing it. The timbres, textures, and spatial placement of sounds (left-right, close-distant, louder-softer) reinforce the gestures of the melody and rhythm. Understanding the influence of sound production in the texture and structure of the piece is as vital as a musician playing in tune with fellow performers. Consider, again, the difference between *hearing*, often considered a passive function of the ears, and *listening*, which is an act of interpretation. Abramo (2014, pp. 85–86) takes these terms in an illuminating direction, writing: "*Hearing* is equated with *meaning*; it is hearing what the sound signifies and to pin down and fix its meanings and make those signifiers permanent. Conversely, a *listener* searches for *sense*". Hearing is not passive, but instead a basic function during which we catalog sounds we hear (a guitar, a voice, a verse, a chorus). Listening is an advanced function in which we build a plausible meaning of the sounds—a *sense* of what they could represent. *Critical* listening, then, is listening which attempts to *make sense* of multiple elements in a piece of music.

Making Sense of Musical Elements

Students may notice a number of details in a given recording yet find difficulty articulating what they hear. Finding ways to translate concepts into easier terms is helpful. The Beat, Bass Line, Melody, Fun Stuff rubric that I used with third-grade students is analogous to Moore's (2012, pp. 19–21) concept of "textural layers", of which he defines four: and Explicit Beat layer ("nominally unpitched"), a Functional Bass layer, a Melodic layer, and a Harmonic Filler layer. Instruments can move among layers—guitar may first function as Harmonic Filler, then take a solo in the Melodic textural layer. The varied instrumentation in the Melodic and Harmonic Filler layers help identify "style" (pp. 27–28).

To learn how each layer represents certain styles, let us compare the layers of a funk song with a Latin pop song. As mentioned earlier, in James Brown's "I Got You", the drums fill the Beat layer, the bass guitar fills the Bass layer, and the voice and saxophone fill the Melody layer. The Harmonic Filler in "I Got You" consists of the rhythm guitar and horn section, mainly short upbeat punctuations. Miami Sound Machine's "Conga" features many of the same instruments, but the piano and Latin percussion influence the sense of style. The piano fills the harmony, with broken chords in a syncopated rhythm, while the percussion supplies active, overlapping rhythms. Even the *a capella* vocal line "come on shake your body baby, do the Conga" serves both a rhythmic as well as a melodic function. Although there are differences in rhythm, groove, and melody in these songs, the roles of the instruments provide more immediate clues to the songs' styles.

Beyond voices and instruments, there are broad and subtle aspects of timbre that one may notice in a music recording. Several authors of audio engineering texts (Moylan 2002; Owsinski 2006; Corey 2017) share similar lists of qualities to listen for in recordings. Moylan's (2002, Table 6.1) list includes Time, Pitch, Dynamics, Sound Quality, and Spatial Properties. Owsinski (2006, p. 10) lists Balance of instruments, Panorama, Frequency Range, Dimension, Dynamics (not only volume, but tempo, meter, and modality as well), and general "Interest". Corey (2017, p. x) lists Timbre, Dynamics, Balance, Distortion, Extraneous Noise, Space, and Panning.

The qualities are analogous, but split in different ways. Dynamics (changes in perceived volume) and space (perceived distance) are the clearest corollaries.

Table 6.1 Comparison of recording qualities

Moylan (2002)	*Owsinski (2006)*	*Corey (2017)*
Time	Balance	Timbre
Pitch	Panorama	Dynamics
Dynamics	Frequency range	Balance
Sound quality	Dimension	Distortion
Spatial properties	Dynamics	Extraneous noise
–	"Interest"	Space
–	–	Panning

Corey and Owsinski separate panning from space/dimension, while Moylan includes these together as "spatial properties". Moylan's "Sound quality" likewise encompasses Corey's "Timbre", "Distortion", and "Extraneous Noise" categories. I have adapted a modified approach to these concepts, listed in Table 6.2.

According to Johnson (2011), students generally describe music in three ways: musical, affective, and associative (pp. 261–262). Musical descriptions include objective terms, like instrument names. Affective descriptions appeal to emotions, and associative descriptions rely on extra-musical images. Students may not know terms like "distortion" or "noise floor", but that does not mean they cannot *hear* these qualities. Moylan (2002, Chapter 4) warns that describing sound through associative and affective analogies is personal and too imprecise for professional audio engineers, but I think that is a fine place for student listeners to begin. As students gain confidence in listening, they will develop knowledge of musical/audio concepts and increase the precision of their descriptions. "It sounds cloudy" becomes "the decay time of the reverb is over five seconds". This is what Kratus (2017) describes as four levels of listening: fluency (ability to identify multiple sounds), flexibility (ability to think across categories), elaboration (ability to notice combinations of ideas), and originality (ability to contribute unique interpretive ideas) (pp. 48–50).

One way in which students can develop their fluency and flexibility in listening is what Green (2008, pp. 6–7) calls "informal learning". Informal learning reflects methods by which pop musicians learn music—by listening and copying. Green proposes "purposive listening", where students listen in order to adopt musical elements into their own practice. By listening and copying, students naturally develop a musical vocabulary, to which they can later apply technical terms (Green 2008, pp. 68–71). For my third-graders, the process of listening and copying was immediate. My residency was short (about eight visits within a month) and the main goal was for the students to create original compositions. The students listened to songs to get a sense of the elements they would need to create in a fully realized pop song. In other words, they needed at minimum a Beat, a Bass Line, and a Melody.

For my college students, the goal of listening was broader. My Introduction to Music Technology class at UF culminated in individual composition études using Propellerhead's Reason software. I required each student to use at least four

Table 6.2 Qualities of recordings

Color
What timbres can be identified?
Dynamics
What is the overall dynamic? When are there changes in dynamics?
3-D space
Where is each element located, left to right, and closest to farthest?
Balance
Does each element "fit" with the others, in terms of volume and frequency space?

timbres and two processing effects, but style and content were otherwise open. The listening lessons I held during the course demonstrated ways in which the students could craft a sophisticated sonic landscape, through panning, reverberation, changes of instrumentation, and so on. My Introduction to Music Technology class at ECU culminated in group recording projects in which the students multi-tracked jazz/rock combos with voice.

One successful collaboration was a cover of Martin and Blane's "Have Yourself a Merry Little Christmas". To create this recording, the students needed to evaluate the technical quality of their microphone placement and mix balance, but also make creative decisions about the arrangement, based on available instruments and aesthetic interest. They recorded a piano-bass-drums rhythm section playing throughout, then overdubbed acoustic guitars and Harmon-muted trumpets during the bridges. These compositional decisions were informed by previous listening lessons, in which we discussed arrangement, timbre, balance, and room acoustics.

Recommendations for Critical Listening by Age

The process of critical listening to popular music is applicable to all levels of music education. Listening, and exposure to music of varying styles, is required by government and academic accreditation agencies, such as the National Curriculum in the UK (Department for Education 2013) and the National Coalition for Core Arts Standards (2014) and National Association of Schools of Music (NASM) in the US (2018). Although the level of detail may be different, a similar process of listening and discussing can be used in K–5 through college.

Critical Listening in K–5

Elementary-aged children may focus on identifying melodies, rhythms, and instrumental timbres. These students can also distinguish different roles of instruments, like Melody and Bass Line. Listening lessons might lead to composition projects, as described previously, but can also reinforce listening during performance. By identifying which instruments have the melody or accompaniment, describing dynamic and rhythmic contrasts, etc., students will better understand their own parts in ensembles. It is useful for students at this age to listen to simpler songs, in terms of duration, number of formal sections, and instrumentation. "I Got You", which features clear textural layers and a predictable structure, may be more accessible for these exercises than "Strawberry Fields Forever", which features changes in instrumentation, texture, and production effects.

Critical Listening in Secondary School (Grades 6–12)

In secondary school, listening exercises offer additional benefits for ensemble performance. For ensembles, Bell (2018) proposes listening to recordings of the piece the group is learning, but only after a few rehearsals. This way the students have developed familiarity with the piece, and can listen for higher-level aspects of the ensemble sound, such as blend and dynamic contrast. The addition of popular music listening lessons—perhaps songs with similar lyrical themes to the performance repertoire—can be a welcome head-clearing exercise. If working on Robert W. Smith's band piece "Africa: Ceremony, Song, and Ritual", one might listen to Toto's song "Africa". In this case, the comparison between these pieces also provides an opportunity to introduce traditional music from African cultures, and to discuss ways in which these pieces appropriate or emulate those traditional styles. Listening to popular music encourages attending to details and to the larger sonic landscape (or "soundbox"). These lessons can reinforce listening to even more aspects of musical performance, such as chord changes, formal structures, dynamics, orchestration, the functional layer of each instrument (melody, rhythm, "fun stuff"), connections between genres, and even the sonic characteristics of the rehearsal and performance spaces (a dry, carpeted music room versus a reverberant gymnasium).

Critical Listening in College Music Courses

Since many schools of music offer an Introduction to Music Technology course, this may be an appropriate venue for listening lessons. One might begin with discussing arrangements, then expand the conversation to include production techniques. Basic audio skills (removing applause, performing amplitude adjustments) are important to all musicians. To better perform these tasks (and better present their work), musicians should know *why* equalization and compression will help a recording sound better on laptop speakers, know *why* a reverberant church sounds optimal for a choir and inappropriate for a percussion ensemble—and so on. These pieces of practical knowledge are well articulated in discussions about pop music because pop music foregrounds timbre and texture.

There is also an opportunity here for college students to engage with ideas such as sonic landscape, style, symbolism, the idea of the text, etc. When students provide affective descriptions or associative connections to songs, they provide an opportunity to investigate the musical content further. "In the Air Tonight" by Phil Collins has elicited descriptions in my classes like "creepy", "intense", and "nightmarish." It is important to think through production techniques that elicit such responses. The song is in a minor key and has a moderately slow tempo, but other factors contribute to the nightmarish affect. For example, the soundbox for this song is quite large—the synthesizer and guitar are heavily reverberated to provide a sense of distance—yet no instruments occupy the high frequency range, revealing a darker timbre.

Even more advanced discussions could take place concerning what elements define styles and genres, the practice of sampling, or other theoretical ideas. Massive Attack's "Unfinished Sympathy" is a rich example in this regard. The song features elements which allude to various genres: a marching hip-hop drum beat, a soul/R&B lead vocal, a jazz-influenced piano, and a classical string orchestra. The song also features a prominent vocal sample from Mahavishnu Orchestra's "Planetary Citizen". In "Planetary Citizen", this "hey hey hey" vocal signifies excitement, but in the new context of "Unfinished Sympathy", it sounds pleading and anguished. Once students have identified a song's textural layers and recording qualities, they can explore interpretive and theoretical questions to deepen their understanding of the song and its relationship to the wider repertoire.

Listening Lessons

In my critical listening lessons, I have followed a laboratory approach as encouraged by Allsup (2016), in which I ask questions to elicit reactions about a song's elements and functions. When appropriate, I use a song to teach a specific acoustic phenomenon or production technique. But my main task, as Abramo (2014) suggests, is to facilitate a *sense* of what the class has heard, and lead students to their own interpretations. Or, as Green (2008, pp. 87–91) describes, the class and I try to determine inter-sonic meanings about musical relationships. We may propose a "delineated meaning"—what a song represents—but only as one possibility among many.

When playing Steely Dan's song "Reeling in the Years", a number of students have responded that the song sounded light and fun. However, the lyrics lament a romantic breakup, with lines like "are you stowing away the tears/have you had enough of mine?" What the song *means* in a poetic sense is up to each individual. What matters for music students is recognizing the elements that make the song sound light and fun—in other words, the inter-sonic meanings. The major mode, the quick tempo, the shuffle groove, and the balance of instruments (across the stereo field and frequency range) all work together to create the sense of lightness.

The ten songs that follow cover a 51-year period of recorded music in a variety of popular genres. For convenience, I have arranged these in chronological order. Rather than provide a detailed analysis for each, I have listed a few guiding questions I have asked my classes. These questions are aimed at the college level; one can use simpler, broader questions for younger students. As mentioned previously, the choice of song is important as well, not only for lyrical content but also for accessibility. I would recommend from this list "I Heard It Through the Grapevine", "Conga", and "Single Ladies" for elementary school. Secondary/high school could add any of the others except "Bulls on Parade", due to profanity. In elementary and high school classes, I recommend listening to just one song in each lesson, to keep the discussion brief and focused. In college classes, one might play several songs during a one-hour lesson. This allows the class to think critically about each song, and also discuss production techniques or musical ideas common to the songs.

Beatles—"Strawberry Fields Forever"—1967

- What is the sonic landscape of this song? Does it change?
- How does the voice sound in the first verse compared to the rest of the song? The voice is mechanically slowed down for verses 2 and 3.
- What happens at the end? The song fades out, fades in, then fades out once more.

Marvin Gaye—"I Heard It Through the Grapevine"—1968

- What is our sense of space, and where are the voices and instruments placed? Let's draw a "soundbox".
- Compare the lead vocal to the backing vocals. Which is closer or farther away? The lead vocal is heavily reverberated, yet it is louder than the "dry" backing vocals.
- How do the strings sound? They may sound thin, since they were equalized to remove their lower frequencies. This helps create balance in the midrange.

Steely Dan—"Reeling in the Years"—1972

- In one word, how does this song sound? Responses may include words like "bright" or "clear". The instruments are carefully stratified in the frequency spectrum, so there is minimal masking or muddiness.
- Besides frequency placement, the instruments are generally played staccato, and sustaining instruments like cymbals are avoided. What affect does this have?

Charlie Daniels Band—"The Devil Went Down to Georgia"—1979

- How does the music represent the two characters? The Devil's solo is quite dissonant, and he is accompanied by distorted guitar. Johnny's solo is more consonant, and he is accompanied by a country band.
- What other sound effects do you hear that help illustrate the story?
- The narrator is quite close, yet heavily reverberated. What effect does that have on the listener. Does that make the story sound scarier, more epic?

Phil Collins—"In the Air Tonight"—1981

- Describe the sonic landscape and what elements lead you to imagine that.
- What effects are placed on the voice, and how does that contribute to the landscape? Slap-back echo and vocoder may make the voice sound larger, robotic, etc.

- Why are the drums saved for the end?
- How "natural" do the drums sound? This gives a chance to discuss drum machines as well as the song's famous gated reverb.

Miami Sound Machine—"Conga"—1985

- Which instruments are in the rhythmic layer, and which are in the "fun stuff" layer? For example, the piano may serve both functions.
- The vocals in the chorus are fast and rhythmic. Are the vocals "only" melody, or do they serve as a rhythmic layer as well?

Massive Attack—"Unfinished Sympathy"—1991

- Listen to the samples used in this song (from "Parade Strut" by J.J. Johnson and "Planetary Citizen" by the Mahavishnu Orchestra). How do they sound in this new context?
- What are some of the genres alluded to here? Hip-hop, classical, marches, jazz?
- What is the orchestra's role? What layers does it operate in? It performs the harmony, the bass line, and also the melody.

Rage Against the Machine—"Bulls on Parade"—1996

- What is the overall emotional affect? Anticipate answers alluding to anger and power.
- What aspects of this recording sound angry or powerful? The distorted guitar, the voice treatment (shouted and processed).
- What is the melodic layer here? Do rap vocals and guitar noises count as melody?

Beyoncé—"Single Ladies"—2008

- Did you notice the short synthesized sounds? What do they add to the song?
- Think about density and thickness. How many Beyoncés are singing in the different sections?
- What is the bass line in this song? Why does it only occur during the chorus?

St. Vincent—"Digital Witness"—2014

- Listen to the rhythm of the drums versus the rhythm of the horns. How do they contrast?
- What kind of emotion is the voice aiming for? It's rather affectless—how does that reflect the lyrics?
- How does the arrangement for the verse contrast with that of the chorus? The verse is dry and staccato, while the chorus features sustained chords and more active percussion.

Conclusion

In this chapter, I presented strategies for using popular music recordings to help elementary- and college-aged students develop critical listening skills. Listening, and exposure to multiple styles of music, is desirable for students (and often required by state educational standards). My approach provides a fresh and fun way to introduce both. While drawing upon critical listening concepts used by audio engineers, I showed that students of all age groups can listen for elements of recordings. Students can use their experience in these lessons to improve their listening in other contexts, such as performing in an ensemble.

To listen critically to music, the listener must attend to multiple dimensions of a given piece. Musicians do not simply create a lattice made from pitch and rhythm, as implied by harmonic dictation exercises. Instead, musicians—in both live and studio contexts—create sonic landscapes with pitch and rhythm, but also with timbre, texture, dynamics, and space. Audio engineers are well-versed in attending to these aspects of music recordings, but students from elementary school through college can also discern these elements. With some guidance, students will be able to "make sense" of a given recording by articulating affective and associative descriptions. As they refine their listening skills and learn new vocabulary, they will better articulate what they can hear in a recording.

Two rubrics for describing elements in a recording will be helpful for students. The first, applicable from elementary age through college, is Beat, Bass Line, Melody, and "Other Fun Stuff". This reflects Moore's categories of Explicit Beat Layer, Bass Layer, Melodic Layer, and Harmonic Filler Layer. These categories help students focus on the role of each instrument in a piece. A more advanced rubric, which I derived from audio engineering texts, is Color, Dynamics, 3-D Space, and Balance. This is especially useful for college students enrolled in introductory music technology classes, since it leads them to think about sound quality in recordings as well as ensemble performance. It is also useful for thinking about the ways in which composers and producers use timbre and texture to articulate formal design and style.

In my listening lessons with 3rd-grade students, I played one song, with an emphasis on learning to differentiate timbre and the roles/layers of each instrument. In my listening lessons with college students, I presented hour-long themed lessons discussing four or five songs. In both cases, I asked questions that alluded to the aforementioned categories: what are the roles of instruments, where are they placed in three-dimensional space, and what is the overall effect? It is important for student musicians of any age to learn how to listen, and to learn ways in which musical elements interact, in order to find their personal "sense" of a song's meaning.

References

Abramo, J. (2014). "Music Education That Resonates: An Epistemology and Pedagogy of Sound". *Philosophy of Music Education Review*, 22 (1), pp. 78–95.

Allsup, R. (2016). *Remixing the Classroom: Toward an Open Philosophy of Music Education*. Bloomington: Indiana University Press.
Bell, C. (2018). "Critical Listening in the Ensemble Rehearsal: A Community of Learners". *Music Educators Journal,* 104 (3), pp. 17–25.
Beauchesne, D. (2007). "Digital Composing and Recording as Arts Integration: The Centennial Place Experience". *Journal for Music-in-Education*, 1 (1), pp. 159–160.
Ceraso, S. (2014). "(Re)Educating the Senses: Multimodal Listening, Bodily Learning, and the Composition of Sonic Experiences". *College English*, 77 (1), pp. 102–123.
Corey, J. (2017). *Audio Production and Critical Listening: Technical Ear Training*. 2nd edn. New York: Routledge.
Department for Education (2013). "National Curriculum in England: Music Programme of Study—Key Stage 3" [Online]. Available at: www.gov.uk/government/publications/national-curriculum-in-england-music-programmes-of-study (Accessed 5 May 2019).
Foundation for Critical Thinking (2019). "Glossary of Critical Thinking Terms". Available at: www.criticalthinking.org/pages/glossary-of-critical-thinking-terms/496 (Accessed 12 May 2019).
Green, L. (2008). *Music, Informal Learning, and the School: A New Classroom Pedagogy*. Burlington: Ashgate.
Johnson, D. (2011). "The Effect of Critical Listening Instructions on Verbal Descriptions of Music". *Journal of Research in Music Education*, 59 (3), pp. 257–272.
Kratus, J. (2017). "Music Listening Is Creative". *Music Educators Journal*, 103 (3), pp. 46–51.
Moore, A. (2012). *Song Means: Analysing and Interpreting Recorded Popular Song*. New York: Ashgate.
Moylan, W. (2002). *The Art of Recording: Understanding and Crafting the Mix*. New York: Focal Press. Available at: www.oreilly.com/library/view/understanding-and-crafting/9780415842808/ (Accessed 5 May 2019).
National Association of Schools of Music (2018). "NASM Handbook 2018–19" [Online]. Available at: https://nasm.arts-accredit.org/wp-content/uploads/sites/2/2019/01/M-2018-19-Handbook-1-7-2019.pdf (Accessed 4 May 2019).
National Coalition for Core Arts Standards (2014). "National Core Arts Standards: Music at a Glance" [Online]. Available at: www.nationalartsstandards.org/sites/default/files/Music%20at%20a%20Glance%20rev%2012-1-16.pdf (Accessed 9 May 2019).
Owsinski, B. (2006). *The Mixing Engineer's Handbook*. 2nd edn. Boston: Thomson Course Technology PTR.
Wishart, T. (1996). *On Sonic Art*. New York: Routledge.
Worthington, D. and Bodie, G. (2018). "Defining Listening: A Historical, Theoretical, and Pragmatic Assessment". In *The Sourcebook of Listening Research*. Hoboken: John Wiley and Sons.

Discography

Armstrong, R. (1975). Planetary Citizen. Mahavishnu Orchestra. *Inner Worlds* [mp3 file]. New York: Sony.
Becker, M. and Fagen, D. (1972). Reeling in the Years. Steely Dan. *Can't Buy a Thrill* [CD]. New York: ABC Records.
Brown, J. (1991). I Got You (I Feel Good). James Brown. *20 All-Time Greatest Hits!* [CD]. London: Polydor.
Clark, A. (2014). Digital Witness. St. Vincent. *St. Vincent* [mp3 file]. Beverly Hills, CA: Loma Vista Recordings.
Collins, P. (1981). In the Air Tonight. Phil Collins. *Face Value* [CD]. New York: Atlantic Records.
Commorford, T., et al. (1996). Bulls on Parade. Rage Against the Machine. *Evil Empire* [CD]. Los Angeles: Epic Records.
Daniels, C., et al. (1979). The Devil Went Down to Georgia. The Charlie Daniels Band. *Million Mile Reflections* [CD]. Los Angeles: Epic Records.
Del Naja, R., et al. (1991). Unfinished Sympathy. Massive Attack. *Blue Lines* [CD]. London: Virgin.
Garcia, E. (1985). Conga. Miami Sound Machine. *Primitive Love* [mp3 file]. Los Angeles: Epic Records.

Johnson, J.J. (1974). Parade Strut. J.J. Johnson. *Willie Dynamite: Music from the Original Motion Picture Sound Track* [mp3 file]. Universal City, CA: MCA Records.
Lennon, J. and McCartney, P. (1967). Strawberry Fields Forever. The Beatles. *Magical Mystery Tour* [mp3 file]. Los Angeles: Capitol Records.
Stewart, C., et al. (2008). Single Ladies. Beyoncé. *I Am . . . Sasha Fierce* [mp3 file]. New York: Columbia Records.
Whitfield, N. and Strong, B. (1968). I Heard It Through the Grapevine. Marvin Gaye. *In the Groove* [mp3 file]. Detroit: Tamla Records.

7

Listening Sessions

Critical Listening in a Social Context

M. Nyssim Lefford

Introduction

Audio engineering students usually develop technical listening skills through autonomous study and practice, by listening to recordings and through ear training programs. However, it is by working in a group that one learns how to listen to others and appreciate their perspectives. Social listening activities build the abilities to learn from listening, identify qualities worthy of attention, and manage complexity—without the support of a teacher's guidance. Students transitioning from university audio programs to professional domains need these skills. Formal education can scaffold this kind of learning with extracurricular *listening sessions* where student-peers spontaneously appraise and critique recordings in a social environment, minimally mediated by a teacher.

Listening sessions build more than listening skills. They create learning opportunities in which students can explore the relationships between content creation and audience perceptions. Furthermore, they are places to develop communication and social skills, which are essential for managing interaction in the work place.

> "What you learn from working with other performers and musicians is invaluable, really, and can only help you grow. I mean, if you spend your whole life focusing on yourself, you're not really learning much." Damon Albarn
>
> (in Mulholland 2011)

Albarn is one among many practitioners who acknowledge the crucial influences of community and interaction on their work. Herein lies a lesson for those teaching tomorrow's music and audio professionals. Creative practitioners, including professional audio engineers, utilize for the betterment of their own work what they learn from interacting with and observing others. They heed feedback from collaborators and audiences. Attending to others' perspectives and opinions requires motivation and prowess, and these traits that are not necessarily innate in all audio students. People are not born wanting critique or knowing intuitively how to receive and utilize feedback well. We are born preoccupied with our own experience, and over time, learn to appreciate the value of understanding the experiences of others. University audio programs have a consequential contribution to make to this learning, and thereby to students' professional development.

Students generally arrive at university expecting and wanting feedback and critique from teachers; but as students' grades are rarely dependent upon peer evaluations, not all students automatically ascribe similar value to the ideas and experiences of their peers. Our curricula can directly and implicitly shape what they come to value. Giving and taking feedback, sharing ideas constructively, contributing to fruitful discussions, these are skills that consummate creative professionals hone, and that our audio engineering curricula may be inadvertently be deemphasizing.

For example, the apportioning of time to topics sends students signals about what is deemed important. To take the case used throughout this chapter, music production, when we dedicate a lot of time to examining attributes of recordings and very little time to discussing the creative and technical processes and social interactions that yield those attributes; students are likely to infer that—in their learning at least—recognizing and labeling attributes is more valuable than reflecting on the processes that yield those attributes. Without additional educational experiences to tip the balance, this value is then likely to be transferred to their views on professional practice. In actuality though, audio professionals work with the intrinsic interrelations of process and product. By seeking out feedback, by inquiring after listeners' varied experiences of attributes in varied context, professionals come to understand that many may agree on the labeling of an attribute, but not on the affect or appropriateness of its occurrence in every circumstance. Professionals grow to appreciate this complexity by focusing on others' experiences.

Reflecting

Even so, our natural instincts to prioritize our own experiences are not completely counterproductive to learning. All learning demands a certain amount of self-directed attention and reflection, a certain amount of self-observation and inwardly directed questioning of one's own thinking and performance. However, Von Glasersfeld argues that learning is not a product of mirroring information, rather it is a process of construction, as concepts are tested out and "adapted" to circumstances and ultimately deemed "viable" given the learner's "range of experience" (Von Glasersfeld 1998: 4). Learners assimilate this new knowledge (Illeris 2017) over time with experience of testing out concepts in context. These concepts are essential attributes of Piaget's constructivism (Piaget 1971). Within the constructivist learning framework, to test out adaptations, for there to be growth and development, the applicability of internalized concepts must be challenged with new perspectives and increasing levels of complexity to manage.

Not Knowing

Jarvis claims that learning begins when we experience "not-knowing". Learners "enter a state of disjuncture and immediately raise questions: What do I do now?" (2009: 27). Jarvis goes so far as to suggests that disjuncture connects to the learner's "biography" or identity, which suggests the ties between receptivity to learning and self are deeply personalized and subjective. Other social constructionists, like Illeris (2017) suggest that Jarvis perhaps takes the "gap" too far, or at least, it may be more accurately understood as dependent on the age, maturity and thus the receptivity of the learner. In any case, not knowing and confronting subjective bias is challenging and potentially intimidating to some students, even at university level. Not knowing inevitably places the learner in a vulnerable position. Educators may be reluctant to address these issues, but receptivity and subjective bias are problems for audio education. Bielmeier (2018) notes that audio industry employers have concerns about "New Hires" who are "not interested in feedback and have an inability to take criticism" (Bielmeier 2018: 4).

Open to Many Perspectives

Professional audio engineers cannot operate well from within a subjective bubble. Professionals interact with clients. They may outsource to or engage colleagues with needed expertise (e.g., mastering engineers). Professionals work in teams, with other audio and non-audio professionals; working alongside, acknowledging, accepting input from and interacting with *others* who have complementary skill sets. To function in this collaborative context, professionals must know how to apply their expertise, use inwardly focused reflections, and also, integrate information about and from others and gleaned from what they observe in others' work. For a professional, this synthesis of concerns and perspectives is—it has become—a natural part of their daily activities.

Educators face the challenge of creating learning environments and experiences that encourage not knowing before knowing, openness to feedback (as an invaluable tool for development) and curiosity in these things. How do our curricula develop these capacities to integrate information from others, and to notice others' perspectives and experiences? Some kinds of learning activities make underdeveloped abilities more readily apparent; for example, activities where engineering and music students work together in a recording session. Engineering students, preoccupied with their own performances in tasks such as placing microphones or adjusting pre-amps, fail to notice the musicians restlessly playing the same thing over and over or noodling, growing tired and bored, which incidentally causes the musicians to retreat into their own subjective bubbles. Professional engineers and producers are sensitive to such issues and take steps to avoid them—precisely because they are counterproductive to collaboration; be it a collaboration in a music or game studio, a theatrical drama, TV news broadcast, etc. Furthermore, professionals, even as they consider the experiences of their colleagues, also consider the expectations of the audience who will ultimately listen to the content created as a result of collaboration. All these interests are being balanced; but in classes, we typically focus on engineering concepts rather than building the skills needed to find an effectual balance.

The Social Context

Managing bored musicians and audience expectations requires social awareness as well as technical skill. A student whose focus is pinned on the recorded artifact easily forgets that "[m]usic is not a thing at all but an activity, something people do" (Small 2011: 2). The role playing, group projects, and assisting in professionally led, in-school sessions commonly found in audio education, while extremely valuable, are (according to Bielmeier's 2018 report) often not enough to instill in students a deep appreciation of others' perspectives, particularly others who do different kinds of work, and have different methods, concerns, and values. Employers surveyed, report that once in a professional environment:

> [audio] students' tone with clients did not match the required tone of the situation (i.e. either too passive or aggressive) and often the student was not able to assume a proper role within the studio hierarchy. This type of nuanced interpersonal skills is developed over time and can only be honed with situational learning environments (i.e. internships).
> (Bielmeier 2018: 6)

In the professional sphere interpersonal problems or a lack of social awareness can be cause for dismissal. In school there are comparatively few consequences for not playing well with others.

Business structures and workplace practices are evolving, and thus strict hierarchies may soon be less strict than they once were. Nevertheless, audio production continues to be a field where individuals with varied skills and talents work together in one way or another. For professional audio working environments to maintain efficacy, financial viability, and the integrity of work products, professionals in management positions, new hires, and educators must all acknowledge that someone with 30 days of experience is simply not in the same professional position as someone with 30 years of experience. The experienced and less-experienced cannot bring the same knowledge and abilities to their work, and therefore professionalism requires an appropriate apportioning of responsibilities, credibility, and opportunities. New hires need to be able to find their niche.

But even that is no guarantee of their effectiveness within that position or as they grow professionally. Professionals at higher levels of experience and expertise are able to think critically about the concerns of varied stakeholders around them, as well as, the technical concerns associated with their responsibilities. They are able to fold this information into engineering actions. Hence, superficial awareness of others is not enough to flourish professionally over the long term. Another layer of reflection, reflections on others, is needed to manage how the technical impacts the social and how the social impacts technical solutions. The Conceive Design Implement Operate (CDIO) initiative in non-audio engineering education (www.cdio.org)[1] has for decades prescribed methods for encouraging creativity, awareness, and foresight in engineering students, predominantly by building critical thinking skills not only about the implementation of solutions but also about the ramifications of those solutions for society and the environment, among others. Audio engineering needs practitioners with similar critical thinking abilities.

Critical thinking involves communication. Communication is the means by which pertinent information is gathered and by which ideas are tested, by which solutions are posed and efforts coordinated. In music production, for example, collaborators are in constant communication, back and forth between the engineer and producer and among the producer and the engineer and the musicians; all of who have different perspectives. Very often they communicate about what they hear in what is being produced. They listen together. Listening together provides an anchor for collaboration and the exchange of ideas. Audio education can use the same activity to develop students' communication and critical thinking skills.

Music Production as an Example

This chapter uses music production as an example of listening together, but the proposed approach is adaptable for most kinds of critical listening. Still, we should consider that an interest in music production can in some audio programs be something all students have in common. Regardless of their primary professional ambitions, students usually appreciate recorded music as a cultural phenomenon, and most have opinions about it. Additionally, current practices in bedroom producing or DJing can obscure the fact that music production has historically been a collaborative activity. All this makes it rich material for teaching about the perspectives of others. Listening together and critically, with the appropriate scaffolding (Belland 2017), is an invaluable means for teaching students about participating in sound making in the broadest sense, about being part of a professional community, interacting with others, and communicating to creators and audiences.

Listening Sessions

Not just any kind of listening will develop the skills described here. Most teachers play audio examples to demonstrate and illustrate. That has its purpose, but in the classroom, students are fixated on acquiring new concepts and their own comprehension. Classroom listening, because it is so highly structured, does not encourage students to notice, much less reflect on, the experience of sharing ideas among peers. For that, some of the structure that makes a class class-like needs to be removed. Dedicated listening sessions, outside of a classroom setting that allow students and teachers to discuss informally what they hear can meet these criteria. They provide opportunities for students to learn about what others hear, and to integrate that with what they know.

The purpose of a listening session is to create a collaborative listening experience. Like a typical studio-based music recording session, this is a time-bounded activity in which collaborators work together to produce a deliverable. Each collaborator brings to this work expertise and an individual perspective. Perspectives and expertise are integrated into the work product through interaction. A music recording session produces a recording. A listening session produces insight into the experience of listening to recordings. My own "listening sessions" are extracurricular and usually last about two hours (on occasion we have run over).

Audio students from all years of study and faculty are invited to attend and to bring recordings to play for each other or to just listen. As space permits, students and faculty from other relevant programs are also invited. Then, very simply, we play recordings and discuss them. Occasionally, there are themed sessions (focused on genres that students may not normally encounter on their own, such as 20th-century electro-acoustic or avant-garde music, for example), but usually anything goes: commercially produced recordings, mixes they are working on, game/film soundtracks, anything. These listening sessions can deliver an educational experience that more traditional educational activities cannot.

Play, Discuss, Repeat

Play, discuss, repeat. Though easy to implement, listening sessions are not at all simple in terms of pedagogic aims and objectives. The primary aim is to listen and get everyone talking spontaneously about what they hear, sharing different perspectives and impressions. Explaining to each other out loud is as valuable as asking questions. Not only does articulating one's meaning nurture reflection and critical thinking, externalizing knowledge makes ideas available to others for elaboration and critique. Questioning facilitates deconstruction, and by discussing what is heard and experienced, students probe the (probable) rationale behind engineering and production choices and when, what, and why certain techniques are deemed affective. This can subsequently inform their own problem solving.

In addition to hearing new ideas, at these sessions, everyone hears new music or sound recordings or has opportunities to hear something new in something familiar. When listening sessions are driven by student-selected content, they usually include some of the newest commercial recordings, making them one of the few educational activities that can keep pace with real-world developments. However, regardless of whether the content is new or just new to the listeners, listening sessions add to both the students' and the faculty's "aural libraries" (Schippers 2013). Relatedly, according to McIntyre, in music producing, to function professionally, producers need to understand their "domain", which "includes the body of songs they use as a template to make judgements in the studio. The more a producer understands the domain the stronger their knowledge will be and the greater their ability to produce work in a studio situation" (McIntyre 2008: n.p.). The more examples students catalog in their aural libraries, the more possibilities they carry into professional situations.

Furthermore, listening sessions can broaden perspectives on music recording by providing a forum to discuss topics that often slip through audio engineering curricula; for example, the history of recording (i.e. earlier practices or equipment), movements in music or art; social theory; or other historical developments that have impacted music production and communication. Audio as a profession responds and adapts to cultural as well as technological shifts. Professional audio engineers are always working within a broader context—a social, economic, and cultural context. That context imposes norms, expectations, and values above and beyond any held among the immediate community of (audio) practitioners (Lave and Wenger 1991); and to these, audio professionals are answerable. Periodically, norms, expectations, and values are challenged, and at these times media plays a critical role in social change. Rock, punk, and rap have all sparked social change, and somewhere in the mix of media and culture were professionals engineering influential content. Audio curricula offer few chances to discuss the engineer's responsibilities in these matters.

As If

To sufficiently prepare students for this complexity and other real-world challenges, the constructivist approach to learning (Piaget 1971) advocates learning by doing with others. The *doing* and *with others* are equally important. Learning with others who are learning similar concepts leads to proximal development (Vygotsky 1978). Co-learners naturally compare their understanding, and that adds naturalistic complexity. Learners thereby consider multiple perspectives and approaches . . . when there is time and inclination to do so.

Audio educational programs typically emphasize the importance of doing or practical training, for example, recording sessions with fellow students and student musicians. These in-school, "authentic activities" provide opportunities to assume professional-like responsibilities in "learning contexts that model proficiency" (Barab and Hay 2001: 72). In these situations, students test practical skills and knowledge and get them wrong, and as a result, learn from those mistakes. Over time, (ideally) the complexity of these tasks increases and facilitation and supervision decreases, "fading scaffolding as students develop competence" (Barab and Hay 2001: 72) and strengthening students' independence. Collins describes such activities as "cognitive apprenticeships" (Collins 2006).

Obviously, there is a social component to in-school recording sessions. In these situations, students interact with student musicians and with peers. However, this interaction almost always happens in the context of trying to solve challenging technical or production problems at the edge of the students' abilities, under time pressures and under the scrutiny of examination. Students do need to develop coping skills and learn to manage stress. However, the combination of factors with which they must contend, can leave little time to think about the professional quality of social interactions; and indeed students are not always encouraged to prioritize quality in social interactions.

Professional environments demand the more sophisticated capacities to manage technical or production problems, under pressure *while also* negotiating social interactions. In 2010, Tough surveyed experienced engineers to identify "core competencies". Competencies related to communication skills ranked amongst the most important, including abilities to "[d]emonstrate the ability to be an effective listener towards co-workers and clients" (Tough 2010: 7). Bielmeier (2016, 2017, 2018) has also investigated the skills of audio program graduates; and reports, as of 2017, that many audio industry employers as well as recent graduates/new hires feel that necessary social and communication skills are lacking in graduates. A unique benefit of learning about audio technology in an academic context is that it—potentially—offers students opportunities to practice, build and reflect on both technical and social skills in proximity to others.

That said, the interrelations of technical and social factors are not easily demonstrated in the academic environment. In the professional environment, novices witness professional interactions all the time. It is difficult to model this behavior authentically in academic contexts. No matter how professional the behavior of teachers while interacting with other teachers, visiting practitioners, and students, school is not a professional production environment. When teachers assume the role of chief engineer in authentic activities, they do not set down their responsibilities as teachers. Visiting practitioners also make concessions in the educational environment, for mistakes and inappropriate actions. They answer questions they would not answer in professional contexts, and give explanations for things that they might expect an intern to learn through observation.

The academic environment's inherent tolerance enhances learning overall; but sometimes at the cost of other kinds of lessons, for example, in professionalism and professional discourse. Professionals discuss what they hear, among themselves, in particular ways, and in ways conducive to the work at hand. For teachers, the work at hand in the classroom is teaching. Teachers field all sorts of questions while probing student comprehension. Listening sessions fall somewhere in between a professional and an academic environment. A listening session is an "authentic" activity in that professionals do have these kinds of discussions, but often under conditions that involve time pressures and responsibilities that structure interactions. In an educational environment, the pressures of time and to produce a work product may be removed to create conditions under which students can attend solely to their verbal interactions, but still as if professional to professional.

Collaborative Production of Intelligibility

Another issue that limits how the real world is modeled for students is time. Only so many authentic activities can be crammed into one academic program. That number no matter how large cannot possibly reflect the diversity of real-world scenarios and circumstances that graduates will encounter in professional environments. Yet to understand the need to adapt and how to adapt in professional contexts, graduates need to grasp that recording/production/working scenarios are similar in some ways and unique in others; that problems and potential emerge from local circumstances and conditions. There is no single set of rules that yields an intelligible recording, no check list of correct actions. It is not a quiz. One cannot memorize correct solutions. Sound engineering is a "situated" practice; meaning the engineers' actions are

> ad hoc responses to the actions of others and to the contingencies of particular situations. . . . [S]uccessful interaction consists in the collaborative production of

> intelligibility through mutual access to situation resources, and through the detection, repair or exploitation of differences in understanding.
>
> (Suchman 1987: n.p.)

A professional's ad hoc solutions are contingent on local discussions through which understanding is shared with others. Listening sessions that offer opportunities to experience differing understandings, offer opportunities to explore how perspectives and recordings are similar in some ways and unique in others. As discussions around a recording, or situational resource, evolves, as students share their perceptions and experiences, they practice partaking in the collaborative production of intelligibility.

Learning itself is situated (Lave and Wenger 1991). The learning environment shapes what can be learned and how. One reason to move listening sessions out of the classroom into more informal contexts is to let the situation afford a different kind of learning potential. The structure of a lecture prioritizes the communication of information from teacher to student. Students ask questions and make comments from which the group as a whole learns, but the nature and quality of discussion is not the central focus of attention. In a listening session, thoughtful sharing to benefit the group as a whole is the priority. These conditions may motivate students to develop communication and interaction skills they cannot exercise in class. For example, instead of the teacher validating or refuting claims, in a collaborative group, concepts for which there is weak evidence or little agreement are abandoned in favor of stronger arguments.

> "You can read all the textbooks and listen to all the records, but you have to play with musicians that are better than you." Stan Getz
>
> (Martin 1986: n.p.)

Valuing Others' Ears

There are many ways students develop critical listening skills, including ear training programs as well as sharing examples in the classroom with a teacher drawing attention to attributes practitioners consider important. Both these activities help to inform the students' perspectives and develop abilities to identify similar attributes in other examples. Beyond this, professionals also have abilities to independently identify other important attributes, and also decide for themselves what makes them important. For example, professional engineers consider the listening audience's experiences, and strive to make decisions that meet or surpass their expectations. To understand those expectations, professionals, and therefore students, need an appreciation for what others hear; not just teachers and not just ear training programs, but a wide variety of others.

In the professional world, there are no facilitative, supportive teachers to direct listeners' attention, to identify who can provide suitable feedback or whose views should be appreciated. Listening sessions are a half-way point, an opportunity for all members of the learning community to share and compare understanding, and thereby direct each other's attention to things to consider and hear, all without the validation of an appointed teacher (tasked with evaluating student performance). Volunteering information about what is heard fosters students' independence in thought. Contributors have to explain their experiences, and work to understand the experiences of others to make those explanations relevant and meaningful to the context. In the process, they access a breadth of perspectives and compare.

No longer responsible for directing attention or the de facto chief authority, the teacher's role shifts to facilitating discussion and inquiry. Resnick proposes:

> [Teachers can] play a role as catalysts, sparking ideas, as consultants, helping people navigate the learning process and providing new strategies to avoid frustration, as connectors to bring people together—peers who can be helpful to each other—and also as collaborators themselves. We want to blur the boundaries so teachers become learners and learners become teachers.
>
> (in Kamenetz 2017: n.p.)

Communication and Vocabulary

To teach each other, we have to share a vocabulary. Recording classes provide students with terminology. They learn words to describe microphone types and techniques, signal processing effects, audible artifacts, the basic physics of sound, among other things; but this too is insufficient preparation for entering the professional sphere. In his study of graduates' competencies, Bielmeier revealed that "according to employers, almost half of New Hires learned the ability to apply and interpret a technical vocabulary as related to audio while on the job" (2018: 4). Moreover, a functional lexicon learned in one working context, though valuable, is also insufficient for the purposes of communicating and sustaining professional interaction across a variety of scenarios.

Anyone who has studied a second language knows that memorizing vocabulary or using it in a test is nothing like using those same words in a conversation with a fluent interlocuter. Porcello (2004) observes that, during professional recording studio interactions, novice intern audio engineers and experienced professional engineers and producers use different kinds of descriptors. Furthermore, learners of any kind of spoken language are slow and cautious. They have to think carefully about formulating what they are going to say next, and there is a general lack of confidence about pronunciation and correct usage. As soon as one is able to sustain basic interactions in a second, spoken language, it becomes apparent that fluent communication requires knowledge of colloquial expressions, commonplace idioms, and vernacular conventions.

There may be no fundamental language barrier when learning audio-speak, but it still takes time and practice to develop confidence using terms appropriately in context. Also, audio students are training their ears to identify things worth speaking about even as they are learning audio-speak. Like all language learners, they need practice speaking about audio spontaneously and without preparation. If that practice happens with faculty or more experienced students in the room, audio-speak learners have opportunities to pick up some of the vernacular.

However, communicating professionally is not only a matter of confidence. To have meaningful and not merely semantically correct exchanges, interactors have to share conceptual understanding. They need to find reference points about which they can speak around. When audio students interact with other audio students and audio faculty, common references are not hard to find. They all share knowledge of the curriculum and local environment. Professionals, meanwhile, have to interact professionally with non-engineers, with others working within another conceptual framework. As Nina Simone reminds us, "Jazz is not just music, it's a way of life, it's a way of being, a way of thinking" (in Brown 2010: 20). Successful jazz engineers understand jazz. They can communicate in jazz-ese and they understand how to approach engineering through a jazz-adapted conceptual framework.

But no matter how experienced or musically literate, everyone finds it difficult to describe sound with words. Groove. Authentic. Edgy. Relevant. Music production is filled with fuzzy, decidedly non-technical, concepts. Even seasoned professionals work at describing the sounds they hear or want to hear, and moreover identifying what about sound is important to describe to others in a particular working situation. These skills are worth practicing because it is hardest to describe, to steer attention toward, or come to consensus about sounds that do not yet exist, that might come to exist if work proceeds in a certain way. Professionals are able to find words for what is audible; and they are also able to describe hypothetical listening experiences, experiences that a production might bring about if certain decisions and actions are taken.

What if . . . Thinking

For the classroom, teachers often choose example recordings that unambiguously demonstrate key concepts and canonical and concrete examples. Through exposure to such materials, students acquaint themselves with norms and conventions. Professional creators need knowledge of the typical to map out their domain (McIntyre 2008). Listening session can contribute to domain knowledge when students (and faculty) bring familiar favorites and classics. Re-playing the canon invites re-examination and deeper consideration.

But there is no requisite set of examples. Aa professional's aural library (Schippers 2013) is never filled. Listening sessions expose both students and faculty to the unfamiliar, to new artists and genres and unheard recordings. This not only starts new mental categories in the aural library, but, more fundamentally, it illuminates the breadth of what is possible in terms of manipulating recorded sound, the many taking-off points for creativity. Listening sessions open doors for imagining *what if* . . . a sound could sound like. . . . Students who have physically heard potential and discussed it with others are better prepared to realize those possibilities in future recordings. Charlie Parker suggests, "Music is your own experience, your thoughts, your wisdom. If you don't live it, it won't come out of your horn" (in Woideck 1998: vii).

Safe Risk

Yet, not all students display an inclination to delve into the unknown. Communicating about new things in new ways inevitably pushes students to the edge of their experience, which they must surpass in order to grow. Listening sessions gently, socially pressure students to communicate about what they hear and think. However, assuming the listening session is extracurricular, unlike an exam, students retain control over how and how much they participate. At their own pace, they can observe, listen, and get acclimated before they risk contributing to discussions. The value of contributing cannot be overlooked. As Pete Seeger reminds us, "The easiest way to avoid wrong notes is to never open your mouth and sing. What a mistake that would be" (in Dunaway 2008: 95).

It is also worth mentioning that learning to participate is as important as learning about others' perspectives. To make this sharing environment work, listening sessions have to feel and be truly safe. Participants must feel like they can use whatever vocabulary they have to make their points. This should be a place/time where everyone can show that they know a lot about some aspect of sound and/or sound recording—a genre, a particular artist, or a favorite album/song or instrument. That kind of mastery usually falls outside standard coursework, but it can shine in a listening session.

Achieving these pedagogic aspirations places a lot of responsibility on the teacher moderating these sessions, even if in this context the teacher does not play a traditional teacher role. Here, the teacher/moderator sets the tone that makes any constructive contribution to discussions welcome, while also encouraging participants to elevate the level of discourse; to think critically about what is heard and commented on rather than fallback on preconceived notions and opinion; and to ensure that evaluations and criticisms feel fair. To make it safe and creative, it may be necessary to mediate negative critique. The teacher-moderator may need to actively support those who have ventured minority opinions. It is worth remembering that professional creativity and excellence happen in environments that feel safe and trustworthy, where it is okay to venture ideas and test things out. Students that experience that kind of environment in school are more likely to carry those expectations into the workplace, places such as music production studios where musicians, engineers, and producers bring varied perspectives to collaboration. Each participant in a music production has a subjective view, and (ideally) each makes way for other ideas; to the benefit of the collaborative work and the shared goal of producing a recording. The collaborative effort in listening sessions similarly tempers the subjective perspective. Since all participants share a mutual interest in learning from listening together, they provide some basic structure for learning how to have a constructive dialogue.

Ultimately, sounding like a professional requires acting like a professional. Professionals can operate outside their comfort zones. They know how to give and take feedback even when they do not want it. Professionals know how to offer, accept and work with differing opinions, even when these opinions are not in line with their own thinking. It can be scary to voice a contrary opinion among your peers, and in the real world, there are no facilitative, supportive teachers to broker compromises among differing opinions. In listening sessions, students learn how to present and accept alternative views constructively. If on the first try someone gets a little obnoxious, since these sessions are outside of class and outside of a professional environment, there can be a lightheartedness about it all, a gentle rebuff without rejection and without serious consequence. Everyone can work on what they need to work on, developing at different stages and speed, and finding a place for their perspectives within a community.

Community

> "You create a community with music, not just at concerts but by talking about it with your friends." David Byrne
>
> (in Cateforis 2013: 62)

Unlike the classes which are divided by year and/or specialization, listening sessions can bring everyone together. This can cultivate a sense of community in a program/school. As individuals, among peers, students see that they can safely have different ideas and interests, show their uniqueness; and at the same time, be part of a bigger whole. It is not a big leap from this to learning to value other people's views and appreciate knowing and working with others because they are different. *Others* are essential for one's own development and professionalism. An interest in people who do related work is good for their careers since audio professionals work in a field built on collaboration.

Finally, to be absolutely pragmatic, more and more, people's livelihoods depend on networking. To find work, people need social and communication skills that enable them to get to know people and their interests, and furthermore, abilities to surmise how their own skills complement other peoples' skills. Often, students are understandably so engrossed in their own development that they rarely think about taking advantage of the connections they can make while at school. However, today's student is tomorrow's producer in a position to hire a mixer. Getting students together to listen together in school (but outside of class) plants the seeds for a burgeoning professional community.

Conclusions

Listening sessions conjointly build critical listening, critical thinking, and communication skills. The process of externalizing the experience of what they hear in the presence of peers, encourages participants to (re)examine what is perceived, as well as, appreciate the experiences of others. By describing their perceptions, assessments, and opinions they develop their professional vocabularies while simultaneously contributing to others' development. These exchanges also demonstrate the inherent difficulties of communicating about recorded sound and offer opportunities for students to explore the values we ascribe to particular attributes. Listening sessions fill their "aural libraries" (Schippers 2013) with examples; and as they participate in deconstructing what they hear, they uncover the critical thinking behind engineering and production choices—when, what, and why certain techniques are deemed affective. The social component of the environment requires them to contend with others' perspectives and opinions, strengthening interaction skills while building a community of future professionals. To prepare students for professional work, audio engineering education, particularly as it is taught in the university context, needs to socialize future engineers even as it fosters technical aptitude. Listening together is one way to accomplish this.

Note

1. CDIO is a self-governing consortia of higher education institutions whose mission is to further excellence in engineering education, and thereby further engineering's potential to contribute positively to industry and society at large. CDIO offers standards and guidelines for engineering curricula and programs, standards and guidelines that reflect feedback from industry, engineering practitioners, as well as academic leaders.

References

Barab, S.A. and Hay, K.E. (2001). Doing science at the elbows of experts: Issues related to the science apprenticeship camp. *Journal of Research in Science Teaching*, 38(1), 70–102.

Belland B.R. (2017). Instructional scaffolding: Foundations and evolving definition. In *Instructional scaffolding in STEM education*. Cham, Switzerland: Springer.

Bielmeier, D. (2016). Audio recording and production education: Skills new hires have and where they reported. *Journal of the Audio Engineering Society*, 64(4), 218–228.

Bielmeier, D. (2017). Audio education: Audio recording production students report skills learned or focused on in their programs. Audio Engineering Society Convention 143. Audio Engineering Society.

Bielmeier, D. (2018). Future educational goals for audio recording and production (ARP) programs: A decade of supporting research. Audio Engineering Society Convention 145. Audio Engineering Society.

Brown, L. (Ed.). (2010). *John Coltrane and Black America's Quest for Freedom: Spirituality and the Music*. New York: Oxford University Press.

Carl Woideck, C. (1998). *Charlie Parker: His Music and Life*. Ann Arbor, MI: University of Michigan Press.

Cateforis, T. (2013). *The Rock History Reader*, Second Edition. Oxfordshire, UK: Routledge.

CDIO (n.d.). Retrieved from www.cdio.org/ [Last retrieved on 25 May 2019]

Collins, A. (2006). Cognitive apprenticeships. In *The Cambridge Handbook of Learning Science*, Sawyer, R.K. (Eds.). New York, NY: Cambridge University Press, 47–60.

Dunaway, D. (1981, 2008). *How Can I Keep from Singing? The Ballad of Pete Seeger*. New York, NY: Villard Books.

Illeris, K. (2017). Peter Jarvis and the understanding of adult learning. *International Journal of Lifelong Education*, 36(1–2), 35–44.

Jarvis, P. (2009, 2018). Learning to be a person in society. In *Contemporary Theories of Learning: Learning Theorists . . . in Their Own Words*, Illeris, K. (Ed.). Oxfordshire, UK: Routledge, 21–33.

Kamenetz, A. (2017). How to make every grade more like kindergarten. *National Public Radio*. Retrieved from www.npr.org/sections/ed/2017/09/18/550448626/how-to-make-every-grade-more-like-kindergarten [Last retrieved on 25 May 2019]

Lave, J. and Wenger, E. (1991). *Situated Learning: Legitimate Peripheral Participation*. Cambridge: Cambridge University press.

Martin, M. (1986). Stan Getz interviewed by Mel Martin. *The Saxophone Journal*, 10(4), Winter, Reprint.

McIntyre, P. (2008). The systems model of creativity: Analyzing the distribution of power in the studio. *Journal on the Art of Record Production*, 3.

Mulholland, T. (2011). Damon Albarn's shapeshifting career. *The New York Times*. Retrieved from www.nytimes.com/2011/07/26/arts/26iht-albarn26.html [Last retrieved on 15 August 2018]

Piaget, J. (1971). *Psychology and Epistemology: Towards a Theory of Knowledge*. New York: Grossman.

Porcello, T. (2004). Speaking of sound: Language and the professionalization of sound-recording engineers. *Social Studies of Science*, 34(5), 733–758.

Schippers, H. (2013). Practitioners at the centre: Concepts, strategies, processes and products in contemporary music research. *Research and Research Education in Music Performance and Pedagogy*, 11, 1–8.

Small, C. (2011). *Musicking: The Meanings of Performing and Listening*. Middleton, CT: Wesleyan University Press.

Suchman, L.A. (1987). *Plans and Situated Actions: The Problem of Human-Machine Communication*. Cambridge, UK: Cambridge University Press.

Tough, D. (2010). Shaping audio engineering curriculum: An expert panel's view of the future. Audio Engineering Society Convention 129. Audio Engineering Society.

Von Glasersfeld, E. (1998). Cognition, construction of knowledge, and teaching. In *Constructivism in Science Education*. Dordrecht: Springer, 11–30.

Vygotsky, L. (1978). Zone of proximal development. In *Mind in Society: The Development of Higher Psychological Processes*. Cambridge, MA: Harvard University Press.

8

Engineering and Recording

Jeremy J. Wells

Introduction

> A graduate [of my "school for soundmen"] . . . should be able to study a score and build up in his imagination a perfect image of the sound of this score. His ear training shall then enable him a) to find out whether or not a recording harmonizes with the image in his imagination; b) to define exactly the differences between his image and the sound of the recording; c) to indicate what should be corrected and how; d) if necessary suggest or even carry out such mechanical improvements of all acoustic, physical or mechanical devices as might be necessary for the case in question.
>
> (Schoenberg, 1940, cited in Feisst, 1999, pp. 104–105)

The ideas expressed in the preceding quote are both important and problematic. Important because they are a starting point of the "Tonmeister concept" (Borwick, 1973), but problematic because of the implicit prejudice against (sound) women. Sound recording, along with the broader subject area of music technology, enjoys popularity (albeit with a profound gender imbalance) as a subject for study at degree level and it appears in many different kinds of programs (Boehm, 2007). It is often viewed as a multidiscipline or interdiscipline, yet it is typically taught within a traditional departmental (or faculty) structure. In such structures the arts are usually separated from engineering, technology, and science, and this segregation is also implicit in the mono-disciplined categories of bachelor's and master's degrees.

Thus, sound recording may be a component of BMus (the UK denotation of bachelor of music), BEng (bachelor of engineering), BA (bachelor of art), or BSc (bachelor of science) degrees. There may also be accreditation by a body, such as the Institution of Engineering and Technology (IET) for UK BEng and MEng degrees in electronic engineering. The term "recording engineer", along with specialisms such as "mix engineer" and "mastering engineer", is commonly used in the documentation that accompanies commercial releases of sound recordings. Although, in the era of downloading and streaming, such documentation has become rarer, it still exists.[1] Yet, there is discontent about this use of the term engineer, particularly amongst some chartered engineers, since many of people who receive this designation on recordings are not chartered (see, for example, the justification for a recent petition to the UK government to legislate to protect the term "engineer" (Robson, 2012)).

There is an opportunity for the nature of both engineering and recording to be explored as part of education and training in the area of sound recording and music technology. If those students and professionals who undertake recording understand how it intersects with the discipline of engineering, and engineers can identify elements of their discipline within sound recording, existing as it does at the interface between musical performance and audio engineering (the design and creation of audio recording, processing, and reproduction tools), then both can be better understood. For some who aspire to a career in sound recording, there are notions of fame and cachet associated with this profession which are not shared by other areas of engineering. This may well be a misperception but it offers a chance to explain engineering as connected to and/or distinct from sound recording, as students transition from (possibly romanticized) perceptions toward concrete realities of recording industries.

This chapter presents a review of the manifestations of engineering within typical professional recording activities, as perceived by audio professionals, educators, and students. The ideas presented are based on existing literature and a number of interviews and focus group discussions conducted by the author at various points over the last seven years.[2] These qualitative data represent views of sound recording practice from those in upper-secondary education making decisions about career and higher (tertiary) routes through those in higher education, professionals at various stages of their careers, and on to those who have retired after decades of work in the industry.

Although commonalities emerge, there are also significant disconnects in perceptions of what sound recording entails and the extent to which the skill set of the engineer is embodied in the discipline. Some of these suggest different epistemologies of sound recording and audio production, and the relationships to those found in engineering are explored. This chapter does not set out to determine or impose a single definition of engineering or a single idea of how sound recording and music technology education should be delivered. Rather, it seeks to present a range of views from a variety of sources, ranging from those who classify themselves as engineers to those who consider themselves musicians, from professionals through educators to students. The purpose is to encourage and inform debate amongst educators in particular, but also professionals and students.

Recording and Engineering as Acts and Disciplines

> Discipline (1): training which produces obedience (= willingness to obey) or self-control, often in the form of rules, and punishments if these are broken, or the obedience or self-control produced by this training.
>
> Discipline (2): a particular area of study, especially a subject studied at a college or university (Cambridge Dictionary, 2019).

Patrick Flandrin, a prolific and influential figure in time-frequency analysis and processing of signals, has suggested that signal processing can be viewed as a three-body system, as shown in Figure 8.1 (Flandrin, 2011). Perhaps a useful starting point for the discussion in this chapter is to adapt this to recording of musical sound[3] and this is shown in Figure 8.2 (Boehm's "triad of music technology" does something similar (Boehm, 2007, p. 9)).

It can be seen immediately that making a recording involves three traditional disciplines, maybe more (which is certainly the case if psychoacoustics belongs to psychology rather than physics). Of course, this cross-disciplinary nature of recording is reflected in the fact that different people undertake the tasks of musical performance, operation of sound capturing equipment and design, development, and manufacturing of that equipment. But to what extent do these individual roles, and particularly that of the person(s) who captures (records) the performance, reside within these single, traditional disciplines and to what extent do they exist across them? Also, to what extent are there separately identifiable tasks within the role of recording? What are the differences in the engineering and production of a recording? Which term is more appropriate for identifying the specific role of

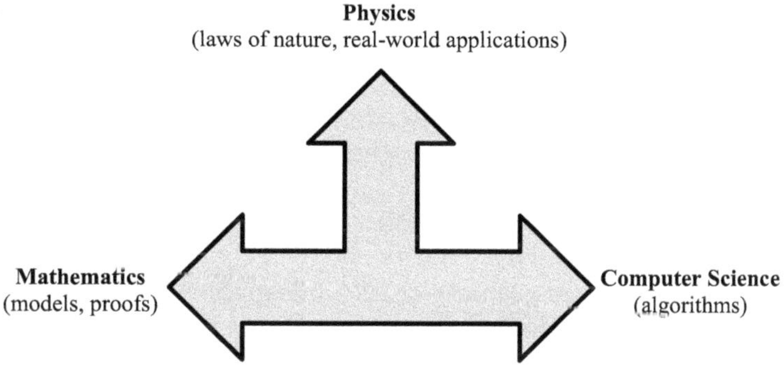

Figure 8.1 The "three-body system" of signal processing, from (Flandrin, 2011).

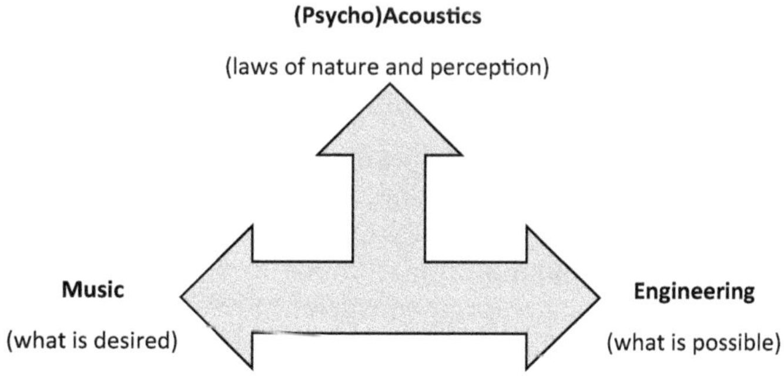

Figure 8.2 A "three-body system" of musical sound recording, after (Flandrin, 2011).

capturing and storing on permanent media the sound made during a performance? It will be one or more of: recordist, craftsperson, operator, engineer, technician, or the German *tonmeister* (Borwick, 1973).

It may be for many who work in sound recording that their role is self-defined or is adapted to the training and experience that they have. However, at the very least, a consideration of these questions is important for education and training as well as understanding how those who work in this industry identify themselves and how they are perceived (and, therefore, valued). Although ideas of role and identity in the recording studio have been considered before, such as the position of the audio engineer on the "aesthetic–technical" continuum (Ojanen, 2015) or the timeline of role development (Hepworth-Sawyer, 2017), the specific purpose in this chapter is to explore how these relate to the more general discipline of engineering.

One does not have to search far to find controversy relating to the use of term "engineer" in relation to recording. As an example, the following is an abridged version of an online exchange about the Wikipedia entry for "Audio Engineering":

Person A: An Engineer is one who has a degree in engineering, and in my opinion, those licensed to practice engineering. . . . Audio technicians (what this chapter refers to as audio "engineering") operate the equipment and mix the sound. While that is definitely a skilled trade or craft, it is not engineering.

Person B: I'm assuming you have done absolutely no work in the music industry, correct? Look on every single CD, cassette, or LP that you have in your house, the people who recorded and mixed the album are credited as engineers.

Person A: I understand they are credited as "Engineers" on record labels. I don't care that this may have been done by the recording and entertainment industries for 60 years, it is still a misapplication of the title to people not qualified as engineers.

Person B: I'm totally okay with a making a note that Audio Engineers are not certified, but I would hate to say that we are really "technicians" who have been mistakenly labeled by the music industry for years. We are not simple operators of the equipment, it is a creative field.

Person A: You do not design anything. You are not creating new equipment and technologies. Yes, you do a creative job, but it is still not engineering. (Anonymous, 2008, n.p.)

This exchange epitomizes one of the issues which arises when combining the terms "recording" and "engineering". One view of engineering is that it is something which is only achieved by formally trained, and then incorporated or chartered, individuals who use quantitative methods to solve problems and achieve outcomes. The alternative view expressed earlier in relation to recording, is that it is the application of some kind of skill in using technology in order to produce an artifact—the sound pressure waves generated by the performance are captured and martialed by the recordist so that they can be recreated (or at least re-evoked) for the listener. The recordist, in actually making the recording, solves the "problem" of how it should be done. This immediately raises the issues of evaluating the quality of the outcome. In many traditional engineering situations, the outcome can, in some way, be quantified, and relatively straightforward comparisons can be made with other possible outcomes or with a requirements specification. The quality of a recording is, certainly to some extent at least, subjective; as Jörg Wuttke[4] states, "there is no adequate basis in pure physics for judging the sound of a recording" (Wuttke, 1999, p. 1). Taste as the final arbiter of the aesthetic appeal of a recording suggests a difficulty in evaluating outcomes using the quantitative methods which are a hallmark of engineering.

That said, it will usually conform to some broad requirements. For example, only in rare circumstances would a recording for audio CD in which both left and right channels are identical (i.e. it is in mono rather than stereo) be desirable or acceptable. Given that stereo reproduction has been the norm in audio production and distribution for the last half-century, a sound recording that does not take advantage of the capability for the presentation of spatial information via differences between its two channels is likely to do so because of a production error, except in a small number of niche cases. This is quantifiable as a requirement for a "non-zero energy difference signal" and can be checked with tools such as a phase scope (Lissajous display) but can also be evaluated by listening with two ears.[5]

Taking a broader view of the outcome, "if nothing else, it has to conform to the spec of the producer" (Wells & Fisher, 2012, n.p.). The following is often quoted from Lord Rayleigh, the noted physicist after whom the unit of acoustic impedance is named, in support of audition as the ultimate quality measure: "Directly or indirectly, all questions connected to sound must come for decision to the ear, and from it can be no appeal". However, it is less common to quote the text that immediately follows that statement, which clearly makes the case for quantitative approaches to the creation and evaluation of audio:

> But we are not therefore to infer that all acoustical investigations are conducted with the unassisted ear. When once we have discovered the physical phenomena which constitute the foundation of sound, our explorations are in great measure transferred to another field lying within the dominion of the principles of Mechanics. Important laws are in this way arrived at, to which the sensations of the ear cannot but conform.
>
> (Rayleigh, 1878, p. 1)

Of course, recordings can be non-artifact (in the physical sense) and exist only to be heard, if considered as just signals; they will not be driven over like bridges, expected to start on cold mornings like cars, required to perform the myriad different tasks of a smartphone or personal computer and so perhaps quantification of them is completely unnecessary.[6]

Negotiating Between What Is Possible and What Is Desired

More than 30 years ago Douglas Lewin identified "Two Cultures" of arts and sciences which dominated universities, and proposed engineering as a third culture.

He argues for engineering to be considered as a third culture, not simply an extension of science (which it predates) and observes:

> Though today engineering artifacts draw extensively on scientific knowledge, it has not always been so, and up to the end of the nineteenth century the methods of manufacture and natural science were quite distinct. . . . Thus engineering can be said to predate science, which is still relatively young, and man's development has been, and still is, determined essentially by his capacity to make artifacts and improve upon his environment rather than the systematic accumulation of knowledge.
>
> (Lewin, 1983, p. 131)

The distinction that Lewin makes between the two is that science often (although not always) deals with systems which are "closed", isolated from the real world, whereas engineering is concerned with open systems which must serve and operate within real-world situations. The idea that engineering, in a recording situation, delivers "what is possible" given what is known about laws of nature (science) and "what is desired" according to aesthetics (art) is depicted in Figure 8.1. However, before becoming too comfortable with the idea that we have identified a convenient slot for recording within the discipline of engineering, there is still Lewin's delineation of the "designer craftsman" and "designing engineer" to consider. In the dialogue quoted previously the correspondent who was resistant to the term "recording engineer" was happy to concede that the act of sound recording was a "skilled trade or craft" (Anonymous, 2008).

Lewin elevates the status of engineering above this in stating that "the design of artifacts such as a communication system, computer software system . . . etc. require[s] a high level of intellectual skills rather than simple craft skills" (Lewin, 1983). Writing in the same year as Lewin, Gordon Rogers states that "craft is the power to produce a preconceived result by consciously controlled action: the craftsman always knows what he wants to make in advance" while the engineer will innovate and have to choose from a number of solutions to a problem (Rogers, 1983, p. 58).

All of this begs the questions, what is the person who makes recordings of music required to do and does what they are required to do encompass the innovation and informed decision making that characterizes engineering? The existence of performance venues with microphones permanently rigged in one place (often positioned with reducing visual obstruction in mind, rather than optimizing sound quality), connected to recording systems with a set of step-by-step instructions to enable recording by a minimally trained person, is certainly an indication that not all recordings are engineered: in this situation no one is exercising choice or innovating, there is just the "preconceived result". Of course, this is a narrow example and it does not imply that the operator does not have knowledge or expertise that could be utilized in situations where they have greater agency.

The agency discussed previously may not have been entirely banished, just displaced: it could be argued that the overall system, as embodied in the microphone positioning, the choice of system components (in this specific example the microphones, preamplifiers, etc.) may have been engineered to best match a specification: a system capable of making the best recordings given that the microphones must not visually obscure performances, their position must be permanently fixed, the operator of the system might have little or no training, etc. But an individual who has knowledge of acoustics is aware of the capability of the equipment at his/her disposal and understands the musical intentions of performer(s) and producer and exploits these in adapting to the recording situation at hand is designing a recording, making choices, and may be innovating. As Wuttke notes:

> A person who understands something about the way microphones work will be well able to master the most diverse assortment of recording tasks. "Cookbook" approaches are certainly easier, and they can be in step with the latest fashion trends, but they are not much help in the constantly changing conditions of live recording. Recording is not easy; being a recording engineer is truly a profession.
>
> (Wuttke, 1999, p. 8)

Engineer is not the only title (rightly or wrongly) attached to the process of making sound recordings. Borwick describes the "Tonmeister concept", which he adopted as the head of the first degree in music and sound recording to be offered at a UK higher education institution, thus:

> A literal translation from the German produces "sound-master" . . . implying that a tonmeister is someone skilled in the arts of sound recording, transmission and reproduction. The complication, and the special delight, is that it calls for a strange mixture of artistic flair and technical knowledge. How you combine the seemingly incompatible aptitudes of art and technology in a single individual and in what proportions, has always been a matter for discussion—and even of heated argument.
>
> (Borwick, 1973, pp. 2–3)

This specialist term is very useful, not least because it emerged (half a century ago) from a discussion about the skills and sensibilities that someone making recordings should possess. To the composer Schoenberg, cited at the beginning of this chapter, the tonmeister would be "trained in music, acoustics, physics, mechanics and related fields to a degree enabling them to control and improve the sonority of recordings, radio broadcasts and sound films" (Maconie, 1984, p. 53). In fact, music aside, this is a list of subjects that might be expected in many engineering courses. A final example, and useful summary, of the difficulties in achieving a consensus on what the act of recording is or should be is the publication and subsequent debate on a public forum of a study of approaches to recording electric guitar. The article attempted to provide a scientific explanation of the results achieved using common recipes for mic'ing guitar amplifiers:

> The typical starting approach at the guitar amp: Shure SM57 microphone, slightly off center of one of the cones of a driver, up close and almost touching the grille cloth. Oh, and angle the microphone a little. Ask veteran engineers why this microphone placement strategy is so common and a range of justifications follows, from seemingly

> scientific explanations, to vague guesses, to an honest, "I have no idea. I've always done it that way. Everyone does".
>
> (Case, 2010, p. 80)

The debate that this stimulated is perhaps more illuminating, for the purposes of the present discussion, than the paper itself (various contributors, 2010). A key contributor to that debate, the late Drew Daniels—someone who made much-admired recordings as well as inventing audio devices for the likes of Fender—was clear that when making recordings, "if you want to know how to [mic'] a guitar amp, do it a thousand times". As was pointed out to this author by one of the correspondents in the study reported later in the next section,[7] what Daniels is proposing is pure experimentation and therefore science (in the sense, of Karl Popper, where knowledge can only come via experience) rather than engineering (Wells, 2012b). Yet Daniels assumes this position in opposition to an article which claims to be about the "science" of recording, but which was published in an engineering journal. It is worthwhile pausing to consider whether the discipline (in both senses of the word) of engineering unnecessarily constrains playful experimentation which might yield unexpected yet aesthetically rewarding results.

Considering examples such as these it is perhaps unsurprising that there is confusion over which disciplines are actually present in typical recording scenarios. Of course, there are many kinds of "typical" recording scenario and their nature will, at least to a certain extent, determine how engineering may, or may not, be present. For the hobby sound recordist, the prime motivation may well be enjoyment and self-satisfaction. For a commercial studio undertaking work for a record label with an international profile and a multi-million-pound turnover, the imperative to obtain a particular type and quality of sound is likely to be the most important: working optimally to meet the (explicit or implicit) requirements of a client. One view is that "engineers, in order to function as engineers, must have a boss, or at least a client" who will provide, or aggregate, value judgments on which success will be judged (Goldman, 2004, p. 166). If we adopt this view, then the hobbyist cannot be an engineer (unless they can simultaneously be the client), but the person working in the commercial studio can. However, if we do adopt this view then, again, we have to consider the possibility that engineers may be constrained in their ability to engage in experimentation with unpredictable, but possibly fruitful, results.

Engineers as People

Another approach to understanding engineering is to consider the people that do it. The under-representation of women in STEM (science, technology, engineering, and mathematics) subjects is a widely acknowledged issue and one which is particularly acute in engineering. In 2017, the Institute for Engineering and Technology estimated that 11% of engineering and technology staff working in the UK were female (The IET, 2017a). At that time women made up 47% of the total UK workforce (The Office for National Statistics, 2017). While focusing on music technology degrees, rather than just those where engineering is explicitly present in the title or the award type (e.g. BEng), Born and Devine found a similar (10% female to 90% male) gender imbalance in enrollments in these programs in the UK, compared to a 55% to 45% female to male split in what they refer to as "traditional music degrees" (Born & Devine, 2016).

Faulkner identifies three themes that emerged from her ethnographic study of five engineering workplaces in the UK and elsewhere: gender inauthenticity (engineers are assumed to be men and, for women, going against this norm in choosing a degree and/or career in engineering is an explicit choice which it is felt needs to be explained because it is not "normal"), belonging ("engineering workplace cultures function, to varying degrees, as men's spaces which women engineers (and some men) have to 'fit in' to or remain on the margins of"), and technical-social dualism (the assumption that men are technical and not social, women are social and not technical, and that engineering is entirely technical and non-social) (Faulkner, 2006, p. 14). It is the strong perceived connection between men and technical prowess which Faulkner also challenges in her work, and the exaltation of a technicist view of engineering, in which the importance of technology is overemphasized at the expense of other elements of the discipline, which most embody what she refers to as the "masculinities" of the discipline. These masculinities of engineering are seen to alienate and exclude women early in life, where technologies in the school classroom "become gendered, in their material use, their symbolic meanings and their ideological function" (Armstrong, 2011, p. 2).

The epistemological stance of engineers is also considered a defining aspect of the discipline. When investigating the "personal epistemologies" (defined as "the stances an individual takes on issues related to knowledge and knowing") of members of a civil engineering faculty, Montfort et al. identified a close, but certainly not exact, alignment with a post-positivist viewpoint: that there is an objective truth about the physical world, but the sheer complexity of that world means that complete apprehension of that truth is not possible and only partial views are accessible which are dependent on the specific circumstances of the observer (Montfort et al., 2014, p. 388). A particular view that emerges is one where the validity of knowledge (its truth) is indicated (but not determined) by its utility in problem-solving situations (ibid., p. 403).

Pertinent to the subsequent discussion in this chapter is the idea that epistemologies develop during an engineering education and/or career, particularly as problems move from the simplified fundamentals of the first year or two of an engineering degree program toward more complex problems which are less well aligned with a narrow set of methods and knowledge. In the discussions which are described in the next section, it was often observed that more fixed and idealized views of recording were expressed by those who were not yet professionals, particularly those in the early stages of their sound recording studies. Professionals and students nearing the end of their degree programs expressed a more pragmatic, goal-oriented approach. This is relevant to learning objectives in degree programs at both module and program level. For example, should they parallel this epistemological development, or should they anticipate its final stage from the outset of students' studies by having situation-focused rather than abstract-theoretical teaching and assessment?

Views on Engineering in Recording and Recording as Engineering

The work described in this section was undertaken to gain a greater understanding of how the relationship between the discipline of engineering and the activity of sound recording and music production is understood by professionals and students working in this area. In particular, it aimed to interrogate the ideas that recording can be seen as an engineering activity or, if not, whether there are any elements at all of engineering in recording activities. Qualitative data was gathered from more than 200 participants, all based in the UK, via interviews and focus group sessions, each of which was led by the present author, recorded and transcribed (Wells, 2018).

For focus group sessions the initial part of the conversation was open, with responses to the question "Is recording engineering?" invited. Then more specific lines of inquiry in the following three areas were opened, in turn: the identity of studio personnel (e.g. "When you are working in the studio, what are you? What do you consider your role to be?"); the knowledge and competencies required for the roles that are identified (e.g. "What do you need to know? What do you need to be able to do?"); the nature of employment ("Will you be salaried or freelance? How you will maintain and update skills and knowledge?"). For interviews the questions were tailored to the individuals and their positions.

Although gender was not an explicit focus at the outset of the project, the data gathered, and the makeup of the focus groups, exhibited such a strong, normative gender bias that this was explicitly addressed in later data gathering (which is ongoing).

For this study, all the qualitative data gathered up to and including December 2018, along with the literature cited previously on gender and epistemology, was coded to identify and link responses to explicit project questions but also to capture additional themes that emerged during the free-ranging discussions. The key themes, how they emerge in the gathered material and how they relate to existing philosophies of engineering and sound recording, are discussed in the rest of this section.

The Qualities and Identities of the Engineer

On the qualities of the engineer, and the extent to which they are required in sound recording scenarios, a wide range of opinions exist but there is a broad consensus on certain attributes, which are considered an important part of the engineer's identity, whether within a recording context or not. The first of these is problem solving. As already noted, the notion of the engineer as someone who uses quantitative methods to solve problems is broadly accepted. For the recording studio this includes being able to cope when things go wrong, as well as being able to determine an optimal approach from a multitude of possibilities. This is something that is described during one of the interviews as:

> An ability to deal with situations which are outside of the normal operation of either what that person is used to, or the way that the equipment should be behaving.
>
> (Wells & Fisher, 2012, n.p.)

Some professionals (and, interestingly, students on non-arts degrees with a live sound component) explicitly identify the design of bespoke systems for particular recording or live-sound situations that exist outside of regularly used and purpose-built studio spaces as definite engineering activities within the paid roles that they aspire to. A professional participant, whose employer describes them as a recording and mastering engineer and record producer describes a typical situation as:

> You set everything up, you get all your boxes out, you plug everything in, turn all your computers on and you have to make sure it all works and quite often it won't, and then you have to trouble-shoot the problem. That is pure engineering skill, I'd say.
>
> (Professional participant)

It certainly matches one of the qualities of the Incorporated Engineer (IEng) described by the UK Engineering Council: "The theoretical knowledge to solve problems in developed technologies using well proven analytical techniques" (The IET, 2018, n.p.).

However, no explicit examples of "the theoretical knowledge to solve problems in new technologies and develop new analytical techniques", an ability required of the Chartered Engineer (CEng) (The IET, 2018, n.p.) are present in the data. In the UK, the academic standards for IEng include (amongst other equivalent qualifications) an accredited bachelor's or honors degree in "engineering or technology" and, for CEng, an accredited bachelor's degree (with honors) in addition to appropriate further learning to master's level or an accredited integrated master's (MEng) degree (The IET, 2018). In these definitions, "accredited" is a key term. Accreditation is defined as:

> [A] mark of assurance that the programmes meet the standards set by a profession. In the UK, the Engineering Council sets and maintains the standards for the engineering profession and sets the overall requirements for accreditation.
>
> (The Engineering Council, 2014)

Working Practices and Roles

When considering how changes in working practices have changed sound recording work, both students and professionals identify automated systems and the reorientation of tools (e.g. Apple, formerly Emagic, *Logic*) toward hobbyists and non-professionals, as examples of how traditional means of doing engineering in recording situations are being restricted or obviated. However, there is not a consensus on whether this implies a reduced, and reducing, role for engineering in such situations: "The thing is that recording is getting easier and easier and easier and things are being made a lot better and work a lot better" (student participant). One view is that cheap automated approaches have always been used:

> The first recordings such as those old Blues artists who would walk down the street in Memphis and go into what was really a record shop and in the back they'd have somewhere that they could cut a side. This was do-it-yourself, at home recording. . . . This notion that there is some great change because of technological forward movement is quite frankly a lot of nonsense.
>
> (Professional pop recording respondent)

Even then, such shifts are inevitable if we accept that engineering "applies well-established principles and methods, and uses proven components and tools" (Robinson, 1998, p. 227). This implies that the development and use of new technologies and techniques within the recording studio means that the locus of engineering is constantly shifting.

Much of the discussion around the existence and role of the engineer in the modern recording studio is cast in terms of the overlap with, or distinction from, the role of the producer. This is important in educational settings, not least because many degree programs in this area have either "production" or "engineering" in their title. A crude depiction of the general consensus is that those with engineering responsibilities in the studio have a responsibility to ensure that equipment works and is properly operated, and to design and implement systems specifically required for the successful capture of sound in a given situation, whereas those with "production" responsibilities are there to ensure that what is captured is musically satisfactory.

Unsurprisingly, professionals are able to give very specific examples of the interactions between these two roles. One perspective on where this demarcation might be, from a participant working in the classical (Western Acoustic Art Music) recording industry, is that between responsibility for sound and responsibility for performance: an audio balance between the elements—instruments, reverberation, etc. that is appropriate to the music, which is the job of the engineer, and overseeing the gathering of a set of convincing and/or editable performances that will be heard through that sound, which is the job of the producer:

> The producer's role is more about the nitty-gritty of the score whereas the engineer's job is about balancing the sound. It's an aural artistic judgement rather than a "that bar's in 5/8 but I didn't hear the fourth quaver from the piccolo;"—that's the producer's job. Some engineers probably couldn't read the score properly but still could get a really good sound.
>
> (Professional classical recording respondent)

This implies a listening ability, encompassing both technical and aesthetic quality, as an essential part of the recording engineer's skill set. It is hard to think of another engineering discipline which has such a strong reliance on a solely sensory and perceptive ability: to apprehend, understand, and act on nonverbal non-discrete data received by the auditory system. This is also an ability that is core to Schoenberg's description of sound recordists cited at the start of this chapter. Another term that is commonly used throughout the data to differentiate the producer from the engineer is "creativity". The following responses ((all from student contributors and labeled SR) give examples of how this term is typically used:

> I think "producer" is a bit more of a creative sort of thing.
>
> (SR 1)

> The roles are still split up with producer doing all of the creative things.
>
> (SR 2)

> I think that creativity is completely void of engineering, it's like a Venn diagram where there's no crossover.
>
> (SR 3)

Yet, within the community, engineers describe themselves as "creative problem solvers" (The IET, 2017b). A professional described themselves as "at the facilitator end of a creative sub-sector, so we're actually doing the work that puts together the ideas of the creative people" with another correspondent adding "which is engineering". This aligns with the view of engineers as "using scientific knowledge as tools in order to achieve goals that are external to the fields of science or engineering" (Montfort et al., 2014, p. 390).

It does seem that the notion of "creativity" has neither a clear or agreed definition. Many who see themselves as engineers also see themselves as being creative, whereas those who do not identify as engineers see one of the defining characteristics of the discipline as being that it is separate from creativity. A likely explanation is that creativity is seen by some as a preserve of imaginative and problem-solving activity, where the outcome has a subjectively aesthetic quality, rather than a quantifiably optimal one. To clarify this issue, perhaps the latter might be identified as "ingenuity", albeit still a particular form of creativity.

Listening certainly emerges, across all participants, as a key activity and ability in recording activity:

> The best training [that] I had was editing, because you just listen for hours and hours and hours and hours and hours. Just headphones on: listening, listening, listening, listening to tons and tons of different recordings.
>
> (Professional)

The only example of an explicit disagreement in a focus group is on this topic: an engineering student claims that you can balance something "quite well" by using level meters and spectral displays, the response from a professional participant is:

> I'm sorry but that doesn't play . . . what you need to do is to learn to use these [points at ears]. The prime thing, before anything else is you need to learn to use these.
>
> (Professional)

Engineers as Human Beings

The social element of recordings is frequently observed by both student and professional participants, memorably summarized as "You could be the best engineer in the world, but if you've got crap social skills then no one will want to work with you" (student), and elaborated by a professional/educator as:

> [A] lot of the job is keeping musicians happy. It's not musical or technical, it's hand-holding or personal skills; and I'm certain that people who do go in and greet musicians and make them happy . . . actually get better recordings, that sound better than the other people, because the people who are playing are playing better, in a more relaxed way.
>
> (Professional/Educator)

While many of the participants seem to regard this social element as particular to recording, it echoes Faulkner's study of civil engineers discussed in the previous section: Engineering is a social activity.

Most participants, student and professional, agreed on and accepted the precarious nature of much employment in the industry. While, as both an audio professional and educator, it is difficult not to be aware of the gender imbalance in sound recording, this author was still shocked at the utter pervasiveness of the assumption that sound recordists are male. Throughout all of the discussions conducted for this research the male pronoun was *always* used by participants when talking about recording personnel, whether "engineer" or "producer". In this sense it matches, perhaps surprisingly or at least to a surprising extent, the "gender inauthenticity" of the female engineer which Faulkner describes. While this assumption is representative of the current reality, if it is accepted that such a gender imbalance does a disservice to both engineering as a discipline and engineers as a workforce, then it must be seen beyond, or disrupted, in some way. A female professional participant stated:

> I felt that I had to double-prove myself. You've got to be better than the man would have to be in the same job because you've got to get past the fact that you're a woman before you get the opportunity.
>
> (Female professional)

The application of theoretical knowledge in real-world recording scenarios is challenging and one of the more marked differences between professional and student participants relates to this challenge. Professionals tended to emphasize the importance of making sure that something is captured, overtaking risks to obtain a particular desired result:

> The first thing that anyone who has [technical expertise and knowledge] learns is what they can't do, or where they can't go or where they're on dangerous ground . . . the most important thing is you get that recording. If you can get that recording sounding fantastic then that's the next step up, if you can help get the artist into a position where they can perform it fantastically and it sound fantastic then that's another step up but fundamentally you have to get the recording.
>
> (Professional)

When considering education and training for recording roles, the need for practice at making recordings was broadly acknowledged as an essential component of courses. This helps to make the connection between typical simple, abstracted problems designed to illustrate underlying scientific principles, and the complexity and often compromised nature of the actual recording activities which such principles and knowledge inform. In particular, work placements were identified as a point of maturation in the sound recordist:

> I am absolutely mouth open, jaw-dropped by the difference in people when they come back from their industrial year. There's such a big change: they've grown up and are employable: you would employ this person tomorrow because they understand what they're doing and they've got an attitude where you can have a discussion with them about the industry somehow in a different way, just more of a "world" view of things.
>
> (Sound recording professional)

This implies an epistemological change: a fundamental shift in the way that goals are identified and prioritized, and how knowledge and technique can be employed to realize them.

Summary

This chapter has considered the nature of engineering and the extent to which it is embodied in typical sound recording activities and personnel. To this end, a wide range of literature has been surveyed and ideas about engineering compared with views of audio-specific professions gleaned from both existing literature and new interview and focus group sessions conducted specifically for this study of recording and engineering. The coverage of the debate is not exhaustive and is largely focused on the UK and, to a lesser extent, the rest of Europe and the US.

There is much more to the discussion about engineering and the extent to which it is found in recording skills and practice, particularly in other cultures and parts of the world, and an entire book could be devoted to the topic. What is presented here is deliberately open-ended. The intention is not to form a single, inflexible view of what engineering and sound recording are, and to dictate how that should inform educational program design. Instead this chapter has sought to present a range of viewpoints from a number of different voices, as well as existing studies and scholarship, in order to inform policy and program development in audio education. However, some key questions that are important to audio education are listed at the end of this section.

Many commonalities with generally accepted ideas and features of engineering exist: the use of scientific knowledge and quantitative methods to solve problems (although these are not exclusively the approaches to problem solving in the studio, with trial and error being a common alternative), the creation of bespoke systems adapted for specific recording assignments (e.g. microphone arrays tailored to obtain a particular sound in a particular scenario), and the change in aptitude and "recording world view" that is afforded by moving from detached and simplified problems in relevant science and technology studies to regularly making recordings in real-world situations. There is also an obvious and strong gender imbalance, in both the student and professional population. Differences exist too, notably the extensive reliance on a particular sensory and perceptual ability (audition) to evaluate outcomes and the view of sound recording as a social activity[8] (whereas for other areas of engineering, such as research and development, there is a view that this is under-emphasized, although this is challenged in the literature (Faulkner, 2006, p. 7)). However, despite this awareness of the social aspect of recording the significant gender imbalance, the entrenched idea dating from at least Schoenberg of the "sound *man*", remains.

In considering how, both positively and negatively, engineering currently relates to audio education and, in designing programs, how the positive elements can be better embedded and the negative mitigated against, a number of questions emerge which are organized into three main themes and discussed next.

Theme One: Both engineering in general and music technology specifically have a profound gender bias, with approximately 10% representation of females in both in the UK. If we accept that this is a situation that needs to change (and it is hard to find any reasonable argument to support the maintenance of the status quo), then audio education has a role to play in this disruption. In order to challenge the perceived gender inauthenticity of engineering, an alternative to the current gender-imbalanced reality needs to be presented.

The Audio Engineering Society in the UK, in order to address what is estimated to be 5% representation of women in the society's global membership, has launched a number of initiatives to present a picture of "what might be" rather than the present situation. These include ensuring that panels at events, including those aimed at students, have at least 50% female representation, making educational bursaries available specifically for female and nonbinary applicants to courses related to audio, and discouraging participation in all male panels (Lopez, 2018). Born and Devine present a number of ways in which the male gendering of technology (partly due to "I-methodology" in the design of technology) and "indirect discrimination" create a different presentation to, and treatment within, music technology programs (Born & Devine, 2016).

A first step to dealing with these is to acknowledge them within teaching and develop explicit policies for monitoring and correction of inadvertent bias. It is very unlikely that one solution can bring significant change. As the authors state:

> [It] would be unwise to search for a single explanation for, or solution to, the gendering of contemporary digital music and sound art. Rather, we witness a confluence of mutually mediating forces and processes . . . the assemblage-like character of these relations means that real change can begin anywhere.
>
> (Born & Devine, 2016, pp. 7–8)

The solutions are unlikely to be the same across individual institutions, but an awareness and acknowledgment of these "mediating forces and processes" is essential everywhere to make progress.

Theme Two: A dichotomy between artistic "creativity" and engineering "ingenuity" is evident, and this is a source of tension between those who see themselves as being (traditionally qualified and/or chartered) engineers and studio-based sound-recordists, technicians, and/or artists. The strong demarcation implied by ternary distinctions between "arts", "science", and "engineering" degrees (made explicit in their titles: BA, MSc, MEng, etc.) reinforces this and, arguably does not serve sound recording well if it is genuinely interdisciplinary. Architects of education programs should consider how both creativity and ingenuity are presented in teaching materials, assessment activities, and learning outcomes.

To what extent can interdisciplinarity (engineering within musically orientated activity) and multidisciplinarity (engineering and music instruction alongside each other) be included in programs to produce graduates with transferable and adaptable skills? How can constraints, such as those of accreditation, be negotiated so that programs form a coherent development of skills and knowledge in sound recording and related areas, while mapping and instilling core elements of the disciplines that they combine? What are the elements of the discipline of engineering that will contribute to the most useful, adaptable, and, yes, creative audio professionals?

This author has previously given examples of recording activities which embody, at different levels, engineering (Wells, 2012a). In aural training consideration should be given to how both technical and stylistic listening skills are delivered and assessed. Exercises and contexts that successfully address quantitative analysis by human audition are likely to be different from those that encourage qualitative stylistic analysis. For example, the former can be successfully achieved in highly repetitive tasks involving a single listener, whereas group listening and discussion, informed by a breadth of works, may well be more suitable for the latter.

Theme Three: A significant difference between professional and student contributors to the study reported in this chapter is opinion on how solutions are arrived at in the studio. This parallels an epistemological change in engineers reported in the literature: idealism and simplicity are replaced by pragmatism and an appreciation of the complexity of recording systems and situations. In one sense, the latter is problematic for the Schoenbergian Tonmeister concept, based as it is upon scientific and engineering fundamentals, which are often taught in isolation from each other (consider the module titles of many such programs) and must, certainly at the outset, make use of simplified models and, therefore, assessment tasks. This is typically mitigated by ongoing practical work and assessment in sound recording. There can be a tension between these two in the experience of this author.

For example, a rigid adherence to simple abstracted understandings of acoustics, psychoacoustics, etc. can lead to the development of mis-ordered priorities and bad habits, whereas experiential and observational learning of studio craft leads to a rigid and nontransferable set of practical skills. Pure observational learning, "sitting with Nellie", is described by one engineering writer as "fine for learning how to operate a bread-wrapping machine" but "that anyone can learn the principles of modern audio equipment just by watching someone using it is obvious nonsense" (Watkinson, 1996, p. 9). Having both theoretical and experiential learning is, of course, the natural solution, but it can create dissonance in students, particularly early on in learning. If this dissonance is not explicitly addressed in teaching then students may select from a false binary choice between "theory" and "practice", rather than developing a sophisticated synthesis of the two.

Taking the electric guitar example cited earlier in this chapter, one should consider how such a synthesis might be arrived at and presented to students. What are the learning outcomes that would best embody negotiating the two approaches and arriving at a satisfactory synthesis? How can theory be supported by practical experience, and how can practical experience enhance understanding of theory, rather than there being a conflict between the two? It might be argued that, although a synthesis of disciplines is implicit in Schoenberg's statement at the opening of this chapter, there are other aspects of his ideas that are outdated; the superior status of the score over the performance and the recording, for example.

Conclusion

This chapter has explored the relationship between sound recording and engineering via a survey of relevant literature from a variety of fields, as well as types of source, and qualitative data from focus groups and individual interviews conducted by the author in the UK from 2012 to 2018. A plurality of views exists, and the purpose of this chapter has not been to impose or champion any one particular viewpoint. The intention has been to present the range of opinions as they relate to sound recording and engineering in the UK, as well as in Europe and the United States, so that the debate about the nature of sound recording and engineering can be broadly informed. It is the belief of this author that these debates should inform program and module level design of audio education provision, and be present in the articulation and delivery of learning outcomes. There may be many different definitions of sound woman and sound man, but all can have an awareness of the skills and knowledge that they have and how their new, emerging, interdisciplines relate to existing disciplines such as engineering.

Notes

1. In 2017, physical media sales still accounted for 30% of global recorded music revenues, or 36% if synchronization and performing rights income is excluded (International Federation of the Phonographic Industry, 2018).
2. Much of this work was supported by the Royal Academy of Engineering, via their *Ingenious* Public Engagement Fellowship program.
3. The specific area of sound recording that this chapter addresses includes the capture of musical performances, or the capture of sound for subsequent assemblage into a musical artifact, for simultaneous (i.e. live broadcasting) or subsequent reproduction. This area of audio recording is what is intended by the term "musical sound" and subsequent uses of the term "recording" refer to this type of recording.
4. Emeritus Technical Director, Schoeps Microphones.

5. A Lissajous display (typically referred to as a phase scope in audio production) is a two-dimensional representation of the amplitude of two signals. In typical audio applications, it is essentially a cartesian plot of left versus right amplitude. The width of the plot is a crude indicator of correlation between the two signals, with a single, thin line indicating high correlation. The relative amplitudes are indicated by the angular orientation of the plot: If it lies along the left axis then there is zero amplitude in the right signal and vice versa.
6. Of course, this is not true of the media and their associated formats. The Compact Disc was a triumph in the provision of a physical medium and storage format that preserved the integrity and fidelity of the recordings they stored over a significant time span, and number of playings, compared to the preceding vinyl disk.

   However, the focus in this chapter is on the signals that are stored on such media rather than the media themselves.
7. This correspondent referred to themselves as "an assistant professor in virology, but continue to produce EDM [electronic dance music]. I like to think I'm the only person who's worked at biosafety level 4 and has a [record] label deal!".
8. Here, participants are typically referring to scenarios in which the client and other workers (e.g. performers) are present. However, it should be remembered that there are numerous roles that can be undertaken in more solitary conditions (e.g. editing, mixing, and mastering).

References

Anonymous, 2008. *Talk Page–Audio Engineering* [Online]. Available at: https://en.wikipedia.org/wiki/Talk:Audio_engineer [Accessed 28 September 2011].

Armstrong, V., 2011. *Technology and the Gendering of Music Education*. Farnham: Ashgate.

Boehm, C., 2007. The Discipline That Never Was: Current Developments in Music Technology in Higher Education in Britain. *Journal of Music, Technology and Education*, 1(1), pp. 7–21.

Born, G. & Devine, K., 2016. Gender, Creativity and Education in Digital Musics and Sound Art. *Contemporary Music Review*, 35(1), pp. 1–20.

Borwick, J., 1973. *The Tonmeister Concept*. New York: The Audio Engineering Society.

Cambridge Dictionary, 2019. *Discipline* [Online]. Available at: https://dictionary.cambridge.org/dictionary/english/discipline [Accessed 14 May 2019].

Case, A., 2010. Recording Electric Guitar: The Science and the Myth. *The Journal of the Audio Engineering Society*, 58(1/2), pp. 80–83.

The Engineering Council, 2014. *The Accreditation of Higher Education Programmes: Third Edition* [Online]. Available at: https://www.engc.org.uk/ahep [Accessed 27 May 2019].

Faulkner, W., 2006. *Genders of/in Engineering*. Swindon: Economic and Social Research Council.

Feisst, S.M., 1999. Schoenberg and the Cinematic Art. *Music Quarterly*, pp. 93–113.

Flandrin, P., 2011. *Fourier + 200*. Paris: s.n.

Goldman, S.L., 2004. Why We Need a Philosophy of Engineering: A Work in Progress. *Interdisciplinary Science Reviews*, 29(2), pp. 163–176.

Hepworth-Sawyer, R., 2017. Mixing as Part-History, Part-Present and Part-Future. In: R.H.A. Hodgson, ed. *Mixing Music*. London: Routledge, pp. 270–278.

The IET, 2017a. *The IET Skills Survey 2017* [Online]. Available at: https://www.theiet.org/media/1350/skills17.pdf [Accessed 15 May 2019].

The IET, 2017b. *Thinking About Engineering as a Career* [Online]. Available at: https://www.theiet.org/students/studying-engineering/why-engineering/creativity-within-engineering.cfm [Accessed December 2018].

The IET, 2018a. *Chartered Engineer (CEng)* [Online]. Available at: https://www.engc.org.uk/ceng [Accessed 27 May 2019].

The IET, 2018b. *Incorporated Engineer (IEng)* [Online]. Available at: https://www.engc.org.uk/ieng [Accessed 27 May 2019].

International Federation of the Phonographic Industry, 2018. *Global Music Report 2018: Annual State of the Industry* [Online]. Available at: https://www.ifpi.org/downloads/GMR2018.pdf [Accessed 19 May 2019].

Lewin, D., 1983. Engineering Philosophy: The Third Culture. *Leonardo*, 16(2), pp. 127–132.

Lopez, M., 2018. Audio Engineering Society Joins Forces with UN Gender Equality Campaign HeForShe. *Pro Sound News Europe*, April [Online]. Available at: https://www.psneurope.com/business/audio-engineering-society-joins-forces-with-un-gender-equality-campaign [Accessed March 2020].

Maconie, R., 1984. Sound Recording in a Music Department. *International Journal of Music Education*, 4, pp. 51–56.

Montfort, D., Brown, S. & Shinew, D., 2014. The Personal Epistemologies of Civil Engineering Faculty. *Journal of Engineering Education*, 103(3), pp. 388–416.

The Office for National Statistics, 2017. *Labour Force Survey Summary: People by Economic Activity for Those Aged 16 and Over and Those Aged from 16 to 64 (Seasonally Adjusted)* [Online]. Available at: https://www.ons.gov.uk/employmentandlabourmarket/peopleinwork/employmentandemployeetypes/datasets/employmentunemploymentandeconomicinactivityforpeopleaged16andoverandagedfrom16to6 4seasonallyadjusteda02sa/current [Accessed 15 May 2019].

Ojanen, M., 2015. Mastering Kurenniemi's Rules (2012): The Role of the Audio Engineer in the Mastering Process. *Journal on the Art of Record Production*, (10).

Rayleigh, L., 1878. *The Theory of Sound*. 1945 ed. New York: Dover.

Robinson, J., 1998. Engineering Thinking and Rhetoric. *Journal of Engineering Education*, July, pp. 227–229.

Robson, D., 2012. *Make Engineer a Protected Title: HM Government E-petition* [Online]. Available at: https://petition.parliament.uk/archived/petitions/6271 [Accessed 14 May 2019].

Rogers, G., 1983. *The Nature of Engineering*. London: The MacMillan Press.

Schoenberg, A., 1940. *School for Soundmen*. Unpublished manuscript. Arnold Schönberg Center Vienna.

Various, 2010. *AES Journal Forum* [Online]. Available at: https://secure.aes.org/forum/pubs/journal/?ID=17 [Accessed 15 May 2019].

Watkinson, J., 1996. Visions of the Invisible. *Studio Sound*, December, p. 9.

Wells, J., 2012a. *Engineering in Recording*. Sheffield: s.n., pp. 159–166.

Wells, J., 2012b. *Is Recording Engineering? Your Views* [Online]. Available at: http://jezwells.blogspot.com/2012/04/your-views.html [Accessed 15 May 2019].

Wells, J., 2018. *Is Recording Engineering?* [Online]. Available at: www.jezwells.org/engineering_recording/ [Accessed 15 May 2019].

Wells, J. & Fisher, D., 2012. An Interview with Dave Fisher. *Journal on the Art of Record Production*, 7.

Wuttke, J., 1999. *The Microphone: Between Physics and Emotion*. s.l.: The Audio Engineering Society.

9

Environmental Sustainability in the Audio Engineering Curriculum

Ben Mosley

Introduction

There is little doubt that environmental sustainability is becoming increasingly important globally. Audio activities are often seen as a tiny part of global industry. However, audio has great reach, with audio equipment and components a fundamental part of all computing, communication, and entertainment. As well as being the primary part of music and radio, audio-based activities are also key aspects of film, games, and television. Through activities such as mass production of audio equipment in the far east which is then shipped across the world, worldwide travel of people, the manufacture and use of a variety of materials for acoustic treatment, energy use, and the disposal of waste products, the audio industry has a great potential to do harm to our environment. However, with such wide reach and influence, it has great potential to influence and inspire changes in practice throughout a wide cross section of global industry.

Audio educators must consider how they will nurture environmentally aware graduates who have the skills and understanding to make themselves more employable in the short term and potential leaders on environmental sustainability in the future. A wider awareness of the environmental impact of their activities will contribute toward producing a more rounded and employable graduate. We will face great challenges with respect to our planet and environment in the coming years and we need to prepare our students with the awareness and skills to face up to and overcome these challenges.

> We face awesome environmental challenges: climate change, food production, overpopulation, the decimation of other species, epidemic disease, acidification of the oceans. Together, they are a reminder that we are at the most dangerous moment in the development of humanity. We now have the technology to destroy the planet on which we live but have not yet developed the ability to escape it. Perhaps in a few hundred years, we will have established human colonies amid the stars, but right now we only have one planet, and we need to work together to protect it.
>
> (Dr. Stephen Hawking 2016)

Aims and Scope

The chapter aims to explore the wider context of sustainability in the audio industries current practice and policy. It will then investigate how environmental sustainability can be embedded in the curriculum of an audio course through projects, assignments, classwork, and discussions, highlighting good practice and discussing the effectiveness of different activities in developing awareness and skills in this important area. It will take a practice-based approach and provide readers with a range of resources and ideas to implement in their own teaching.

The chapter considers how environmental sustainability can be integrated with traditional audio engineering course exercises, such as energy requirement calculations, energy source considerations, and materials choices. It will discuss how students can be encouraged to apply traditional engineering skills, such as calculation, design, and measurement to produce better products and solutions and reduce environmental impact. Finally, the chapter will look toward the future of sustainability in the audio industry and audio education settings and consider possible directions and developments in this area.

This chapter does not seek to debate the validity of climate change science, although there is tacit acknowledgment that this is a serious issue for our planet. The focus here is twofold. First, the chapter works on the principle that we should not leave the planet in a worse state than when we arrived here, both individually and as a species. Second, the chapter recognizes that education has a wide remit to produce good citizens who are culturally and socially aware.

Background

The Wider View

> Environmental sustainability features prominently in the press, academia, government and industry. It is often considered as one part of a wider concept of sustainable development, alongside social and economic sustainability.
>
> (Goodland 1995)

The United Nations defines sustainable development as "development that meets the needs of the present without compromising the ability of future generations to meet their own needs" (UN 2015). In 2016, members of the United Nations signed up to 17 Sustainable Development Goals and 169 targets as part of its 2030 Agenda for Sustainable Development. These goals are wide in scope and focused on improving people's lives throughout the world but include a range of statements relating to the environment and infrastructure that have important implications for our everyday activities both in industry and our personal lives.

Morelli defines the more specific concept of environmental sustainability as:

> [M]eeting the resource and services needs of current and future generations without compromising the health of the ecosystems that provide them, . . . and more specifically, . . .

> as a condition of balance, resilience, and interconnectedness that allows human society to satisfy its needs while neither exceeding the capacity of its supporting ecosystems to continue to regenerate the services necessary to meet those needs nor by our actions diminishing biological diversity.
>
> (2011: 6)

At the time of writing, 184 parties have signed up to the Paris Agreement which enshrines a commitment to hold the increase in the global average temperature to well below 2°C above pre-industrial levels, pursue efforts to limit the temperature increase to 1.5°C above pre-industrial levels, foster climate resilience and to make finance flow toward low greenhouse gas emissions and climate resilient development (UN 2018).

The Intergovernmental Panel on Climate Change's most recent research suggests that:

> human activities are estimated to have caused approximately 1.0°C of global warming above pre-industrial levels, with a *likely* range of 0.8°C to 1.2°C. Global warming is *likely* to reach 1.5°C between 2030 and 2052 if it continues to increase at the current rate.
>
> (IPCC 2018: 6)

Climate change is often the headline issue and most discussed aspect of environmental sustainability. However, environmental sustainability covers a wide range of issues, such as emissions to air, releases to water, releases to land, use of raw materials and natural resources, use of energy, energy emitted, generation of waste, and use of space (BSI 2015).

One could argue that the so-called traditional engineering sectors (mechanical, chemical, built environment, etc.) are a long way ahead of audio engineering and the Creative Industries in terms of considering environmental sustainability as a key part of their practice. Many of the organizations promoting environmental sustainability in the Creative Industries have only become active in the last ten years, whereas the United Nations Sustainable Development Goals and organizations such as Engineers Without Borders and Practical Action can trace their roots back to the early 1980s. Possibly the main reason for this is the recognition of the great potential for environmental harm and long-term impact of the activities of these sectors and vice versa; the ability of these sectors to engineer solutions that mitigate long-term environmental impacts. The impacts of the activities of the audio and the Creative Industries are perhaps not as obvious. However, the audio industry features many activities that have the potential to affect our environment, such as travel, energy use, and product manufacture.

Environmental sustainability is a significant issue for business and industry already and is likely to become more and more pressing in the coming years. If your view of education is that it should prepare students for industry, then a strong understanding of environmental sustainability issues makes them more employable and improves their opportunities for success. If your view of education is to improve the intellectual, cultural, and moral capital of a society then there is little argument that environmental sustainability is a vital part of any educational experience.

The University Sector

Embedding sustainable development teaching in university curricula faces many challenges. Dawe et al. (2005) found four major barriers to the successful embedding of education for sustainable development into many of the subject disciplines in UK Higher Education and an overcrowded curriculum, perceived irrelevance by academic staff, limited staff awareness and expertise, and limited institutional drive and commitment. Ralph and Stubbs (2014) highlight a lack of understanding and awareness of sustainability issues, resistance to change, a lack of staff commitment, engagement of all levels of stakeholders in the decision-making process and initiatives as important factors in developing sustainability in the university environment.

All too often educators become so focused on the core curriculum that they may neglect the wider context, and sustainability is often a victim of this. Dmochowski et al. (2016) note that although many universities offer courses that address topics of sustainability within the existing curriculum, the connection to sustainability is not always clear in course documents and that connection is not as strong as it could be.

Current Practice in the Audio Industries

Julie's Bicycle

Julie's Bicycle is a London-based charity that supports the creative community to act on climate change and environmental sustainability (Julie's Bicycle 2018). They provide a range of Creative Industry Green Tools to "allow organisations to calculate and understand the impacts of their venue, office, tour, production, event or a festival's environment impacts". They also offer Creative Green Certification that recognizes and celebrates environmental best practice where organizations are awarded between one and five stars based on their commitment, understanding, and improvement. It is currently the only environmental certification scheme in the UK designed specifically for the creative and cultural industries (Julie's Bicycle 2018). Julie's Bicycle's webpages host a wide range of documents and resources covering industry research, guidance, and data.

Albert

The BAFTA Albert consortium members include all3media, BBC, Channel 4, Endemol, IMG, ITV, Kudos, NBC, Universal International Television, Sky, Shed Media, Twofour, and UKTV (BAFTA 2015). Established in 2011, it is a think-tank on sustainability for film and television in the UK, working to raise the profile of sustainability in the moving-image media production industry. The organization champions sustainable production techniques and provides tools, guidance, and direction (BAFTA 2015).

One of the key initiatives of the Albert consortium is the albert+ standard. Based on original research and development by the BBC, albert+ is a mark of sustainability that indicates that the program has taken steps to manage and reduce its environmental impact during production. Although Albert focuses on the production process for TV programs there are many parallels with live and studio music production. Considerations such as travel, energy use on location, waste disposal, and equipment supply chain are all pertinent to music events (BAFTA 2014).

Albert provides a checklist for production companies to work to covering general issues, in the office, on location, transport, sets, props and wardrobe, in the studio, and postproduction and show launch. Many of the issues in this checklist can easily be applied to audio and music projects The Albert checklist encourages production teams to write a simple statement of intentions and goals; nominate a senior individual responsible for sustainability; and to ensure all cast, crew, and supply chain are aware of any plans around sustainability (BAFTA 2014).

The Broadcasters

The largest UK broadcasters are a useful starting point for guidance on policy and practice in terms of environmental sustainability, as most broadcasters address environmental sustainability through their corporate responsibility agendas. The BBC has a strong sustainability policy under the banner of "Green Broadcasting" which sets targets for 2022, against a baseline of 2015/16 (BBC 2018). Sky's corporate responsibility agenda is titled "The Bigger Picture". This is a very wide-ranging agenda of which environmental sustainability is only a small part. The corporation set goals for 2020 and makes a commitment to continue to monitor these as part of its annual Bigger Picture Impact Report (BSKYB 2018). Channel 4 also place their environmental policy within their corporate responsibility stating that "we want to ensure that the way we conduct our day-to-day activities reflects our commitment to reducing any negative impact we may have on the environment" (Channel 4, 2017: 109). ITV has its own dedicated corporate responsibility webpages and lists a series of broad aims related to environmental sustainability (ITV 2017).

These policies are broadly similar in their aims. All of them mention energy use and CO_2 reduction for buildings and technology. They also target reduction of unnecessary travel and waste production. Other notable aims include zero waste to landfill, removal of single use plastics, and the raising of awareness both within the employee base and through editorial content in broadcasts. The BBC and SKY policies set very clear targets whereas the ITV and Channel 4 polices are based on broader aspirational statements. They are all useful policies to inform classroom discussions, especially where students are involved in media production based activities.

Others

Arts Council England publishes a biannual environmental report titled "Sustaining Great Art and Culture" in partnership with Julie's Bicycle. This document presents data and evidence relating to sustainability performance across the creative and arts sector. Data and performance can be tracked form the first publication in 2012/13. Other organizations of note that are actively promoting sustainability in the Creative Industries include the Sustainability in Production Alliance (SiPA) and Creative Carbon Scotland. The Institute of Engineering and Technology (IET) have sustainability written into their rules of conduct for members which suggests that sustainability should be a part of all members' daily professional practice (IET 2018). The Institute of Acoustics (IOA) have a sustainability task force in place that is running workshops and seminars to progress discussion in the area. This has yet to result in any formal guidance but it is promising to see these discussions taking place. The IOA's code of conduct includes a statement with regards to the environment, stating that members should not "needlessly pollute the environment except when legally authorized to do so" (IOA 2018).

Relevant International Standards

BS ISO 20121:2012

This guidance document titled "Event Sustainability Management Systems—Requirements with Guidance for Use" was released in 2012 and was developed from BS 8901:2007, which itself was developed as a result of the organizing team for the London 2012 Olympics recognizing the importance of sustainability as part of the event and a desire to formalize and share good practice in this area. The document provides a very generic framework for event sustainability with a strong emphasis on documentation and tracking with clear and strong management of the whole event process from planning and procurement through to evaluation. There is, however, some useful guidance for the audio side of events, with energy efficiency of equipment, choice of renewable energy sources, transport, logistics, and noise levels all being identified as key issues. It can be of particular interest for educators and students involved in audio for live events.

BS EN ISO 14001:2015

"BS EN ISO 14001:2015, Environmental management systems—Requirements with guidance for use" (BSI 2015) specifies the requirements for an environmental management system that an organization can use to enhance its environmental performance. This standard is intended to support the long-term development of an organization's environmental sustainability policies. It focuses on leadership, management, policy, and monitoring. It provides much useful guidance on how to run an organization in an environmentally sustainable way. It is a widely known international standard that has been in operation in various forms for 20 years. The British Standards Institution offers a certification route against BS EN ISO 14001:2015 for organizations.

Resources

UNESCO

The United Nations Educational, Scientific and Cultural Organization (UNESCO) has been a world leader in promoting sustainability since its inception in 1946. UNESCO's Teaching and Learning for a Sustainable Future pack is by far the most complete source of education guidance and materials (UNESCO 2010). This is a 27-module series of activities, worksheets, and materials covering a wide range of issues under the banner of global sustainability.

Engineers Without Borders UK

Engineers Without Borders UK is an organization with the mission to "lead a movement that inspires, enables and influences global responsibility through engineering" (EWB-UK 2018: 100). It is part of a wider network operating from various countries throughout the world. Each year, Engineers Without Borders UK works with one of their partner organizations to produce a series of engineering challenges based on real-world problems the partner's community faces. These challenges span engineering disciplines including water and sanitation, energy, the built environment, transport, waste management, information communications technology, and local industry. Students are asked to design a solution to one of the challenges appropriate to the economic, social, and environmental context in which it will be used (EWB-UK 2018). Universities can formally engage in this challenge by making it a mandatory part of their curriculum. The outcomes of the student activities are published by Engineers Without Borders UK. Prizes are on offer and at the time of writing the challenge is running in 31 universities across the UK and Ireland involving over 6,500 students.

Practical Action

Practical Action is a registered UK charity established in 1965 with the mission statement to "build the technical skills of poor people in developing countries enabling them to improve the quality of their lives and that of future generations". They publish a range of resources for use in the classroom based on real-world problems in energy, sustainability, and agriculture on their webpages. Of particular interest are their STEM challenges, which, although designed for use at school level, outline some really interesting exercises for STEM students at University level (Practical Action 2018).

Global Dimension in Engineering Education

The Global Dimension in Engineering Education (GDEE) is "an initiative that aims to increase the awareness, critical understanding and attitudinal values of undergraduate and postgraduate students in technical universities related to Sustainable Human Development (SHD) and its relationship with technology" (GDEE 2013). They provide a range of teaching materials, case studies, and resources. The focus is on engineering but there are a range of critical thinking exercises that are valuable for those in more creative areas.

Teaching Sustainability

Based on the guidance, policy, and resources just summarized, key issues that require consideration when planning teaching environmental sustainability are:

- Energy—consumption, efficiency, and renewable sources
- Waste products—reduction of waste and increase of recycling
- Travel—minimizing emissions and reducing
- Awareness—targets, leadership, documentation, planning

The teaching resources from the organizations just discussed fall into two broad categories. First, exercises that raise the awareness and understanding of students in the area of environmental sustainability. Second, skills-based exercises that develop new skills or apply existing skills to sustainability problems. All exercises are characterized by a clear context strongly related to real-world problems in sustainability. Guerra suggests that:

> An engineering curriculum for sustainability must embed education for sustainable development learning principles and create opportunities for students to learn and practise critical, holistic and systemic thinking, apply it to real-life situations and enhance lifelong learning, responsible decision-making and professional practice.
>
> (2017: 437)

Guerra's suggestion appears to be an appropriate approach to take based on examples from the area.

Classroom Exercises

The exercises and activities discussed next have all been delivered as part of the BSc (honors) Audio Engineering course at Leeds Beckett University in the UK. The course is a three-year undergraduate program and has between 15 and 20 students per level. This course is very much at the technical and engineering end of the audio education spectrum. It features acoustics, programming, electronics, system design, game audio, spatial audio, and audio recording. This is reflected in the technical nature of most of these exercises. However, they can be easily modified to take into account the experience and specialisms of any cohort studying audio-related courses. Exercises have been adapted where appropriate to the student's study level, study environment, and the particular module that the exercise is embedded into.

Awareness-Based Exercises

Surroundings

This short exercise can be carried out as a worksheet or as a discussion session. It is intended to draw attention to a student's work environment, how this environment consumes energy, and how the energy consumption could be minimized. Responses should be discussed, recorded, and summarized for student's further use.

Students should be asked to take a little time to look around their current surroundings and then make a list of the three main consumers of electrical energy. For each of these, they should write down ways of reducing their energy consumption. Students should also consider how is the room heated and write down how they think this heat energy is generated, what are the main areas of the space where heat energy is wasted, and how the space could be made more heat efficient.

Daily Activities

This short exercise can be carried out as a worksheet or as a discussion session. It is intended to draw attention to a student's own daily activities and their impact on the world around them and can be repeated at various intervals during a course in order to encourage students to monitor how this impact might change over time. Responses should be discussed, recorded, and summarized for students' further use.

Students should consider their activities today/yesterday and name three things they have done that have consumed significant energy and how they could have reduced the carbon footprint of these activities? They should also list the waste items that they have generated and whether these are likely to be recycled given the nature of their disposal. It is also interesting at this point to look at what recycling options are available around the university/school/college and how these might be improved.

Policy Documents

This exercise encourages students to engage with key policy documents. It develops students' reading and critical evaluation skills. It can be carried out as homework or could be an assessed activity for a module, resulting in an extended critical report on a specific document or range of documents. Students should be asked to find and download the environmental sustainability policy for either their university/college/school, a broadcaster (BBC, ITV, SKY, Channel 4, etc.), or a relevant professional body (IOA, IET, etc.).

For the chosen document, students should summarize the main points of the policy, discuss and critique the policy, and suggest improvements and changes. Students can also consider how easy it was to find the document and how this document might impact on their own work and/or daily activities?

Materials—Acoustics

This exercise works best in a recording studio environment and draws students' attention to the materials and manufacturing processes required to produce the acoustic treatment for the recording studio/critical listening environment. It can be run as a discussion or an assigned activity. Students should make a list of the main materials used for acoustic treatment in the studio and find out how each material is manufactured. Students should consider the sustainability of these materials and look into alternative materials from sustainable sources.

Materials—Electronics

This is a case study exercise that is easy to scale from a short classroom-based studio exercise to a longer assignment piece. Students should choose an audio device from their own recording studio or live sound rig and note down the main materials that are used in the manufacture of this device. They should then explore where the materials were sourced from, how sustainable the materials are, and whether there are any sustainable alternatives or ways of producing these materials in a more sustainable way. The manufacturing processes involved in the production of the device can also be explored along with sustainability issues in packaging and shipping the device.

Skills-Based Exercises

Basic Calculations—Solar Panels, Batteries, Power Usage

Sustainable system design is a great way to add context and practical examples to discussions of energy, power, voltage, current, and efficiency. A good exercise involves exploring running a car audio system from a solar panel and battery. The problem: A 100-Watt solar panel is connected via a battery charger to a 12 Volt, 2200mAh battery. This battery is used to power a car stereo that draws an average electrical power of 50 Watts.

Questions could include how much current does the stereo draw, how long can the battery run the stereo before it is completely empty, and what is the maximum current from the solar panel? All these questions are simplifications of reality but do allow further discussion of power and efficiency curves.

Laptop Energy Usage

Anyone teaching in audio cannot fail to notice the prevalence of Apple Macbook usage within the student community. This provides a great starting point to consider the sustainability of the devices we use every day. Apple (and of course other computing manufacturers) publish environmental data for all their products, which are easy to find with a simple internet search.

These data sheets include power figures for various different usage scenarios that can be used to establish energy consumption. Asking students to consider their own weekly patterns of laptop usage (actively charging/idle/sleep/etc.) and carrying out basic calculations to work out their total energy consumption (in kWh) is a really worthwhile exercise. It allows development of skills in traditional engineering calculations (power, energy, kWs, kWhs) with a raised awareness of the student's own daily and weekly energy consumption. Cost can be factored into this by requiring students to search for common energy tariffs and look at electricity costs per kWh. An extension to this is to repeat the calculations for a commonly available electric heater (for example, a 2kW electric oil filled radiator). This can be really eye-opening for students in terms of the huge energy requirements of devices used for heating.

The Apple data sheets also include information on greenhouse gas emissions, material efficiency, packaging, and recycling. These provide valuable discussion points for a group classroom exercise. For example, a 2017 15-inch Macbook Pro quotes 77% of its greenhouse gas emissions purely from production, 4% from transport, 1% from end-of-life recycling, and 18% from customer use during its normal lifecycle (Apple 2017: 1). Discussion points could include what do we mean by greenhouse gas emissions and how could we reduce these emissions? Students should be encouraged to explore why "production" is the highest contributor here and how the figures may differ for other devices in daily use.

Larger Projects

Major Projects—Ethics and Environmental Impact

At final year and postgraduate levels, students should be capable of integrating environmental sustainability awareness into their work, rather than as separate structured exercises. Good practice here includes requiring students to consider the environmental impact of their project work as part of the ethics approval process and asking students to devote a section of their final written work to discussing and evaluating the environmental impact of their work. Grading criteria and rubrics can be used to give credit for discussion of environmental impact in their work. Where students are working on extended creative productions (such as films, music recording projects, and games) they should be encouraged to engage with the Albert checklist in order to mirror industry practice in broadcast.

At second-year undergraduate level and below, large directed projects can be used to encourage students to engage with environmental sustainability through problem-based learning alongside developing their skills in audio. The two projects detailed next set students a design challenge that requires research, calculation, and design work while considering environmental sustainability throughout. Encouraging students to work together on these tasks also develops peer learning and team working skills.

The Off-Grid Sound System

This project requires students to produce a design for a sound reinforcement system for an outdoor event that runs from purely renewable energy sources and works well as an in-depth assignment. The project works best where students are given clear guidelines and a specification to meet, including requirements for channel counts, outboard, monitor feeds, and target sound pressure levels.

The main challenge for this project is calculating the energy requirements of the sound system and designing a sustainable energy generation system that can reliably supply this energy. This requires a lot of detailed calculation work using specification sheets and published information and also many assumptions on issues such as efficiency, weather conditions, and worst-case scenario using often incomplete information. Students will have to carry out a lot of reading into energy efficient audio equipment and off-grid energy generation and make justified choices for both aspects of the system.

The Sustainable Recording Studio

In this acoustics project students are asked to produce an acoustic design for a recording studio control room using purely sustainable materials. Many audio courses include some element of recording studio design in the curriculum and it is an easy extension to integrate environmental sustainability into this process.

Students should be given clear guidance on target predicted reverberation times, early reflection levels, and operational room response (EBU Tech 3276 (EBU 1998) can be of use here). General control room design philosophy will obviously be explored here but there should be special focus on the sustainable credentials of the materials used to implement absorption, diffusion, and insulation in the design.

This is an in-depth project requiring a great deal of reading and calculation work, resulting in a detailed design and report. This project can be made even more in-depth by adding the requirement for live rooms and target sound reduction index between the live and control rooms and between the control room and external environments. Another possible extension includes specifying the equipment for the control room and predicting power requirements and energy consumption. This can lead to cost prediction calculations and exploration of green energy tariffs to supply the energy for the studio.

Evaluation

Student engagement overall with these projects and exercises has been very good. Students are keen to learn about this area and show reasonable initial awareness of the subject area. Students were aware of the wider climate change debate and many of the actions that they should be taking in order to reduce their impact on the environment. Where students were less aware was in areas such as energy production, waste recycling and specific policy details. The prevalence of environmental sustainability in the news has a part to play in this and students generally agree with regards to the importance of the subject. From discussion with the students it would appear that, despite environmental sustainability being part of the UK national curriculum, few of them had explored the subject at school and, where they had, it was not a memorable experience.

The classroom exercises in awareness and skills are best kept as short sessions of around 30–45 minutes. The discussion sessions after these exercises looking at the students' responses and exploring the key issues were by far the most valuable part of the sessions and produced a wide range of interesting views and perspectives. Students were able to demonstrate their analysis and communication skills. The discussions also ensured that key ideas were shared with the rest of the class, as most of the exercises were individual or small group tasks.

During the classroom activities it was found that the older students were more aware and prepared to engage with the discussion. This is to be expected and, fortunately, in the cohorts where these exercises were trialed, there were some strong "older" voices which helped to drive the discussion. This worked well because some of the younger students were not as confident and not as aware of sustainability as an issue so were able to learn from their peers. Students also really appreciated context and real-world examples for these activities and the responsibility for providing this does fall on the educator to a certain extent. Student feedback on these discussions were overwhelmingly positive, ranging from "thanks for talking about this" through to "this has inspired me to look at my choices".

The larger projects (sound system and studio design) were delivered as part of a choice of second-year undergraduate project briefs. The projects seem quite straightforward and interesting for the students at first glance, however, the task soon opens up a wide range of issues and challenges that develop the students' skills in multiple areas. Students often got frustrated with the lack of environmental data available for equipment and materials and struggled with many of the assumptions that were required in order to calculate energy usage. One student working on the "green" festival stage project admitted to being very lost and unsure about the project in the initial stages but once they had got through this period and completed some of the key calculations they commented that they had "never learned so much" during a project.

Despite the challenging subject (and possibly because of it) these projects produced some innovative work. Students have produced fully pedal powered solutions and hybrid systems using biomass generators. One student even mapped the UK's average hours of sunshine and average windspeeds to produce regional and seasonal variations for the maximum and minimum solar and wind energy available for a "green" festival stage. Where students were asked to integrate sustainability into major project work the engagement was not as good. Students often saw this as just "another hurdle to jump through" when all they wanted to do was get on with their own project work. One way to address this was to integrate environmental sustainability into the standard research ethics process for the institution. This gives it more formal weight from the student's perspective.

Future Direction

These are early days for sustainability in audio courses and early days for the author's attempts to develop this area. It would appear that sustainability does not feature in many audio courses and there is much work to do. It is clear, however, from the general direction of industry and government policy that it is a vital area that needs to be integrated into all audio-based courses.

Useful future work in this area would be to develop a repository of audio-specific classroom exercises and project-based activities that are open for educators to access and use, similar to those identified earlier in this chapter by UNESCO, GDEE, and Practical Action. These resources could be developed into a structured Environmental Sustainability in Audio curriculum that could form the basis of a module within a university program or a short course to run alongside a university program. Certification for this course could be offered by an educational or industry body.

With regards to working toward a set of general guidelines and recommendations for embedding sustainability into an audio course it is clear that the various guidance documents considered earlier have an important part to play. These documents can be used as discussion points and to encourage students to understand the key issues. Context is really important. Students are much more likely to engage if the exercises and activities are closely related to their own practice. Educators should try and relate examples to audio and music wherever possible. Embedding environmental sustainability within wider projects has also been shown to be a successful approach in improving student awareness of this topic. Encouraging students to take ownership over their actions and highlighting the employability benefits of a strong understanding of sustainability issues are also very important in improving student engagement. It is also worth making clear at this point that a "holier than thou" attitude is not helpful. Educators should try to explore sustainability alongside their students and acknowledge that nobody is perfect.

Making significant changes to university curricula can be challenging but Ralph and Stubbs (2014) concluded that "people, in particular committed individuals, was a key success factor for operational and teaching areas" (p. 86) when integrating environmental sustainability into university courses. Developing this feature in higher education courses is a complex and long-term ambition and may require a paradigm shift in education. It is not only a matter of transforming institutional responsibility but also curriculum reorientation and teaching to better serve the needs of current and future generations. This will take time and will require a resilient and inclusive approach (Albareda-Tiana et al. 2018).

Conclusion

Much of the writing on embedding environmental sustainability in higher education recognizes that there are many challenges but also stresses that this is an issue that cannot be ignored. Educators must rise to these challenges in order to prepare our graduates for an industry that is placing great importance on sustainability in its day-to-day operations. The examples discussed here are just a snapshot of the current approach in the audio, music, and Creative Industries and many of the organizations involved have strong plans for the future. Educators across all sectors must integrate this into their practice. Policy and guidance from the

UN, UNESCO, ISO, BSi, and EU bodies provide a clear mandate for embedding sustainability in education. The fact that major broadcasters and creative industry bodies are clearly and dynamically addressing this area means that educators must respond in kind.

There are many useful resources out there to support educators in developing teaching materials in sustainability. Very little of this is specific to the audio sector but much of it can be easily adapted. The examples provided in this chapter have been shown to work well and it is hoped that these will be of help to fellow audio educators.

There is a huge amount of insightful science and writing on sustainability and climate which cannot possibly be discussed here. However, the following quote summarizes our current position as a species very well.

> We have overrun it in a way that is unprecedented, no other creature in the world has had the effect on the planet that the human species has, and so we ought to be aware of what we've done and recognize the responsibility that we now have in our hands.
>
> (Sir David Attenborough to the United Nations Katowice Climate Change Conference, December 2018)

References

Albareda-Tiana, S., Vidal-Raméntol, S., and Fernández-Morilla, M. (2018) Implementing the sustainable development goals at University level. *International Journal of Sustainability in Higher Education*, Volume 19, Issue 3, pp. 473–497.

Apple (2017) *15-inch MacBook Pro with Thunderbolt 3: Environmental Report*. Cupertino: Apple Inc.

BAFTA (2014) *Programmes That Don't Cost the Earth*. London: BAFTA.

BAFTA (2015) *The BAFTA Albert Consortium* [Online]. London: BAFTA. Available at: <http://wearealbert.org/about/the-consortium> [Accessed: 23 November 2018].

BBC (2018) *BBC Environmental Sustainability: Targets and Performance 2017/18*. London: British Broadcasting Corporation.

British Standards Institution (BSI) (2007) *BS 8901:2007. Specification for a Sustainable Event Management System with Guidance for Use (Withdrawn/Superseded)*. London: BSI.

British Standards Institution (BSI) (2012) *BS ISO 20121:2012. Event Sustainability Management Systems: Requirements with Guidance for Use*. London: BSI.

British Standards Institution (BSI) (2015) *BS EN ISO 14001:2015. Environmental Management Systems: Requirements with Guidance for Use*. London: BSI.

BSKYB (2018) *Sky Bigger Picture Impact Report 2018*. London: BSKYB.

Channel 4 (2017) *Channel 4 Annual Report 2017*. London: Channel 4.

Dawe, D., Jucker, R., and Martin S. (2005) *Sustainable Development in Higher Education: Current Practice and Future Developments—A Report for the Higher Education Academy*. York: Higher Education Academy.

Dmochowski, J., Garofalo, D., Fisher, S., Greene, A., and Gambogi, D. (2016) Integrating sustainability across the university curriculum. *International Journal of Sustainability in Higher Education*, Volume 17, Issue 5, pp. 652–670.

Engineers Without Borders (EWB) UK (2018) *Engineering for People: Design Challenge*. London: Engineers Without Borders UK.

European Broadcast Union (EBU) (1998) *EBU Technical Document 3276, 2nd Edition, Listening Conditions for the Assessment of Sound Programme Material: Monophonic and Two—Channel Stereophonic*. Geneva: European Broadcast Union.

GDEE (2013) *About GDEE* [Online]. Available at: <http://gdee.eu/index.php/about-gdee> [Accessed: 12 December 2018].

Goodland, R. (1995) The concept of environmental sustainability. *Annual Review of Ecological Systems*, Volume 26, pp. 1–24.

Guerra, A. (2017) Integration of sustainability in engineering education: Why is PBL an answer? *International Journal of Sustainability in Higher Education*, Volume 18, Issue 3, pp. 436–454.

Hawking, S. (2016) This is the most dangerous time for our planet. *The Guardian*, 1 December.

IET (2018) *Explaining the IET's Rules of Conduct* [Online]. Stevenage: The IET. Available at: <www.theiet.org/membership/career/ethics/rules/index.cfm> [Accessed: 12 December 2018].

Intergovernmental Panel on Climate Change (2018) *Global Warming of 1.5°C: Summary for Policymakers*. Geneva: IPCC.

IOA (2018) *Code of Conduct* [Online]. St Albans: IOA. Available at: <www.ioa.org.uk/membership/code-conduct> [Accessed: 23 November 2018].

ITV (2017) *Corporate Responsibility Summary Report 2017*. London: ITV.

Julie's Bicycle (2018) *Julie's Bicycle: About Us* [Online]. London: Julie's Bicycle. Available at: <www.juliesbicycle.com/about-us> [Accessed: 14 December 2018].

Morelli, J. (2011) Environmental sustainability: A definition for environmental professionals. *Journal of Environmental Sustainability*, Volume 1, Issue 1, Article 2.

Practical Action (2018) *Practical Action Schools* [Online]. Rugby: Practical Action. Available at: <https://practicalaction.org/schools> [Accessed: 1 December 2018].

Ralph, M. and Stubbs, W. (2014) Integrating environmental sustainability into universities. *Higher Education*, Volume 67, pp. 71–90. Dordrecht: Springer.

UNESCO (2010) *Teaching and Learning for a Sustainable Future*. Paris: UNESCO.

United Nations (UN) (2015) *Transforming Our World: The 2030 Agenda for Sustainable Development*. Geneva: United Nations.

United Nations (UN) (2018) *Paris Agreement*. Geneva: United Nations.

10

Breaking the Sound Barrier

The Importance of Interdisciplinary Audio Curricula

Elsa Lankford

Introduction

Audio programs span a wide range of disciplines and are situated in many university departments. Audio programs can be found in diverse areas, including the sciences, fine arts, performing arts, music, communication, computers and technology, and liberal arts. Audio curriculum crosses boundaries, incorporating practical and vocational skills, musicianship, art and experimental sound, and the cultural and historical study of sound. An interdisciplinary audio curriculum evaluates and contextualizes this disciplinary work. Society has collectively been enthralled with live, recorded, and broadcast sound; therefore, the industry needs professionals (with both formal and informal education) to build the technical and creative skills necessary to make these productions as engaging as possible. From live concerts to podcasts, there are different ways to approach teaching methods, styles, and programs.

This chapter considers how audio programs can look beyond their disciplinary borders for increased opportunities staffing, classes, and projects. I define an audio program as an educational major, department, or unit within a department, where students learn to record, edit, and mix sound and media, of which music technology is a subset, and students possess varying levels of musicianship. A discipline is defined as the division of subject matter through which knowledge is organized and transmitted (Davis 1995). A discipline is not a university department, as disciplines transcend universities (Dressel et al. 1970). A department is defined here as the basic scholarly and administrative unit in academia that houses a fairly autonomous community of scholars and is responsible for teaching and researching within a specialized field of knowledge (Andersen 1977). Departments are the link between discipline and the university (Trow 1977). An interdisciplinary class is one that involves the subject matter, faculty knowledge, and expertise of multiple disciplines or professional specializations (Davis 1995). Interdisciplinary courses can be taught by one or more instructors in related departments and the class may or may not incorporate civic engagement or service learning with a community partner.

Following a brief introduction of interdisciplinary education and its history, I highlight a study on audio curricular offerings across the United States and its level of disciplinarity. I examine and detail several examples of interdisciplinarity in audio education—from a class project to an entire program. As curriculum might begin with, but not end with, a syllabus, I conclude with the future and challenges of further work in audio education across disciplinary boundaries, learning outcomes, as well as assessment tools and academic workload.

Disciplinary and Interdisciplinary Education

Disciplines are essentially the building blocks that make up a university (Klein 1996). Disciplinary material is taught at the departmental level. Through and inside departments, money and resources are allocated, curriculum located, and educator workloads are constructed. Most disciplines were created organically, split off from larger departments or created by faculty, administrative, or student demand. There are benefits to disciplinary education and departments. They create an ability to deliver and deploy knowledge, offer a hierarchy of authority; such administrative positions include the provost, dean, and department chair. Disciplinary education also engages faculty interaction with related subject matter, provides scholarly and faculty mentorship, as well as scholarship appraisal by colleagues for promotion and tenure (Andersen 1977).

Boundaries between disciplines can be permeable, particularly through associations with related disciplines (Klein 1996). Disciplinary lines are created accidentally, intentionally, or even arbitrarily; each discipline incorporates its particular aesthetics, ways of working, analyzing, explaining, thinking, and worldview (Miller 1982). Furthermore, interdisciplinarity consists of the interaction of disciplinary elements, not the addition thereof (Graff 2015).

Brief History of Disciplinary and Interdisciplinary Higher Education in the United States

Interdisciplinary education includes programs in the liberal arts (e.g. women's studies, environmental studies, cultural studies) and the sciences (e.g. public health, environmental science, and bioinformatics). Before the mid-1800s, most colleges and universities in the United States were interdisciplinary, as they had been from their start in the 1600s and 1700s. During the late 1800s to early 1900s, curriculum and faculty were then divided into departments (Brubacher and Rudy 1958).

The creation of academic departments in the United States occurred for several reasons. The university system of scholarship in Germany used a hierarchy of departments and graduate education that top universities in the United States emulated. This German university style emphasized a research-intensive method (i.e. learning for the sake of learning), which the United States later adapted into pure science and research (Veysey 1965). The newly emerging structure, supported by Harvard University leadership under Charles W. Eliot, the opening of Johns

Hopkins University in 1876, and the desire for utilitarian education by students and businesses, were the main drivers for change (Andersen 1977). In the mid-1860s, the Morrill Act created and funded public universities across the United States, many of which had multiple schools and departments. By the 1890s, departments and majors solidified in the American university system, and most university students were able to choose their paths of study (Dressel et al. 1970).

A push for a common curriculum in primary and higher education helped lead the way for interdisciplinary pedagogy. Maria Montessori developed curriculum for children in Rome in the early 1900s (Gnanakan 2012). John Dewey, a leader in the Progressive Education Movement, wrote, in his seminal *Democracy and Education* in 1916, of the importance of education to cultivating critical thinking and preparedness. Dewey found these skills lacking in a more traditional disciplinary education. Noting that many students were "unwilling to be disciplined", Dewey (1980: 140) expressed concern about students who were not connecting to the disciplinary material found in most schools and universities.

The 1960s and 1970s were a watershed era for interdisciplinary studies (Klein and Newell 1998). Interdisciplinary research grew in popularity through federal government funding. Accordingly, environmental activists, along with members from various civil rights and social movements, joined to create new interdisciplinary centers, studies for research, and programs to provide solutions to societal, medical, and scientific problems. Philosopher and educator Paulo Freire's *Pedagogy of the Oppressed* (1970) introduced critical pedagogy to developing and developed countries; Freire's notion of critical pedagogy rejected the teacher-student hierarchy (Hill 2018).

Freire and others in the critical pedagogy movement looked to education as an intensely empowering and political process for creating change. Henry A. Giroux followed with the Radical Education movement, starting in the 1980s and continuing through the present day. Radical Education is defined as a practice of interdisciplinary education that questions disciplines and works toward a mission of democratizing society (Giroux 1988). Between 1975 and 2000, there was a nearly 250% increase in the number of interdisciplinary fields, with many of the new majors consisting of non-Western, ethnic, and women's studies (Brint et al. 2009).

Multidisciplinary and Interdisciplinary Education

There is a common misconception that combining elements of two or more disciplines, for a class or a program, makes it interdisciplinary. Without the integration of disciplines, the work is multidisciplinary. Multidisciplinarity includes exposure to multiple disciplines, but no integration. Allen F. Repko (2007) defines multidisciplinarity as a subset of interdisciplinarity, involving disciplines that are close to each other; interdisciplinarity "identifies relevant insights, creates or discovers common ground among them, and uses it to integrate them" (p. 12). Multidisciplinarity can be thought of as collaboration; students and faculty from different disciplines work together on an issue while maintaining their disciplinary identity. Also, multidisciplinarity need not be construed negatively. It can be seen as one step in a progression to interdisciplinarity, or the end goal for a particular class or program.

Julie Thompson Klein (2015) defines interdisciplinarity as the "integration of information, data, methods, tools, concepts, and/or theories from two or more disciplines or bodies of specialized knowledge" (p. 15). In *Crossing Boundaries*, Klein (1996) describes interdisciplinary work as the breadth, the comprehensiveness, while disciplinary work is depth or competence. To create deeper and more meaningful knowledge, interdisciplinary education needs to be informed by integrated disciplinary expertise. Transdisciplinarity is a process of reaching beyond disciplines to synthesize information and create new frameworks (Klein 2015), The differences between interdisciplinarity and transdisciplinarity are not always clear and the terms are sometimes used interchangeably.

Roots of Interdisciplinarity

Barry and Born (2013) divide interdisciplinarity into three modes that help form the base of an interdisciplinary program or process. These three modes are *integrative synthesis*, which is the integration and synthesis of two or more disciplines; *subordination*-service, where one discipline is subservient to another; and *agonistic-antagonistic*, which involves interdisciplinarity and trandisciplinarity evolving from opposition to the disciplinary limits. This classification can be used to examine existing interdisciplinary pedagogy.

Sometimes interdisciplinarity is top-down, emanating from the goals of a particular university. Sometimes it starts from the roots, with faculty from multiple disciplines working together or even students wanting to work on projects or problem solving that lives between departments. Raymond C. Miller (1982) described how an interdisciplinary program can start as a curriculum used to piece together and build on knowledge, where neither students nor faculty are satisfied and want more. An interdisciplinary program, class, or project can be created to elevate an existing program, to make it more whole and satisfying for the creative, technical, and professional needs of the students.

Interdisciplinarity often begins randomly, when faculty from different departments and academic tracks bump into each other in meetings, at social events or academic conferences, and even the hallway. A project across multiple classes can easily grow out of a chance encounter. This proximity factor coupled with an open mind makes interdisciplinary projects possible.

The physical location and accessibility of programs, classrooms, faculty offices, and studios is crucial. Recent scholarship examines the importance of spatial proximity for businesses (Zeller 2004), academic interdisciplinary medical centers (Rekers and Hansen 2015), and a newly merged interdisciplinary audio, music, and drama program (Nixon et al. 2015). In examining over 40,000 published papers and 2,300 patents all out of MIT from 2004–2014, Matthew Claudel (2017) used network analysis to determine that spatial proximity for collaboration was even more important than departmental or institutional structures.

Within a classroom, integration of disciplines can happen at the student or faculty level. Interdisciplinarity also happens on an individual basis, where students bring in experiences and knowledge from other disciplines, and the integration happens within that student. It is then up to the student to do the integration. It can occur within the class from a student perspective, where students from multiple disciplines are in the same room and interacting on projects with a chance to learn from each other. It also emerges at the faculty level, with one or more faculty members synthesizing disciplines to students from a singular or multiple disciplines. Traditionally, disciplinary integration happens in general education classes or in introductory classes. There is an advantage to having the interdisciplinarity later in upper-level courses, when the students acquire knowledge about their major and its boundaries, and have perhaps more to gain by testing those boundaries (Mahoney and Brown 2013).

Benefits and Challenges of Interdisciplinary Education

Interdisciplinary education is both challenging and beneficial for departments, colleges, and students. There can be additional risks and increased work for students, administrators, and community partners to make interdisciplinary collaboration possible. Interdisciplinarity also exposes the scope of a discipline as well as its limitations (Carlisle 1998). Maintaining sustainable interdisciplinary projects and pedagogy can be difficult, as university and department policies, values, and economics change (Barry and Born 2013). Faculty workloads often limit the feasibility of long-term interdisciplinary projects or courses; this is especially true particularly if the class is offered in another department. It is helpful to have workload and/or financial support for faculty doing interdisciplinary work as such endeavors require time for careful planning and implementation.

Many of the challenges for interdisciplinary education stem from bureaucratic university systems that are departmental and/or disciplinary based. Standards for scholarship and faculty research output are typically delivered within departments. This leaves interdisciplinary work traditionally undervalued and not counted properly for faculty workload (Biancani et al. 2018). Interdisciplinary faculty may not feel part of a department. Departments are where academics frequently meet, both socially and professionally, on university campuses. Many academic and professional organizations and conferences are disciplinary-based, which can prove problematic for scholarship, networking, and educational opportunities for faculty who cross disciplines. On the other hand, the benefit here is that professional organizations may create opportunities to broaden their scope and grow their ranks.

Pedagogy that crosses disciplinary boundaries may expand into territory already occupied by another department. This creates curricular turf battles that can prohibit future cross-disciplinary work. In the field of audio, where so many academic programs can make a claim, the question that can be easily be asked is “Who ‘owns’ audio?” It is more difficult to assess the quality of the pedagogy, focus, and pedagogy associated with non-disciplinary education. Disciplinary education, on the other hand, connects deeply with subject matter, rather than superficially, which is more common in interdisciplinary education. The upside is that such challenges can be addressed with multidisciplinary peer evaluators—who help to build bridges and create opportunities in higher education.

Interdisciplinary education does pose some risks. Such coursework introduces a chance of failure for a faculty member's class (untenured faculty with poor teaching and student evaluations are particularly vulnerable), and ought to be taken into consideration. Partnerships, organizations, and projects can frustrate all involved or fall apart. For example, for quite a few years, I had an assignment in my required upper-level audio production course where my students worked with a middle school in Baltimore City. As an interdisciplinary service learning project, my students taught and worked with the 10- to 14-year-olds in helping them learn how to record and produce music and sound stories. This was a great opportunity for everybody involved, including bringing the students to campus for a recording/mixing session each semester in our studio. However, when the middle school changed leadership and direction, it quickly became apparent that the partnership was no longer feasible.

As with a student group project, an interdisciplinary education partnership may not be equitable or equal, where one of the faculty members has to shoulder more of the load. There is a lot of communication, planning, and additional work needed to make interdisciplinary projects succeed for the faculty member(s) as well as the students. Because students are taking multiple classes, this time factor should be taken into account and communicated clearly to the students at the start of the semester; this is especially true if the class involves service learning and external partners. There are risks outside of the primary disciplinary department as well. Curricular agreements with other departments might change due to personnel and policy. Other departments that are treated as subservient through the subordination-service model of interdisciplinary education might grow tired of that role.

There are definitive benefits to interdisciplinary education as well. Interdisciplinary programs may have access to additional institutional funding. Interdisciplinary understanding and problem solving can help create links and deepen disciplinary knowledge (You 2017). It offers faculty a chance to explore other topics and to work with students in a way that can re-energize their disciplinary courses. For audio faculty in particular, it gives them a chance to move away from simply teaching students to use audiovisual equipment, to a more critical learning focus that places the medium of sound (and related areas) in a broader perspective. Existing research can be found about music educators similarly using interdisciplinary and transdisciplinary courses as a way to place their work in larger contexts (Tobias 2015).

Working with somebody in another discipline can be helpful in rethinking established ideas and make some curricular changes (Müller and Morton 2018). Innovations in pedagogy can help both faculty and students. The faculty member can be reinvigorated in a topic that might be getting stale and on reports asking about innovative teaching techniques. When interdisciplinarity involves an internal or external organization the exercise can help students gain a broader perspective (Augsburg 2005), learn to work better collaboratively, and improve critical thinking (Augsburg and Henry 2009). For teaching-focused institutions, interdisciplinary curriculum can be used to fulfill university missions to help students become productive citizens. It can play an integral role in preparing students to be fully engaged in society, where they can make more intelligent decisions through knowledge acquisition by evaluating multiple sources (Mansilla 2005). It can identify new career opportunities for students and ready them for the challenges to come, before and after graduation. It can help build, strengthen, and diversify their portfolios and help them network with students and professionals outside of their discipline.

Why Does Audio Work Well as an Interdisciplinary Option?

Audio has connections with math, science, technology, music, film, theater, dance, art, media, radio, communications, architecture, sociology, media literacy, storytelling, and urban planning among other fields. Each of these disciplines has its own vocabulary and ways of working. Interdisciplinarity helps to bridge these gaps and expand students' vocabularies. On a practical level, audio professionals will likely be working on or with projects that touch on these other disciplines. Having an understanding of how the fields connect and interact is important.

Researchers and educators are addressing multiple aspects of developing interdisciplinary pedagogy in programs that involve audio. Incorporation of sound into interdisciplinary programs includes an integrated arts and media studies program (Watson 2008) and creative material design and production in European design institutions that importantly introduces sound to designers (Tahiroğlu et al. 2014). Pedagogy and research on interdisciplinary classes with sound design and video games (Olivieri and Wang 2018; Onen et al. 2011; Hamilton 2015) and art and design (Collins and Kapralos 2014) has been published. Similarly, Felipe Otondo (2013) wrote about curriculum and course planning for spatialization of sound in an interdisciplinary class for dance, theater, and music students.

According to the United States Department of Labor, the industries with the highest level of employment for sound engineering technicians are the motion picture and video industries; sound recording industries; radio and television broadcasting; performing arts companies; and independent artists, writers, and performers (Bureau of Labor Statistics 2018). Additional industries listed that hire sound engineering technicians include software publishers, computer systems design, schools, and other information services (Bureau of Labor Statistics 2018). To be competitive, students benefit from having creative audio and technical skills in addition to film and media, broadcasting, performing arts, programming, and information technology (IT) experience.

Classes in music industry, business, and entrepreneurship can help prepare the student for what lies ahead, as a high percentage of those working in the music industry are part-time or self-employed (Brown 2007). Even if an audio program does not offer opportunities to take classes in all of these fields or interdisciplinary classes, students can look to general education classes and free electives, if the major leaves room for additional classes, to go on their own interdisciplinary learning journeys. The broad scope of audio requires the interaction, if not the integration, with other disciplines and colleges. This can be seen in examining audio programs in the United States and Canada. There are particular areas of audio that are easier to connect to other disciplines. I will highlight these, including possible connections, and personal examples in the sections that follow.

Interdisciplinarity Expressed Through Audio Programs in North America

Much has been written about audio and music technology programs in the United Kingdom (Boehm 2006, 2007; Born and Devine 2015; Mosley and Thompson 2014). In examining the interdisciplinarity of existing audio programs, I gathered information on each entry in the Audio Engineering Society (AES) educational listing. I collected basic information about the audio programs in the United States and Canada. The information collected included program name, focus or track (if audio was one of multiple curricular paths in a major), university, department, degree, school, college, geographical, private/public, links to the program and course listings, CIP (Classification of Instructional Programs) codes found within the courses (ex: MUS or AUD), total number of credits, categories of types of classes (business, film sound, synthesizer, recording/studio, radio, music), and comments. I focused on degree programs in non-profit colleges and universities in the United States and Canada. This is not a fully comprehensive list, as there are inevitably audio programs that are not listed in the AES education directory. The end list includes 38 graduate (master's or PhD), 44 associate's (two-year degree), and 119 bachelor's degree programs.

The type of degree programs shows how interdisciplinary audio is, including arts, science, engineering, music, and business degrees (see Figure 10.1). Over three-quarters of the programs are undergraduate two- or four-year degrees. Figure 10.1 does not and cannot showcase interdisciplinary pedagogy, but it highlights the variety of types of degrees that can be employed by students at all levels of learning. Students more interested in vocational learning of audio skills and craft can enlist in an associate's degree program whereas those wanting to teach can pursue an MFA or PhD degree.

Tables and Figures

AA—Associate of Arts (two-year)
AAS—Associate of Applied Science (two-year)
AS—Associate of Science (two-year)
BA—Bachelor of Arts
BS/BSc—Bachelor of Science
BFA—Bachelor of Fine Arts
BM—Bachelor of Music
MM—Master of Music
MFA—Master of Fine Arts
MA—Master of Arts

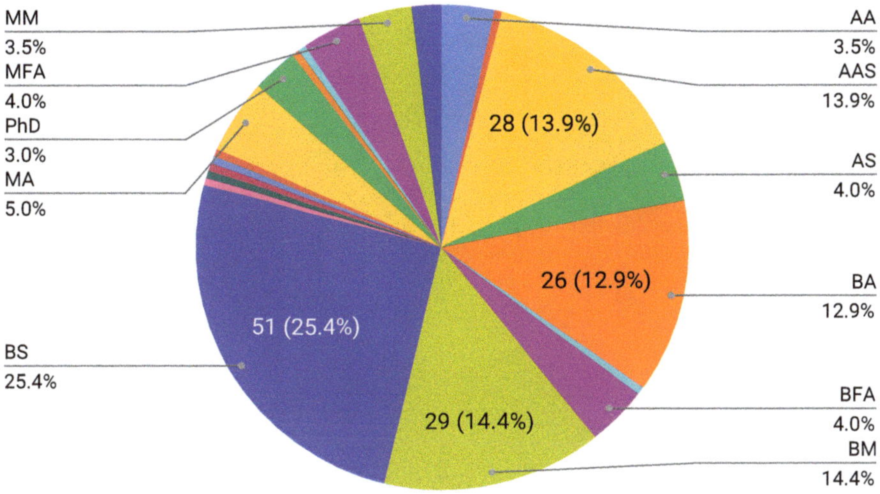

Figure 10.1 Audio degrees in US and Canada.

There were many different program and/or concentration (area of study) names that also help to visually show how audio programs cross disciplines, from acoustical studies to worship arts. Because there were so many variations, I performed a keyword search for some of the most used terminology in program names (see Figure 10.2). The term "music" occurs in over 20% of the program names, followed in order by technology, audio, record, production, sound, art, and engineering. See Table 10.1 for a list of the programs classified by program type.

One of the primary determinant factors of the curriculum and direction of the audio program is where the program is located within the college or university. If the college or school was listed on the audio program website, then I noted the information. For colleges that hold different types of departments, for example a college of the arts and sciences, I classified that in both the arts and sciences categories. As shown in Figure 10.3, by far the majority of audio programs in the US and Canada are located in an arts college or school.

This examination of audio programs was helpful, as the college and department in which the audio program is based will affect what type of interdisciplinarity can occur. For example, my program is part of a film and media department. This means that the easiest place to start with interdisciplinarity, in my program, is by combining film and sound and animation and sound, which we do inside and between classes. This collaboration in the classroom leads to or is inspired by collaboration between the faculty on creative scholarly projects as well.

Audio Engineering as Interdisciplinarity

Audio engineering, one branch of audio education and the profession, requires technical, creative, and communication skills. Students may need to know music

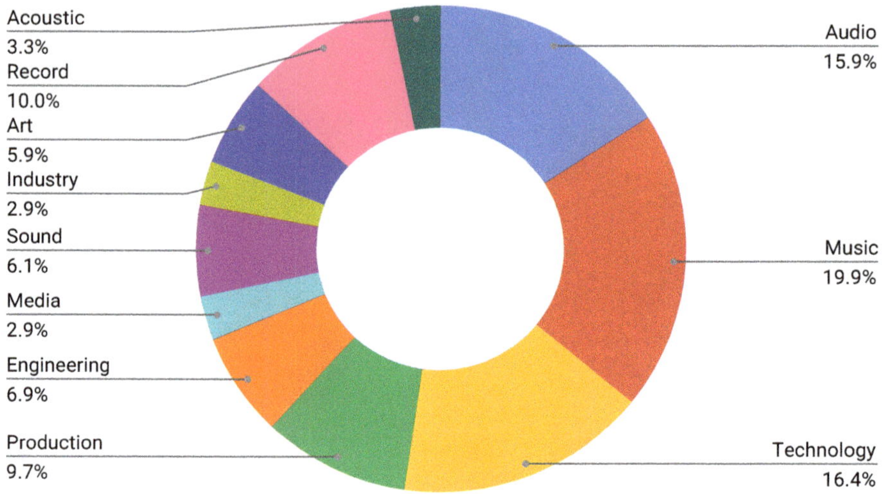

Figure 10.2 Keyword search in audio education program titles.

Table 10.1 Audio degree programs classified by program type

Audio engineering and production	Media arts
Commercial and computer music	Music
Digital arts	Music business or industry
Electronic media	Music production or technology
Entertainment management	Sound design for film, theater, or art
Mass communications	Interdisciplinary computing and music technology

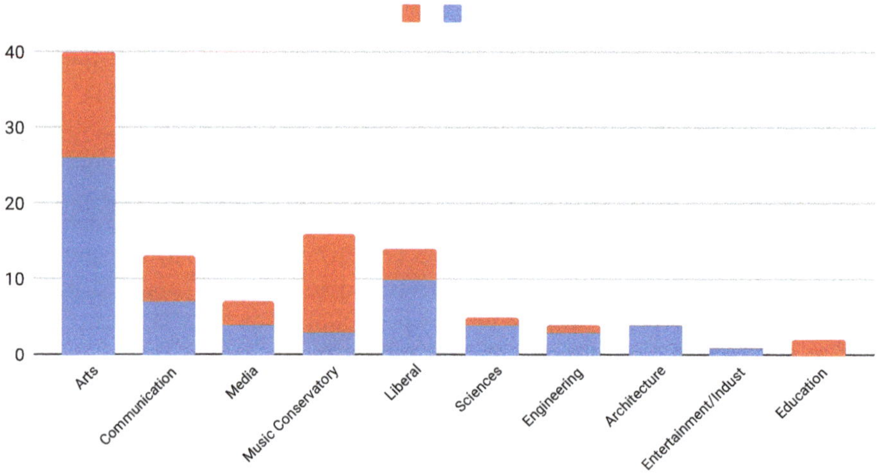

Figure 10.3 Audio program schools and colleges.

theory and performance, and understand rhythm and overall musicianship. They need to be able to use and troubleshoot the equipment and studio, sometimes in spaces and with equipment they have not used before. They need to understand the basics of acoustics, electronics, and their tools, which will differ depending on what type of audio they are working on. They also need to be able to communicate effectively in the studio and in their other creative places of work.

The curricular topics and projects in audio engineering include many of the following: live sound, studio recording and production, sound design or production sound for media or theater, mixing, mastering, synthesis and sound creation, sound and/or music for interactive media, and/or production or engineering for broadcast or podcast. Classes or projects with any of the following departments can be directly beneficial to the creative lines of work just listed: mass communications, music, physics, acoustics, music technology, theater, video, and music business. Depending on the university academic hierarchy, some of those departments may be housed in the same college as the audio program, which might make interdisciplinary interaction easier.

There are multiple resources on the interdisciplinarity of audio engineering, in research and in the classroom. Such research includes a project that utilizes a binaural radio play to help teach spatiality and abstraction to architecture students (Karathansopoulou 2019); a project for architecture students who were introduced to the overlap of sound and architecture and spatialization, where they had to solve an acoustic problem on the campus (Sheridan and Lengen 2003); and a student-run record label at the University of Memphis that gives students skills in entrepreneurship, law, marketing, and music industry (Butler 2007).

Case Study of an Interdisciplinary Audio Program

The history of an academic program tends to remain visible in the curricular requirements and elective classes, even after years and changes. My program at Towson University is a concentration in the Electronic Media and Film (EMF) major that split from Mass Communications department around the year 2000. Because of its Mass Communications history, it was initially geared toward radio, and only began to offer more than one audio class after I started teaching there in 2005. While EMF is a department, it is multidisciplined through its concentrations of film, audio, and film and media theory and criticism. Within the audio program, students can take EMF courses, music classes in recording techniques or music industry, or a theater class in sound design. They can also take an interdisciplinary arts and social justice class, that can count as an elective. This introduces the students to audio through the lens of other fine arts, as the EMF program is under the College of Fine Arts and Communications.

Students within the audio program can choose a path of sonic study including audio, radio, or sound for picture (Lankford and Schwartz 2018). While some audio programs have more prescriptive curricular paths, that students get to choose one or more paths creates interdisciplinarity in the classroom and within each student. Any one audio or radio class will have students with interests in film sound, audio engineering or radio and the possibility of at least two or three student film-makers or animators enrolled. Interdisciplinarity need not always come from the front of the classroom, it can emanate from the students to create more networking and shared and peer learning opportunities.

There are many opportunities for students to work with other students on projects, within and outside of classes. The production sound class records on-set audio for the upper-level film class where they collaborate during several in-class joint sessions, and then on set. Students within the sound design and sound creation classes create sounds and soundtracks for animations, interstitials and ads created in other film and animation classes. Students record and create music and other audio projects to be played on the student radio station, which also has academic connections to a radio stations operations class. The radio station class offers events involving live sound reinforcement where students taking both the live sound and radio class can put their skills learned in the classroom to work. These examples of interdisciplinary and collaborative work between and within classes gives students opportunities to get hands-on experience, build their portfolios and resumes, and connects them more with their faculty and peers. Beyond the purely utilitarian, it pushes them further creatively and adds breadth to what they are learning and creating.

Audio Documentary, Broadcast, and Podcasts as Interdisciplinary Models

Beyond the connections to communication and the arts, audio has a strong link to both the liberal arts and the sciences. For example, consider audio as recorded or live content for narrative nonfiction, broadcast, or podcasts. This work involves a combination of skills that take time and practice to develop, from utilizing research methods to psychology and interpersonal communications. A student producing or creating this content would benefit greatly from access to faculty and students in the liberal arts.

There are multiple resources about the interdisciplinarity of audio documentaries, both in research and in the classroom. A few examples include a collaboration between an anthropologist and radio journalist in Germany on the effect of mining to German villages (Müller and Morton 2018); a biochemistry course that uses podcasts to support teaching material in their courses (Müch-Harrach et al. 2013); courses that teach audio documentary bootcamps in Chicago and St. Louis through the lens of urban communication (Makagon and Gould 2016); a study of the management practices of students at college radio stations (Raymond 2016); and the study of pedagogy and disciplines involving podcast and audio documentaries at Australian and New Zealand universities and the blurred lines between audio, podcasting, radio, and journalism (Wake and Bahfen 2016).

In this type of situation, there are multiple ways in which the integration of disciplines can be handled. The interdisciplinarity can occur with specific persons, where a student brings in knowledge from a sociology or psychology class taken as a general education class and can apply it to the project. The integration could happen at the faculty level, with a shared project between an audio class and a liberal arts course. It could also happen at the program level, where a module of elective classes or a track contains pertinent classes in the liberal arts. With an agreement between departments, this could be a bidirectional interdisciplinary connection, where certain audio classes could house students from multiple disciplines. This can create rich interaction and co-learning experiences in discussions and group projects. The issue with this type of interdisciplinary learning involves potential problems with prerequisite structures and space for non-major students in the class.

Audio documentaries, in particular, are also easy to connect with service learning (or community engagement through experiential learning) which is often associated with interdisciplinarity. These activities can add to the class experience and students' portfolios. This can involve field trips and working with community partners. The partnership needs to be equitable, the community partner needs to know in advance how the student interaction will work, and the partner needs to understand the deadlines. The students need to feel enough ownership of the subject matter and project so that they are all in and fully involved. For as much as service learning is about communication for the students, it is equally, if not more so, for the faculty member(s). The faculty must ensure that both the students and the community partner have a clear understanding of the timeline, decision-making process, logistics, scope, and final product. Personally, I have been involved with interdisciplinary service learning projects that have gone well and were repeated in following semesters, and others where I had to help steer the project to completion.

Case Study of an Interdisciplinary Audio Project

The following example of an interdisciplinary service learning class project grew out of shared interests between three faculty members at Towson University: Matthew Durington in anthropology and sociology, Audrey Falk in family studies, and myself (Falk et al. 2012). We applied for a teaching innovation grant from the university to work on a project together through three different classes, my *Audio Documentary* class, Durington's *Life in the City*, and Falk's *Community Services for Families*.

The Audio Documentary class is designed to teach students to create short National Public Radio (NPR) style audio documentaries. The class also contained an ongoing assignment where the students become personally engaged in some aspect of Baltimore City. Towson University is only two miles north of the Baltimore City line, but in reflective discussions before and after the project was completed, the students envisioned Baltimore City through the lens of the show *The Wire*, an American television show from 2002–2008 shot in Baltimore that highlighted police, drugs, and crime in Baltimore City. The students did not have an understanding of gentrification, and most spent very little time in the city. None had been to Sharp-Leadenhall before, a historically black community in southwestern Baltimore City.

Each of the three classes began by working independently on their aspects of the project. *Community Services for Families* worked with the neighborhood to survey employment needs, *Life in the City* involved ethnography and team and individual fieldwork. The *Audio Documentary* class researched the history of the neighborhood, through seminars, talks, and web and library searches. After several shared seminars between students from the audio documentary and family studies class, the *Audio Documentary* students toured the neighborhood and interviewed residents.

The interviews were arranged by the community leadership with my involvement. The interviews were transcribed and edited into a five-minute audio documentary. The interviews, transcripts, and finalized documentary were handed over to the anthropology students to incorporate into their research. The community was given the files for archives, their website, and for presenting to possible and current funders. Students were encouraged to continue to work with residents of Sharp-Leadenhall who were interested in additional projects with the class. Acquiring a dedicated recorder and equipment purchased with funds from the grant allowed for that opportunity. While equipment for the class was available for checkout through the department, it was important to have dedicated equipment for this project so that the students could have longer access.

Often, university research is initiated for publications and scholarship from the faculty member(s) while "community collaborators are left wondering how they benefit" (Falk et al. 2012: 40). This project was designed to ensure that the community received a meaningful product. The group project to create short audio documentaries about Baltimore City continued for several years after the initial project with encouragement from the neighborhood, with one or two groups each year working with Sharp-Leadenhall's residents. The class continued to do field trips to talk with and interview community members. The community received the recordings and the finished audio documentaries to use for archiving and promotional purposes and the original audio documentary was uploaded to Public Radio Exchange (PRX) for online listening and possible licensing by public radio stations. There were additional outcomes through the other classes, involving archival and employment data for the neighborhood. It was important, for all of the faculty members, that this project have some positive impact for the neighborhood.

There were challenges involved with this interdisciplinary class project. It involved a lot of planning, including helping to write the grant, lining up transportation, coordinating with the other faculty members and classes, and arranging interviews. We were fortunate to have an excellent and willing community partner, who was aware of the restrictions of academia, including the semester schedule. The students demonstrated, through their written reflections, that they appreciated the ability to work with a client.

Case Study on an Interdisciplinary Class in an Honors College

An Honors College class can provide a supportive structure for a faculty member to get interdisciplinary teaching experience. An Honors College is typically an educational entity at a university for exceptional undergraduate students independent of their majors. I have developed and taught a course entitled *Sound and the City* for the Honors College at Towson University. This class investigates urban history through the lens of urban soundscapes, literature, paintings, film, and sound. Additionally, the students learn to critically listen, creatively work with sound, and interpret the urban soundscape using creative media.

When I originally conceived of the idea, I knew it would benefit from an interdisciplinary class environment. Having students from different majors would serve to enrich the learning experience and offer a chance to develop critical listening skills in non-audio students. With a class comprised of different colleges, departments and majors, students can bring different perspectives into class discussions. Having the class live in the Honors College also allowed for more readings, writing, and the ability to do a deeper dive into the material.

The first weeks of class involved readings from soundscape and sound studies experts and getting the students up to speed with terminology, history, and learning to listen. The outputs of the class included a group creative narrative, a research paper on urban soundscapes, and a literature/film survey presentation and discussion. The creative narrative could be in any media that involved the urban soundscape. What the students developed for the final creative narratives exceeded my expectations to include a puppet show, animation, radio drama, theater, music video, and poetry.

During the class, students went on field trips to record the Baltimore City soundscape. The audio from the field trips was used in class to learn how to edit and layer sound. The literature or film survey involved choosing a time period or author and evaluating the realism of the described or sound designed soundscape. In doing so, they had to research what sounds were realistic for the chosen time, and compare and contrast with what the author or sound designer had created on paper or on screen. The research paper had to incorporate at least two primary sources, that could include recorded interviews. The topic was fairly open, on any aspect of urban soundscapes, past or present. Students had access to recorders purchased by the Honors College for this assignment and the creative narrative. While the benefits of teaching an interdisciplinary honors class are clear, the faculty member's workload must be able to handle a class being taught outside of their home department. This can be challenging due to the limited number of faculty slots for audio courses, both full-time and adjunct. Teaching outside of the department might be a luxury that most cannot afford.

Other Interdisciplinary Audio

Some branches of audio are more interdisciplinary than others. Game audio, for instance, requires music, audio, technology, programming, engineering, math, business, law and physics. Those working in game audio need to be adaptable, able to learn technologies that often change, and agile enough to work independently and in teams (IASIG 2011). Students and professionals working in game audio benefit from at least basic music composition skills, computer programming skills, and the ability to tell a story effectively through sound and/or image. Skills and classes on sound for video and film are essential as well, as game audio practitioners typically record and edit sound effects, design sound, and work with dialogue and voice-overs (Onen et al. 2011).

Sound studies also differs from most other audio paths. It involves the study of the "material production and consumption of music, sound, noise, and silence and how these have changed throughout history and within difference societies" (Pinch and Bijsterveld 2004: 636). Film, theater, music, and art, have more established studies or history paths, less about the production of a creative product and more about the study, history, and cultural effects on that media. Sound studies is more oriented toward humanities and the bridge between audio and the liberal arts and cultural studies. Kate Lacey (2018) writes about the interdisciplinarity of a related field to sound studies, radio studies, which crosses into culture, politics, identity, and community, among others.

Student Learning Outcomes and Assessment Tools for Interdisciplinary Classes

Interdisciplinary work is challenging in many aspects, including developing outcomes and methods of assessment. Interdisciplinary classes are also challenging in that the knowledge generated by the class will not be repeated the same way each time (Schneider and Caswell 2003). Interdisciplinarity in audio education can be wide-ranging in scope. Listed next are examples of three outcomes: *communication*, *teamwork*, and *creativity*; and three assessment criteria, developed by Mansilla (2005): *disciplinary grounding*, *integrative leverage*, and *critical stance*. As can be seen in subsequent sections, they can apply to many interdisciplinary audio academic ventures, with a focus on project-oriented student outcomes.

Outcomes

Multiple studies suggest that audio and music students need interpersonal and soft skills that they are not always getting in disciplinary programs (Bielmeier 2013, 2014, 2017; Branscome 2010; IASIG 2011; Onen et al. 2011; Tough 2010). Interdisciplinary projects offer an opportunity for students to build on these skills. Outcomes should be discussed in class at the onset of the interdisciplinary project. They should be reflected upon both in writing by individual students and by the class as a whole, including the faculty member. These outcomes for interdisciplinary audio courses can be divided up into *communication*, *team-oriented* based objectives, and achieving *creative* objectives.

The ability to *communicate clearly and effectively to peers* should be in the list of outcomes for an interdisciplinary audio project, class, or program. Interdisciplinary work offers even more opportunities for communication with team members and possibly external organizations. Especially when solving complex problems, students benefit from group work with teams of students from different disciplines (Carr et al. 2018). Communication involves building on and sharing an existing disciplinary vocabulary, learning vocabulary from other disciplines, and translating between them. This can happen within and outside of groups, in communication to the rest of the class, external communication, and related to tasks and projects. Communication can be further broken down into a related learning outcome regarding *interpersonal communication*, the ability to develop and build relationships.

Teamwork is essential to working with media and in interdisciplinary projects and classes. *Working effectively as a member of a team* is a crucial skill. This includes the ability to follow and lead, listen and talk. Teams also offer important experiences in conflict resolution. While this is a learning outcome that could be assessed in a disciplinary class in audio, group work is not only more common in interdisciplinary learning, it is the most effective way to ensure students do not stay in their disciplinary comfort zones. Working in groups can offer additional benefits to creative thought and processes in collectively building a creative project (Sawyer and DeZutter 2009).

While the concept of team membership may seem nebulous, it can be incorporated into the pedagogy. Early classes could involve team building exercises and in-class projects can help strengthen teams. This is particularly important for teams that will be together the entirety of the semester. A lecture on what interpersonal communication skills are and their importance for the project or class can be incorporated, particularly if there is different vocabulary between students of different disciplines that are on the same team, tying back to the communication outcome sets.

Projects or classes that integrate audio and another fine or liberal art, such as music, theater, video, or anthropology, to name a few, can utilize more specific learning outcomes. *To explore and enact potential solutions to a problem through creative and scholarly projects in performance, visual art, media and communication* is a learning outcome from an interdisciplinary fine arts class on art and social action that I co-developed. Many creative projects emanating from an interdisciplinary class will be narrative based. The ability *to use audio or sound to effectively tell stories in a meaningful way across media* can be applied to fiction or nonfiction stories: audio or video documentaries; narrative creative film, writing, or theater; or music, among others. This outcome gives audio students a chance to work from, and go beyond, their existing disciplinary knowledge. Whether a student is interested in music technology, audio documentary, or postproduction sound for video, they can use what they know and help further the project. A related outcome is *to contextualize narratives, created in other disciplines in the fine or liberal arts, through sonic element(s)*. This helps ground audio students in the fine art or liberal arts world, reminding all team members of the importance of critical listening and the power of sound to their team project.

Assessment

Veronica Boix Mansilla (2005) lists three ways to assess interdisciplinary student projects. The first is through *disciplinary grounding* where the student showcases disciplinary knowledge as in traditional assignments. Disciplinary grounding is how audio educators are already assessing disciplinary audio work, which can include technical quality and creativity. Julie Thompson Klein states that interdisciplinary work should be evaluated on how well the students "accomplish the particularities of their tasks and how well they integrate knowledge" (Klein 1996: 211). This is too general for assessment purposes, but a helpful starting point for interdisciplinary assessment. Interdisciplinary work involves disciplinary work, both of which need to be assessed, as well as their integration.

Integrative leverage is then how multiple disciplines and perspectives are integrated in a new way, involving "interpreting the work with an epistemological eye" (Mansilla 2005: 20). From these assessment criteria, learning outcomes can be derived. Integrative leverage can be a writing prompt within the reflective paper, where the students reflect upon the interdisciplinary knowledge gained and how it affects how they work in audio. This can easily involve some level of research into the interdisciplinary or other disciplinary subjects. Davis uses a simple rubric, of sorts, to define a "continuum of collaboration" for interdisciplinary courses and team teaching, which simply maps the following areas on a range from low to high degree of collaboration: planning, content integration, teaching, and evaluation (Davis 1995: 20).

Finally, the *critical stance* is a measurement of the work "against its goals and its ability to withstand critique" (Mansilla 2005: 20). Assessing the critical stance involves both project management and peer critique. The integration of peer critiques into audio classes is helpful, particularly with creative sound and/or interdisciplinary projects. Peer critiques help students develop their professional and creative vocabulary and their ears. Critiquing the work of their peers also helps students to better identify strengths and weaknesses in their own work (Collins and Kapralos 2014).

Assessment of communication and teamwork can be built into the assignment criteria and the grading rubric. Students can assess their team members based on their communication skills, which is typically a problem with group assignments. Items to assess could involve verbal or written communication with clients, internal or external. The assessment could be from peers, self-evaluation, the faculty member, and even clients, if an evaluation form is created, similar to an internship site supervisor evaluation form.

Reflection in interdisciplinary work is critical, as the student does not always recognize or take stock in the skills and knowledge generated through the interdisciplinary exercise. As part of the reflection, a student should write a self-evaluation after each exercise is complete. Because interdisciplinary work often creates opportunities for students to work outside of the comforts of a safe academic structure, it is very likely that they will have encountered new obstacles. Because the knowledge generated with each iteration of the exercise will likely change each time, the information from this reflective exercise should not simply be filed away, because if anything, it will then only reside with the faculty member and that student. A class discussion can help spread the knowledge further, and allow for further discussion of shared experiences, at least within the class.

Assessment should not only be for the students, the project itself should be assessed and reflected upon by the faculty member for future revisions and iterations. Due to additional work for the students, faculty may want to create their own assessment survey, beyond the reflective paper and the university's evaluation system. This could be as simple as an online form. If working with a community partner or other entity, there are opportunities to receive and act on their feedback as well, about the students, the projects, and the outcome.

Faculty Scholarship and Workload

I would be remiss if I did not address how interdisciplinary scholarship and curriculum in an audio program counts toward the educator's workload. Developing curriculum and making inroads with other departments takes time and interdisciplinary scholarship is not always counted properly in promotion and tenure, merit, or contractual matters.

On this point it is worth highlighting Boyer's (1990) model of the various forms of scholarship, including the Scholarship of Discovery and the Scholarship of Integration. The Scholarship of Discovery involves both global and local knowledge contribution and the Scholarship of Integration involves creation and research across disciplines. These models are typically listed in departmental, college, and university promotion and tenure guidelines and can form the platform of having US and Canadian faculty scholarship count properly toward their workload. Mentorship with senior faculty, both inside and outside of the department, should be sought out early, particularly for junior faculty. Senior faculty, especially who have had interdisciplinary and service learning experience, can help guide and be a support and sounding board.

Where to Get Started with Interdisciplinary Audio

The simplest way to get started with developing an interdisciplinary course or project begins with choosing a topic or theme and then determining if you will be the sole educator. If not, and you want to work with somebody else, then you will need to determine which discipline(s), educators, or organizations would be the best partners to achieve the course or project goals.

A straightforward way to start with interdisciplinarity is the better utilization of existing resources. On-campus radio and TV stations give students and faculty an opportunity to be involved in media production, promotions, and portfolio/resume builders while interacting with students, faculty, and staff in other academic and non-academic departments. In a new service role at my university as the student streaming radio station faculty advisor, I have found this out firsthand. The radio station needs content, coupled with a surge in popularity of podcasts. This offers chances for my students in radio and audio to produce, engineer, program, and record content from students in other majors. This creates many opportunities for cross-disciplinary pollination. That being said, I am careful to point out that they are not performing a free service for other departments, which puts them and their skills in a more subservient position. The key is looking for partnership opportunities, where the radio and audio students are as involved in the creation of this content as they want to be.

Events and conferences on campuses offer a chance for hands-on live sound experience, including as assignments or projects within an otherwise disciplinary audio class. Existing interdisciplinary entities such as Honors Colleges, service learning centers, and student organizations can be hubs for team-teaching and interdisciplinary opportunities. These offer an easy way for an interdisciplinary-curious faculty member to bring audio to other students or bring audio students together with other majors. Off-campus organizations, community centers, and schools also offer engaged service and interdisciplinary learning and teaching opportunities. These can lead to in, and out of, class experiences for students to network, become more well-rounded, and learn experientially and collaboratively. Many universities have service learning centers where staff can help faculty get involved with organizations and entities. There may be service learning fellowships and local, regional, and national service learning conferences and workshops. While service learning is not always interdisciplinary, their paths often cross.

Larger changes to make a program interdisciplinary may be problematic. There may be administrators who are biased against interdisciplinary programs, as that may not be their background (Walzer 2017). It is then up to the department and faculty to educate the administrators on the benefits of interdisciplinary audio education. Colleges and departments interested in inspiring interdisciplinarity should be encouraged to have financial and workload incentives to develop and teach on existing classes. Tenure-track faculty should look for mentorship with more experienced and tenured faculty, in audio or another discipline. Team-teaching in interdisciplinary classes should be encouraged, when possible, to allow multiple disciplines or perspectives to be synthesized. Related departments should be co-located, so as to let students and faculty take advantage of the proximity factor of disciplinary integration and success.

Look for ways to explain the types of projects and classes in one's department to a colleague in a different discipline. Identifying overlap in projects and classes can be a way to weed out redundancies or find common ground for a project. This can be through college or university committee meetings, affinity groups, or even serving on search committees. Interesting projects exist at every university, the key is learning about them, and collaborating with those colleagues in the classroom.

Dedicated equipment for interdisciplinary projects can be very helpful, especially if it is to be used by students outside of the program. This involves additional issues to enact a system for students to access the equipment and for handling repairs or replacements. Students may not be in the same major, and the chance for all to access the equipment or spaces is helpful. For the interdisciplinary projects that I have worked on, dedicated equipment has been or would have been helpful for almost every single one. Not every faculty member will know how to teach an interdisciplinary class, especially when they have had limited exposure to other disciplinary knowledge (Watson 2008). There may be workshops or longer-term fellowship opportunities through on-campus student engagement centers or academic affairs for service learning, civic engagement, or interdisciplinary education that they may be able to access.

Conclusion

By way of conclusion, I want to take a step back to the title of this chapter. Disciplinary boundaries are not necessarily barriers. They provide needed structure for the university and the students, but professors and students should be encouraged to add gates to the fences that our departmental boundaries sometimes create. Audio education mixes well, the audio puns abound, with interdisciplinarity, through the subject matter, creative products, and the media itself. While there are different levels and methods of interdisciplinarity, audio students and faculty benefit from integrating disciplines. Developing disciplinary integration in a program can happen in small steps, using existing resources within the department, college or university, which can start with one project in a single class and a single instructor.

While interdisciplinary audio education is more work and adds risk for the faculty and students, the benefits outweigh the challenges. Through active, engaged, and collaborative learning, interdisciplinary audio education adds breadth to the depth of audio technology, resulting in more creative projects, deeper and longer lasting knowledge, and important career-building skills. Interdisciplinary audio projects can provide experiential learning to additionally deepen the learning experience and give students the opportunity to learn and practice their interpersonal and communication skills.

The existing interdisciplinarity of audio education is evident through the degree offerings in the United States and Canada. No matter how it is implemented, there are several student outcomes that can be applied to projects to help with critical communication and teamwork. Creative and social action outcomes can also be implemented in projects and classes. Assessment and reflection of interdisciplinary projects is challenging, but worthwhile, for faculty and students alike. Proper balancing of faculty workload, funding, and support are necessary for sustainable interdisciplinary projects and curriculum development in audio education.

References

Andersen, K.J. (1977) In Defense of Departments. In Mchenry, D.E. (ed.). *Academic Departments*. San Francisco: Jossey-Bass.

Augsburg, T. (2005) *Becoming Interdisciplinary: An Introduction to Interdisciplinary Studies*. 2nd ed. Dubuque, IA: Kendall/Hunt Publishing Company.

Augsburg, T. and Henry, S. (eds.) (2009) *The Politics of Interdisciplinary Studies: Essays on Transformations in American Undergraduate Programs*. Jefferson, NC: McFarland & Company.

Barry, A. and Born, G. (2013) Interdisciplinarity: Reconfigurations of the Social and Natural Sciences. In Barry, A. and Born, G. (eds.). *Interdisciplinarity: Reconfigurations of the Social and Natural Sciences*. London: Routledge.

Biancani, S., Dahlander, L., McFarland, D.A. and Smith, S. (2018) Superstars in the Making? The Broad Effects of Interdisciplinary Centers. *Research Policy*. 47. 543–557.

Bielmeier, D. (2013) *Why Didn't You Learn This at Recording School: Critical Comments by Employers*. Paper presented at AES 50th International Conference on Audio Education, Murfreesboro, TN, 25–27 July.

Bielmeier, D. (2014) *Apprenticeship Skills in Audio Education*. Paper presented at AES 137th Convention, Los Angeles, CA, 9–12 October.

Bielmeier, D. (2017) *Audio Recording Production Students Report Skills Learned or Focused on in Their Programs*. Paper presented at 143rd International Audio Engineer Society Convention, New York, NY, 18–21 October.

Boehm, C. (2007) The Discipline That Never Was: Current Developments in Music Technology in Higher Education in Britain. *Journal of Music, Technology and Education*. 1(2). 7–21.

Boehm, C. (2006) The Thing About the Quotes: "Music Technology" Degrees in Britain. In ICMC (ed.). *2006 ICMC Conference Proceedings*. New Orleans, LA: ICMC, 682–687 [Online]. Available from: https://quod.lib.umich.edu/i/icmc/bbp2372.2006.138/1 [Last Accessed: 7 May 2019]

Born, G. and Devine, K. (2015) Music Technology, Gender, and Class: Digitization, Educational and Social Change in Britain. *Twentieth-Century Music*. 12(2). 135–172.

Boyer, E.L. (1990) *Scholarship Reconsidered: Priorities of the Professoriate*. New York: Carnegie Foundation for the Advancement of Teaching.

Branscome, E.E. (2010) *Music Career Opportunities and Career Compatibility: Interviews with University Music Faculty Members and Professional Musicians*. PhD Thesis. University of North Texas. Available from: https://digital.library.unt.edu/ark:/67531/metadc28398/ [Last Accessed: 15 May 2019]

Brint, S.G. et al. (2009) Expanding the Social Frame of Knowledge: Interdisciplinary, Degree-Granting Fields in American Colleges and Universities, 1975–2000. *The Review of Higher Education*. 32(2). 155–183.

Brown, R. (2007) Enhancing Student Employability: Current Practice and Student Experiences in HE Performing Arts. *Arts & Humanities in Higher Education*. 6(1). 28–49.

Brubacher, J.S. and Rudy, W. (1958) *Higher Education in Transition: An American History: 1636–1956*. New York: Harper & Row.

Bureau of Labor Statistics (2018) *U.S. Department of Labor, Occupational Employment Statistics* [Online]. Available from: www.bls.gov/oes/current/oes274014.htm [Last Accessed: 15 May 2019].

Butler, T.D. (2007) BlueT.O.M. Blueprint: A Case Study on the Creative Design of a Student-Run Record Label. *MEIEA*. 7(1). 99–118 [Online]. Available from: www.meiea.org/resources/Journal/html_ver/Vol07_No01/2007_Vol_7_No_1_A7.htm [Last Accessed: 11 May 2019]

Carlisle, B. (1998) Music and Life. In Newell, W.H. (ed.). *Interdisciplinarity*. New York: College Board Publications.

Carr, G., Loucks, D.P. and Bloschl, G. (2018) Gaining Insight into Interdisciplinary Research and Education Programmes: A Framework for Evaluation. *Research Policy*. 47. 35–48.

Claudel, M. et al. (2017) An Exploration of Collaborative Scientific Production at MIT Through Spatial Organization and Institutional Affiliation. *PLoS One*. 12(6). 1–22 [Online]. Available from: https://doi.org/10.1371/journal.pone.0179334 [Last Accessed: 15 May 2019]

Collins, K. and Kapralos, B. (2014) Sound Design for Media: Introducing Students to Sound. *Journal of Sonic Studies*. 6(1). 1–15 [Online]. Available from: http://journal.sonicstudies.org/vol06/nr01/a04 [Last Accessed: 12 December 2018]

Davis, J.R. (1995) *Interdisciplinary Courses and Team Teaching: New Arrangements for Learning*. Phoenix: Oryx Press.

Dewey, J. (1980) Democracy and Education. In *John Dewey: The Middle Works, 1899–1924*, Vol. 9. Carbondale: Southern Illinois University Press.

Dressell, P.L., Johnson, F.C. and Marcus, P.M. (1970) *The Confidence Crisis*. San Francisco: Jossey-Bass.

Education Working Group of the Interactive Audio Special Interest Group (IASIG) (2011) *Game Audio Curriculum Guideline* [Online]. Available from: www.iasig.org/index.php/component/edocman/?task=document.viewdoc&id=3 [Last Accessed: 24 July 2019]

Falk, A., Durington, M. and Lankford, E. (2012) Engaging Sharp-Leadenhall: An Interdisciplinary Faculty Collaboration in Service-Learning. *Journal of Public Scholarship in Higher Education*. 2. 31–46.

Giroux, H. (1988) The Hope of Radical Education: A Conversation with Henry Giroux. *The Journal of Education*. 170(2). 91–101.

Gnanakan, K. (2012) *Integrated Learning*. New Delhi: Oxford University Press.

Graff, H.J. (2015) *Undisciplining Knowledge: Interdisciplinarity in the Twentieth Century*. Baltimore: Johns Hopkins University Press.

Hamilton, R. (2015) *Designing Next-Gen Academic Curricula for Game-Centric Procedural Audio and Music*. Paper presented at AES 56th International Conference, London, UK, 11–13 February.

Hill, S.C. (2018) A "Sound" Approach: John Cage and Music Education. *Philosophy of Music Education Review*. 26(1). 46–62.

Karathansopoulou, E. (2019) Atmosphere in Radio and Architecture: Using the Revenge Radio Play in Interdisciplinary Teaching as a Means to Understand Notions of Abstraction and the Tensions Between Materiality and Immateriality in Building Physical and Imagined Spaces. *Radio Journal: International Studies in Broadcast & Audio Media*. 17(1). 113–129.

Klein, J.T. (1996) *Crossing Boundaries: Knowledge, Disciplinarities, and Interdisciplinarities*. Charlottesville, VA: University Press of Virginia.

Klein, J.T. (2015) *Interdisciplining Digital Humanities: Boundary Work in an Emerging Field*. Ann Arbor, MI: University of Michigan Press.

Klein, J.T. and Newell, W.H. (1998) Advancing Interdisciplinary Studies. In Newell, W.H. (ed.). *Interdisciplinarity: Essays from the Literature*. New York: College Board.

Lacey, K. (2018) Up in the Air? The Matter of Radio Studies. *The Radio Journal: International Studies in Broadcast and Audio Media*. 16(2). 109–126.

Lankford, E. and Schwartz, A. (2018) *Case Study: An Interdisciplinary Audio Curriculum*. Paper presented at AES 145th Convention, New York, NY, 17–20 October.

Mahoney, K. and Brown, R. (2013) Devising and Interdisciplinary Teaching: A Case Study in Collaboration Between Theatre and Humanities Courses. *College Teaching*. 61. 143–149.

Makagon, D. and Gould, M.R. (2016) Learning the City Through Stories: Audio Documentary as Urban Communication Pedagogy. *International Journal of Communication*. 10. 1263–1276.

Mansilla, V.B. (2005) Assessing Student Work at Disciplinary Crossroads. *Change: The Magazine of Higher Learning*. 37(1). 14–21.

Miller, R.C. (1982) Varieties of Interdisciplinary Approaches in the Social Sciences: A 1981 Overview. *Issues in Integrative Studies*. 1. 1–37.

Mosley, B. and Thompson, P. (2014) Setting the Tone: Considerations for Educating the Next Generation of Sound Reproduction Professionals. *Proceedings of the Institute of Acoustics*. 36(2). 790–798. Available from: http://eprints.leedsbeckett.ac.uk/1796/ [Last Accessed: 15 May 2019].

Müch-Harrach, D., Kothe, C. and Hampe, W. (2013) Audio Podcasts in Practical Courses in Biochemistry: Cost-Efficient E-Learning in a Well-Proven Format From Radio Broadcasting (translated). *GMS Zeitschrift für Medizinische Ausbildung*. 30(4). 1–10. Available from: www.doi.org/10.3205/zma000887 [Last Accessed: 15 May 2019].

Müller, K. and Morton, T. (2018) At the German Coalface: Interdisciplinary Collaboration Between Anthropology and Journalism. *Energy Research and Social Science*. 45. 134–143.

Nixon, P., Young, T., Brown, E. and Wiltshire, L. (2015) *Interdisciplinary Benefits: Encouraging Creative Collaboration*. Paper presented at 26th UK AES Conference on Audio Education, Glasgow, Scotland, 26–28 August.

Olivieri, V. and Wang, R. (2018) Transdisciplinary Sound Design Pedagogy. *Theater Design & Technology*. 10–22. Fall.

Onen, U., Stevens, R. and Collins, K. (2011) Designing an International Curriculum Guideline for Game Audio: Problems and Solutions. *Journal of Game Design and Development and Education*. 38–47 [Online]. Available from: www.rit.edu/gccis/gameeducationjournal/designing-international-curriculum-guideline-game-audio-problems-and-solutions [Last Accessed: 15 May 2019].

Otondo, F. (2013) Using Spatial Sound as an Interdisciplinary Teaching Tool. *Journal of Music, Technology & Education*. 6(2). 179–190.

Pinch, T. and Bijsterveld, K. (2004) Introduction to Sound Studies: New Technologies and Music. *Social Studies of Science*. 34(5). 635–648.

Raymond, R.K. (2016) Managing College Radio: Understanding American College Stations Through Their Management Practices. *The Radio Journal: International Studies in Broadcast and Audio Media*. 14(2). 193–211.

Repko, A. (2007) Integrating Interdisciplinarity: How the Theories of Common Ground and Cognitive Interdisciplinarity Are Informing the Debate on Interdisciplinary Integration. *Issues in Integrative Studies*. 25. 1–31.

Rekers, J. and Hansen, T. (2015) Interdisciplinary Research and Geography: Overcoming Barriers Through Proximity. *Science and Public Policy*. 42. 242–254.

Sawyer, R.K. and Dezutter, S. (2009) Distributed Creativity: How Collective Creations Emerge from Collaboration. *Psychology of Aesthetics, Creativity, and the Arts*. 3(2). 81–92.

Schneider, B. and Caswell, D. (2003) Using Narrative to Build Community and Create Knowledge in the Interdisciplinary Classroom. *History of Intellectual Culture*. 3(1). 1–12.

Sheridan, T. and Van Lengen, K. (2003) Hearing Architecture: Exploring and Designing the Aural Environment. *Journal of Architectural Education*. 57(2). 37–44.

Tahiroğlu, K., Özcan, O. and Ikonen, A. (2014) Sound in New Media and Design Studies. *Design Issues*. 30(2). 56–66. Spring.

Tobias, E.S. (2015) Participatory and Digital Cultures in Practice: Perspectives and Possibilities in a Graduate Music Course. *International Journal of Community Music*. 8(1). 7–26.

Tough, D. (2010) Shaping Future Audio Engineering Curricula: An Expert Panel's View. *MEIEA Journal*. 10(1). 149–160.

Trow, M. (1977) Departments as Contexts for Teaching and Learning. In Mchenry, D.E. (ed.). *Academic Departments*. San Francisco: Jossey-Bass.

Veysey, L.R. (1965) *The Emergence of the American University*. Chicago: The University of Chicago Press.

Wake, A. and Bahfen, N. (2016) Redefining Radio: Implications for Journalism Education in an Era of Digital Audio Storytelling. *The Radio Journal: International Studies in Broadcast and Audio Media*. 14(2). 231–242.

Walzer, D. (2017) Independent Music Production: How Individuality, Technology and Creative Entrepreneurship Influence Contemporary Music Industry Practices. *Creative Industries Journal*. 10(1). 21–39.

Watson, I. (2008) Developing an Integrated Arts and Media Studies and Curriculum: One Model. *Issues in Integrative Studies*. 26. 138–162.

You, H.S. (2017) Why Teach Science with an Interdisciplinary Approach: History, Trends, and Conceptual Frameworks. *Journal of Education and Learning*. 6(4). 66–77.

Zeller, C. (2004) North Atlantic Innovative Relations of Swiss Pharmaceuticals and the Proximities with Regional Biotech Arenas. *Economic Geography*. 80. 83–111.

11

Sound Reflections

The Purpose, Perspectives, and Place of Audio Education

Paul Thompson and Ben Mosley

Introduction

As formal education at college and university level in sound recording and audio production continue to develop and are seen as preliminary routes into the Creative Industries, it has become increasingly important to assess the place and purpose of audio education within the sphere of formal education and its relationship to the wider Creative Industries. The UK government Department for Culture, Media and Sport (DCMS) defines the Creative Industries based on international industrial codes and includes the following sub-sectors: advertising and marketing; architecture; crafts; design and designer fashion; film, TV, video, radio, and photography; IT, software, and computer services; publishing; museums, galleries and libraries; and music, performing, and visual arts.

As part of this discussion of place and purpose, it is also important to consider the perspectives of the numerous stakeholders, benefactors, and beneficiaries of audio education, which include students, educators, education directors, educational institution representatives, music industry bodies, music industry representatives, and the music and Creative Industries more broadly. Each group has their own perspective on the purpose of audio education and each has something to add to the discussion on the content and delivery of audio education programs. For example, in their *Manifesto for the Creative Economy*, Bakhshi et al. (2013) emphasize employability as the central purpose for audio education and argue that "evidence suggests that most universities haven't been producing the kind of talent that the creative industries demand" (Bakhshi et al. 2013: 103). This, they argue, is evidenced in poor employment outcomes of graduates from creative media specialist degrees as evidenced in only 12% of those graduating from games courses secured employment in the industry within six months of leaving university and other studies (ibid.).

Like all educational systems, HE is not removed from the political and economic landscape and its policies are: "heavily influenced by economic institutions such as the Organization for Economic Co-operation and Development (OECD), the International Monetary Fund, and the World Bank" (Rusinek & Aróstegui 2015: 78). "Employability" and the development of professional and vocational skills has become central to the purpose of HE, which in turn has created tensions between these capitalist ideas and neoliberal and democratic perspectives on education (see Benedict et al. 2015 for a more in-depth discussion). Employability is defined as:

> [A] set of skills, knowledge and personal attributes that make an individual more likely to secure and be successful in their chosen occupation(s) to the benefit of themselves, the workforce, the community and the economy.
>
> (Little 2006: 2)

Others argue that higher education functions to educate, to research and to transfer knowledge (Brennan et al. 2013) with an emphasis on the development of higher order skills: "such as, the ability to think logically, the capacity to challenge the status quo, and the desire to develop sophisticated values" (Chen 2016: 2).

As the political, economic, technical, and social landscape continues to shift within the higher education sector and the Creative Industries too, now is an important time to reflect on audio education's place, purpose, and function. The following chapter begins by exploring the place of audio education within the higher education landscape in the UK leading to discussion of the often-contradictory perspectives involved in defining its purpose and function within HE in the UK and the Creative Industries. The chapter concludes by framing the challenges for audio education in light of these perspectives.

Study Design

Perspectives on the purpose and place of audio education were collected through a mixture of literature and document analysis, semi-structured interviews, focus groups, and email correspondence. Literature analysis was conducted in order to review current trends in educational research and to determine the purpose and the broader benefits of studying in higher education. Document analysis was conducted primarily with government agency publications and industry bodies such as Joint Audio and Media Education Support (JAMES) in order to gain a view on their perspective of HE audio education.

Semi-structured interviews and focus groups were conducted with 30 students studying an audio-related program at Leeds Beckett University. The interviews and focus groups helped to explore some of the themes that were highlighted by government and audio industry bodies and more importantly, some of the educational experiences of the participants. Exploring experiences as well as documents or artifacts: "increases our knowledge of the details of cultural processes and practices" (Cohen 1993: 135), and in this instance, allowed a qualitative analysis of the musicians' experiences.

Finally, email correspondence was a useful method in helping to collect the perspectives and thoughts on audio education from audio industry professionals and audio industry representatives. Questions during the interviews and focus groups covered the respondents' views and experiences of audio education at HE institutions in the UK. Further information was sought on whether the respondents considered audio education to be important to the Creative Industries and whether formal institutions are the best places to deliver audio education. Questions to audio undergraduate students also addressed whether they considered audio education as a route into the Creative Industries and whether or not they felt audio education had prepared them for their chosen career path. Finally, respondents were asked what they thought audio educators should be doing to ensure they deliver an appropriate audio education program in their chosen industry area.

The investigation included 40 participants, 30 of these were students, eight participants were industry professionals, and, in an attempt to capture as much of the spectrum of the audio industry as possible, participants from the areas of acoustics, postproduction/film sound, record production and audio engineering were selected. Two of the participants were industry body representatives. The responses from students, audio industry professionals, and audio industry representatives have been anonymized and included alongside the discussion.

Background and Context

Formal HE audio education programs in the UK began with the University of Surrey's Tonmeister program and have become more established and prevalent through the 1990s and now has a strong presence within various other educational disciplines. Audio-specific programs, such as music technology, audio engineering, sound technology, and music production, have seen extensive growth over the last two decades (Davis et al. 2014). Carola Boehm (2006) noted that the term "music technology" (the most common term linked to audio education) was linked to 351 educational programs, distributed over some 62 HE institutions. As of 2014, this had risen to 80 institutions (Mosley & Thompson 2014). At the time of writing (January 2019) this now stands at 443 courses across 112 institutions according to the UK University and Colleges Admissions Service (UCAS). Many of these programs are delivered in engineering departments, some in traditional music departments, and some in creative arts or media departments. The wide range of discipline areas in which audio education takes place illustrates the broad spectrum of its nature, which includes a complex mixture of musical, technical, and cultural elements.

The HE landscape in the UK has changed significantly in recent years through the implementation of particular government policies, which have influenced the place and the perceived purpose of audio education in the HE sector. Most significant, and certainly the most prominent, are the changes in the way HE is funded. Following the Browne review (2010), student tuition fee contributions in England and Wales were increased threefold in order to pass on the full cost of HE to the student. In the 2015/16 academic year government caps on the number of students universities could enroll were lifted and universities have since been able recruit as many students as they wished.

In 2018, the Teaching Excellence Framework (TEF) was introduced, which seeks to rate institutional teaching quality on a "Gold, Silver, Bronze" scale. The metrics for these ratings are based on student satisfaction with "Teaching on my course", "Assessment and feedback", and "Academic support" plus data for

"Non-continuation" and "Employment or further study". Importantly, much of these recent developments in UK HE policy are driven by a marketization agenda with the specific aim of creating market competition between institutions. Consequently, "the choices made by students [are] now conceived of as customers exercising choice in paying for a product in a market—and no longer as citizens exercising a social right" (Anderson 2016).

The Creative Industries too have undergone significant change since audio education's introduction in the 1990s. The post-digital economy landscape has notably changed since the 1990s with a decline in more traditional or established occupations and the appearance of new areas of audio-based creative economy activity. For example, whereas areas of the recording industry have degenerated, new areas in film, television, radio, new media, and online marketing have created fresh opportunities for graduates with skills in music and audio. Figures released by the DCMS in 2017 show that the Creative Industries in the UK have shown an increase in jobs of 28.6% between 2011 and 2017. The closely related digital sector and cultural sector have shown an increase of 16.1% and 23.6% respectively across the same period. This is far in excess of the overall increase of 9.3% across the whole of the UK over the 2011–2017 period (DCMS 2017). In 2017, the number of jobs in the Creative Industries sector stood at just over two million. The Creative Industries accounted for 6.1% of the total number of jobs in the UK in 2017. This compares with 5.2% of the total number of UK jobs that it accounted for in 2011. These figures suggest a healthy and growing industry with many employment opportunities.

Changes to the school level and post-compulsory education sectors (further education in colleges and small private providers) also continue to influence audio education in higher education, particularly as the Institute for Fiscal Studies (IFS) reported in 2017 that funding for sixth form and further education (FE) in the UK has been cut more quickly and more deeply than other areas of education (Belfield et al. 2017). FE is typically the place where students can study audio education programs before either moving into the Creative Industries or moving on to further study at HE level. If audio education programs are closed due to a lack of funding this influences both the educational and employment opportunities for future prospective students or creative industry trainees. Given the growth of the Creative Industries and the important role that audio plays in this, alongside the number of audio-related courses in HE, audio education can be seen as an interrelated part of the wide range of educational routes that feed into the wider Creative Industries.

Perspectives on Audio Education

The perspectives of the students surveyed for this study on audio education were varied but most students considered audio education to be a route into employment within the Creative Industries. However, one final-year student noted that:

> Initially, before I started the course, I saw audio education as a route to employment but now I've nearly finished the course, I don't think it is necessarily tied to preparing you for employment. It's more about gaining broad skills.

The perspectives of industry professionals captured during this study were useful in providing insight into specific industry views on what they considered the purpose of audio education to be. For example, an acoustic consultant noted that although the majority of the practical skills can be learned while undertaking the role: "formal education afforded me a wider and deeper understanding of the principles and concepts that might not have been developed outside of university". An experienced audio engineering professional surveyed for this study also expressed his positive view toward the benefits of studying audio at a formal HE institution:

> "A formal education that results in the award of a recognised qualification provides a bedrock of competence that can be applied to many areas over time" and "A University education that provides individuals with recognised, respected qualifications as well as delivering motivated self sufficient people with up to date technical, operational and business skills is crucial".

Conversely, a postproduction industry professional explained that: "I feel working in the field is far more beneficial than studying in a classroom for gaining experience and a solid skill set".

Generally, audio education was viewed by all the industry professionals involved in the study as useful in developing critical thinking and problem-solving abilities. A significant benefit of studying audio related programs at formal institutions was highlighted by a recording industry professional who explained that: "educational institutions should (and do) give students a place to experiment and work with good quality resources without financial pressures".

In bridging the two perspectives between education and employability, accreditation bodies provide a much-needed intermediary role between formal education and industry. The principle aim of an accreditation body is to ensure that graduates learn the necessary skills and knowledge on their educational program in order to meet the needs of the industry. Currently in the UK there are two main accreditation options for audio courses, the Joint Audio Media Services (JAMES) and the Institute of Engineering and Technology (IET).

JAMES represents the Creative Industries and accredits educational programs on behalf of, and with the endorsement of, the following (JAMES 2018):

- APRS (Association of Professional Recording Services)
- MPG (Music Producers Guild)
- MMF (Music Managers' Forum)
- PLASA (Pro Light and Sound)
- UK Screen Association
- BASCA (British Academy of Songwriters, Composers and Authors)
- IASIG (Interactive Audio Special Interest Group)

The IET represents the engineering industry and its accreditation is awarded to higher education programs that meet the educational requirements of the UK Standard for Professional Engineering Competence (UK-SPEC), now in its third edition (Engineering Council 2014). It centers on a review of a program's content and delivery and includes areas such as relevance, coherence, challenge, assessment, staffing, quality assurance, and resources (IET 2015).

Additionally, ScreenSkills (formerly Creative Skillset) offer their "Tick" accreditation for "practice-based degree courses that most effectively provide students with the skills and knowledge required by employers in the screen industries". ScreenSkills Tick is a quality mark that is awarded to a wide range of creative industry courses from fashion to media business at many educational levels. There are several UK HE courses that feature audio in areas such as radio production, game audio, and postproduction that have been awarded the Screenskills Tick (screenskills.org 2018).

Although these accreditation bodies serve a valuable role, they do so from very different perspectives, the IET from a technical perspective and JAMES from a creative perspective. JAMES is predominately a music industry body and the IET is an engineering industry body and therefore their accreditation focus and expectation of an educational program's content and purpose is understandably different. At the time of writing, only three UK HE programs have been accredited by both JAMES and IET. One key aim of the ScreenSkills Tick is to identify "the UK's most industry-focused courses and apprenticeships enabling universities, colleges and employers to recruit the brightest talent" (Screenskills 2018).

Because accreditation bodies are industry-focused their perspective on the purpose and content of audio education is unsurprisingly from industry-related training with an emphasis on the practical acquisition of skills or knowledge. Perspectives on audio education are therefore varied and range from an HE perspective that views audio education as a way of developing higher order skills in audio production to an industry perspective that sees audio education as a way of gaining industry-related skills and training.

Purpose of Audio Education

In considering the various perspectives so far, it is clear that the purpose of audio education depends upon the nature of the stakeholder, beneficiary, or benefactor. For instance, the government views audio education as a route into the Creative Industries and prioritizes employability skills. The majority of industry professionals interviewed in this study also emphasize the importance of a curriculum that has its main focus as preparing graduates for work in the Creative Industries. However, one postproduction industry professional noted that he wouldn't have benefited from studying audio at university stating:

> Rarely have I met someone in the industry that has formal education in music. The guys [sic] who usually go far are the people who have been working their way up industry since a young age.

On the other hand, one acoustics professional highlighted how studying formally at an HE institution gave them a deeper understanding of acoustics that wouldn't have been possible through on-the-job training. The perspective on the purpose of audio education is therefore also related to the area of industry. When asked about hiring prospective employees, one studio owner answered:

> It wouldn't matter to me whether a person had a degree or not. What matters is that they are ready and willing to learn and can follow instructions. More often-than-not I'm looking for someone who is ready to be trained rather than someone who thinks they already know about engineering.

From this perspective, the purpose of audio education can be viewed as a combination of both the development of higher order skills and industry-focused skills and abilities.

Employability and the Creative Industries

Despite the additional benefits of studying at university level, including citizenship, civic engagement, increased health and well-being (Brennan et al. 2013), and the development of higher order skills, audio education in HE is often seen as a vocational pathway. This reduces its perceived purpose as "training" and to simply provide the workforce for a specific creative industry. However, this view overlooks the diverse and multidimensional nature of the Creative Industries and audio education more generally. The marketization of HE and the increase in tuition fees paid by students to fund their tuition however has fueled this "industry-ready agenda" (Ashton 2010), which has become a dominant perspective on HE's central purpose in the UK. This agenda has contributed to the centralization of employability within the curriculum (Ashton 2010), so much so that the former UK government's Department for Universities, Innovation and Skills (DIUS) stated that:

> We want to see all universities treating student employability as a core part of their mission. So we believe it is reasonable to expect universities to take responsibility for how their students are prepared for the world of work.
>
> (DIUS 2008: 6)

The DIUS has recently been subsumed into the Department for Business, Innovation and Skills and indicates the current government's stance on where they see the place and purpose of HE; as a central contributor to the workforce of the Creative Industries.

The emphasis on employability within the HE curriculum and the industry-ready agenda within audio education has been "dominated by employer and government concerns about the supply of graduates [and] has received little conceptual or empirical analysis" (Brown 2005: 109). However, it is the creative economy and their industries that shape and define the meaning of "employability" and what it is to be "industry ready" (Ashton 2010). Education providers are therefore expected to add "employability" skills to their everyday delivery of subject-specific education and research. The term "employability" has been described as:

> A set of achievements, skills, understandings and personal attributes that make graduates more likely to gain employment and be successful in their chosen occupations, which benefits themselves, the workforce, the community and the economy.
>
> (Yorke and Knight 2006: 8)

Employability skills include "self-management, team working, business and customer awareness, problem solving, communication and literacy, application of numeracy and application of information technology" (CBI 2009: 8) with an underpinning positive and entrepreneurial attitude (ibid.). It has been argued that gaining employability skills benefits graduates as well as "their employer and the wider economy" (ibid.). Sir Christopher Snowden, the president of Universities UK, however maintains that studying at university shouldn't simply be viewed in employability terms and explains: "I still very much believe there is a significant lifetime benefit to be got from going to university" (Snowden 2014).

The general view of those involved in HE is that audio education is not simply a means to get a job. Rather, the purpose of studying a higher degree includes developing critical skills and knowledge within a particular discipline or subject. The purpose of audio education is therefore to gain subject-specific knowledge and skills and to develop critical thinking and problem solving abilities within the broad area of audio production. The results from this study showed evidence that students are also aware of this. One student stated:

> The principles behind audio education can shoot off into other areas of work that would not normally be associated with music technology. I have learnt time management and problem solving skills that can be applied to other careers outside of music technology . . . it teaches you not just about music technology, but also key life skills, team building, problem solving etc. Skills that may not be easily obtained in other courses.

These critical skills and knowledge therefore extend beyond the technical and include the social aspect of the field—inclusivity and accessibility aren't always addressed as part of these social aspects but these are discussed later in this chapter. While the purpose of audio education depends upon the perspective of the relative stakeholders within the debate, and their views appear at times to be in conflict, some industry professionals who had studied audio education at university highlighted how audio education can complement working in some areas of the Creative Industries as in this example from an audio-technical consultant:

> One and the other should form part of an holistic approach. Theory, discussion, experiments and projects are the key to developing a good foundation in a creative subject. This, by definition, can only take a student so far, and much of the methods of dealing with professional contracts/clients/jobs are adapted or augmented as a result of work-life experience.

Focus on Vocational Skills

One of the criticisms of audio education by industry professionals in this study was that curricula and its delivery don't provide enough real-world examples and that more industry experience should be offered as part of an educational program. This is challenging in a university environment where links with industry have not traditionally been that strong. A comment, from an industry professional interviewed for this study, of "most university courses fall down badly when it comes to providing informal learning by not giving students the opportunities to work alongside established expert professionals" hints at the some of the challenges of connecting students with the industry they are hoping to work in.

This challenge coincides with the need for audio education programs to provide an expansive curriculum that introduces the industry's various aspects (including but not limited to recording, postproduction, film, TV, radio, audio maintenance, acoustics, and audio programming) in sufficient depth. Since different areas of industry have different needs, these must also be addressed in the audio education curriculum. For example, audio maintenance requires greater emphasis on the technical aspects of the curriculum than other areas such as composition for film and TV. There are curriculum challenges in providing technical skills on creative courses:

> Not enough courses seem to adequately cover engineering basics. This includes: Basic soldering skills, understanding signal flow paths, understanding the parameters that must be controlled when assembling complex systems (studios for example) from many interconnected audio processors, developing the ability to fault find in complex audio systems.

There are many more.

There is an inherent challenge here in providing an educational experience that has both breadth and depth. As noted by one student:

> [B]ecause of the scope of Music Technology, the course has prepared me in other career paths through the wide variety of modules that have been taught. This has allowed me to consider my own career path, as well as others through modules that I have enjoyed.

suggests an appreciation that opening up a wide variety of career paths is beneficial in a course. However, another student commented that:

> If it is to be a path to employment then the content needs to be more specialised and more geared towards that area of industry, such as game audio, rather than covering lots of little bits of different areas of the industry,

highlighting the challenge of providing an educational experience with enough specificity.

A comment from industry suggests a need for universities to "offer solid foundation qualifications in years 1 and 2, then offer considerable opportunities to specialise in year 3 and post graduation". Many university courses are built on this type of structure already but it is often not possible to cater for the wide range of career paths and specialisms available. An important consideration, therefore, is ensuring that students have the adaptability to apply the skills that they learn during an educational program to new and challenging situations. The concept of transferable skills is often cited in university-level teaching, but many of the core skills from an audio curriculum are transferable in and of themselves. For example, skills in a music recording studio are applicable in a radio studio, a TV studio, or a film set. Audio editing skills can be applied in games, advertising, TV, radio, app design, and many more platforms.

Some of the views of the industry professionals discussed here and some of the issues identified earlier with regards to poor employment rates for graduates in the Creative Industries (Bakhshi et al. 2013) hint at the notion that a degree is not essential for success in the Creative Industries. Unfortunately, the growth of audio related educational programs in the UK "has been paralleled by the apparent decline in informal apprenticeship systems that have typically provided a gateway to employment in the recording industry" (Davis et al. 2014: 1).

Therefore, formal education must perform some of the roles previously delivered by apprenticeship programs. There is also some tension between the government view that a graduate should emerge perfectly formed and ready for industry from a degree course and the view from many employers here that the willingness to learn and a good attitude are most important. A key aim of all degree programs, and built into the QAA framework for higher Education in the UK, is that of developing students into independent learners. One could argue that this aim addresses many of the employer concerns with regards to attitude and willingness to learn on the job. A degree course cannot hope to produce the perfect graduate every time but it should give all graduate enough skills to place them at a good starting point for a role in the industry.

Challenges

In addressing each of the perspectives on the purpose of audio education, the challenges in developing and delivering the curriculum of audio education are wide-ranging. Audio education programs are expected to develop transferable skills and knowledge that are common to all degree programs. These include the development of higher order cognitive and communicative skills in young people, such as, the "ability to think logically, the capacity to challenge the status quo, and the desire to develop sophisticated values" (Chen 2016, p. 2). Audio education programs are also expected to develop subject specific skills and knowledge so graduates can operate in a rapidly changing industry.

Added to these challenges are the educational expectations to periodically refresh the curriculum in response to new research, new innovations, or new economic landscapes in a way that balances the needs and expectations of each of the beneficiaries and benefactors of audio education. Engaging meaningfully with industry is important here as noted by this response to the question "do you view audio education in universities as important for the future of the creative industries?":

> Yes. Providing the creators and providers of courses work VERY hard to consult with relevant industry representation during the structure of course material. Further, industry representation should be sought as a part of any review process, internal or external, to sit alongside any academic representation. . . . Equally, industry input should be sought when designing and updating project and practical module working briefs thereby ensuring up to date and relevant student work and study is derived.

Educators are typically responsible for the design, delivery, and evaluation of their programs but government (particularly in the UK) increasingly seeks to dictate what universities are expected to do and how the outcomes of their activities are measured. The Research Excellence Framework (REF) and Teaching Excellence Framework (TEF) in the UK are good examples of this as they are used to assess the quality of research and teaching of UK higher education institutions through an auditing processes that is designed to measure how research and teaching meet specific expectations. Government also expects universities and their courses to respond to the needs of industry and industry increasingly has input into the design and implementation of university courses through the accreditation process. However, as suggested by some of the discussion earlier in this chapter, the expectations of industry and government do not always align, leaving universities and educators in an unenviable position.

These challenges cannot be met by educators or institutions alone; overcoming some of these issues requires cooperation with government agencies and operatives within the Creative Industries. One recording industry professional highlighted the perceived value of the Creative Industries has reduced and therefore HE institutions have a critical role to play:

> I believe creative industries as a whole are fundamentally suffering from a lack of value—i.e., a cataclysmic funding gap and a complete cultural devaluing from central government downwards. Creative industries, of which audio work is a part, are neither understood nor valued by a systemic infrastructure that is focused heavily on "quantitative success". It is crucial that universities have the funding, the staff and the ethos necessary to correct this alarming trend.

It was further noted by another of the recording industry professionals that the Creative Industries also have a responsibility to work with those in audio education:

> I do not believe that sufficient creative jobs exist in the UK employment market . . . industry has developed a disturbing ignorance of artistic/creative study. Video game and film audio production were areas I felt I was most equipped to work in, following graduation. My time after university coincided with a business world still reeling from the

> banker-led crash of 2008. Simply put, I would prefer to see more creative workplaces making better use of students and graduates from audio courses.

Previous studies in this area noted that facilitating access to the Creative Industries allows students to appreciate its mechanisms and creative values (Thompson & McIntyre 2013). In this context, it also helps students to prepare for the demands of their career path and work toward gaining specific skills and knowledge in their chosen area of the Creative Industries.

The Creative Industries are often difficult for graduates to enter due to the ways in which outputs and activities are monetized. Work is often transient and freelance in nature and value is often wrapped up in intellectual property and copyright. Operating and succeeding in this environment requires both an understanding of the nature of the industry but also flexibility and a certain amount of entrepreneurial skill. Audio courses often feature some sort of "industry" content but actually preparing students for the rigors of a freelance career is not easy and many courses fall short in this area.

Can We Do More? (Recommendations and Further Work)

Social intelligence and emotional intelligence often feature in the literature regarding graduate attributes (Artess et al. 2017). However, these attributes are often not addressed in the core of audio curricula. The development of these skills is more likely to be dependent on individual student's circumstances or addressed through pastoral support. Audio educators need to be aware that they have some responsibility in educating beyond the core curriculum and to explore where their skills and understanding fit within a wider social context beyond the Creative Industries. In a world that faces great challenges with regards to sustainability, inequality, and political stability our young people need a wide awareness to succeed, prosper, and lead.

Projects that connect students with their local community or bring the community into the classroom such as community educational projects, working with people with disabilities, exploring how audio can support blind and partially sighted people, and how audio experiences can be improved for those with hearing problems can provide opportunities for students to develop these vital social and emotional intelligences. Other graduate attributes such as flexibility, adaptability, positive attitude, aspiration, resilience, and work ethic (Artess et al. 2017) could also be seen to be more closely aligned with a student's sense of self-worth and their general well-being and therefore, mental health awareness and its management could be a useful inclusion within an audio curriculum, helping to focus on the "person" as well as their skills, knowledge, and understanding.

The Creative Industries have long been recognized as underrepresented by minority groups. Recent research has also highlighted the issue of sexism and the challenges women face entering the audio industry (Lanzendorfer 2017). Introducing the social study of contemporary issues alongside core audio education curricula could help to introduce all students to the challenges certain groups of people face in the Creative Industries and further equip students with knowledge and experience that will help them, beyond the classroom, to challenge gender stereotypes for example, or how to address everyday occurrences of misogyny or sexism wherever it occurs.

Another area where there is room for development is the area of sustainable development in audio. Most of the large broadcasters (ITV, BBC, SKY, Channel 4, etc.) have strong sustainability agendas and many audio/visual productions are carried out under the Albert guidelines for sustainable production (Mosley, Chapter 9, this volume, and 2015). Students pursuing a career in broadcast are highly likely to be required to work within these guidelines and policies and educators can help prepare students for these challenges.

Conclusions

Given the varied perspectives on the purpose of audio education, the challenges identified, and the sheer breadth of possible career paths within the Creative Industries, it is concluded that audio education cannot address all of the needs of its stake holders, beneficiaries, and benefactors. Although the "industry-ready agenda" (Ashton 2010) perspective has become a dominant voice in the debate of the purpose of higher education in the UK, the educational perspective that there is a significant lifetime benefit to studying at university (Snowden 2014) is equally as valid. Industry accreditation bodies such as JAMES and IET, therefore, play an important role as functional intermediaries between education and industry in bringing these apparently opposite perspectives together.

In addressing each of the perspectives on the purpose of audio education, the challenges in developing and delivering audio education are wide-ranging. Audio education programs are expected to develop transferable skills and knowledge that are common to all degree programs and develop subject-specific skills and knowledge so graduates can operate in the rapidly changing Creative Industries. What is clear is that these challenges cannot be met by educators or institutions alone. Addressing some of these issues will require cooperation with government agencies and operatives within the Creative Industries.

At the time of writing, the discussion on the purpose and place of audio education is arguably more important than it has ever been. In a world where audio visual communication, broadcast, and entertainment is so prevalent in the post-digital economy the need for graduates that have relevant skills, knowledge and abilities with sound and audio is growing increasingly important. However, this varied and relatively unmapped discipline faces challenges in providing sufficient depth and breadth of curricula that can adequately serve such a diverse and rapidly changing industry.

It is acknowledged that although the data presented here cannot accurately portray the numerous and varied perspectives on audio education or reflect the breadth and depth of audio education at different HE institutions. Rather, the perspectives presented serve to provide a useful starting point to begin a more in-depth discussion on the place and purpose of audio education in UK HE, particularly as formal courses in sound recording and audio production continue to develop as preliminary routes into the Creative Industries. This study has highlighted a number of areas for future work and research however the most important is that audio education could benefit from greater consultation between industry representatives and education providers to help bring understanding to each of their perspectives. Further research is also needed in this area to identify the benefits and disadvantages of more focused audio educational programs that target specific industry areas or current skills shortages. Doing so will afford a more empirically informed perspective on the place and purpose of audio education in UK HE and help to meet the needs of each of its benefactors, beneficiaries, and stakeholders.

References

Anderson, R. (2016) *University Fees in Historical Perspective*. London: History & Policy.

Artess, J., Hooley, T., & Mellors-Bourne, R. (2017) *Employability: A Review of the Literature 2012 to 2016. A Report for the Higher Education Academy*. York: Advance HE.

Ashton, D. (2010) "Productive Passions and Everyday Pedagogies: Exploring the Industry Ready Agenda in Higher Education", *Art, Design & Communication in Higher Education*, 9 (1), 41–56.

Bakhshi, H., Hargreaves, I., & Mateos-Garcia, J. (2013) *A Manifesto for the Creative Economy*. London: Nesta, p. 103.

Belfield, C., Sibieta, L., & Farquharson, C. (2017) "Severe Squeeze on Further Education and Sixth-Form Funding in England", Institute for Fiscal Studies [Online]. Available from: www.ifs.org.uk/publications/13307 [Accessed December 2018].

Benedict, C., Schmidt, P., Spruce, G., & Woodford, P. (2015) *The Oxford Handbook of Social Justice in Music Education*. Oxford: Oxford University Press.

Boehm, C. (2006) "The Thing About the Quotes: 'Music Technology' Degrees in Britain", Proceedings of the 2006 International Computer Music Association Conference, New Orleans, LA, 682–687. Available from: http://hdl.handle.net/2027/spo.bbp2372.2006.138 [Accessed July 2019].

Brennan, J., Durazzi, N., & Sene, T. (2013) *Things We Know and Don't Know About the Wider Benefits of Higher Education: A Review of the Recent Literature*. London: London School of Economics and Political Science.

Brown, S. & Drew, S. (2005) "Developing an Employability Framework: An Institutional Approach", 9th Quality in Higher Education International, Birmingham, UK, January. Available from: www.qualityresearchinternational.com/esecttools/eseconference.php [Accessed July 2019].

CBI (2009) *Future Fit: Preparing Graduates for the World of Work*. London: Confederation for British Industry.

Chen, R.Y. (2016) "Understanding the Purpose of Higher Education: An Analysis of the Economic and Social Benefits for Completing a College Degree", *Journal of Education Policy, Planning and Administration*, 6 (5), 1–40.

Cohen, S. (1993) "Ethnography and Popular Music Studies", *Popular Music*, 12 (2), 123–138.

Davis, R., Thompson, P., & Parker, S. (2014) "Preparing the Music Technology Toolbox: Addressing the Education-Industry Dilemma", *Journal of Music, Technology & Education*, 7 (3), 313–323. doi:10.1386/jmte.7.3.313_1.

DCMS. (2017) *DCMS Sectors Economic Estimates 2017: Employment*. London: DCMS.

Department of Innovation, Universities and Skills (DIUS). (2008) "Higher Education at Work High Skills: High Value Consultation", Report prepared for the Department of Innovation, Universities and Skills (DIUS), London.

Engineering Council. (2014) *The Accreditation of Higher Education Programmes: UK Standard for Professional Engineering Competence* (3rd edition). London: Engineering Council.

IET. (2015) *Guide to Academic Accreditation for Higher Education Institutions*. London: IET.

JAMES. (2018) Joint Audio Media Education Support – Homepage [Online]. Available from: www.jamesonline.org.uk. [Accessed July 2019].

Lanzendorfer, J. (2017) "Why Aren't There More Women Working in Audio?" *The Atlantic* [Online]. Available from: www.theatlantic.com/entertainment/archive/2017/08/why-arent-there-more-women-working-in-audio/537663/ [Accessed July 2019].

Little, B. & ESECT Colleagues. (2006) *Employability and Work-Based Learning*. Learning and Employability Series One. York: The Higher Education Academy.

Mosley, B. (2015) "Sustainable Design and Practice in Reproduced Sound", Proceedings of the 37th Reproduced Sound Conference, November, 37 (4).

Mosley, B. & Thompson, P. (2014) "Setting the Tone: Considerations for Educating the Next Generation of Sound Reproduction Professionals", Proceedings of the 40th Annual Conference of the Institute of Acoustics, Birmingham, October, 36 (3).

Rusinek, G. & Aróstegui, J.L. (2015) "Educational Policy Reforms and the Politics of Music Teacher Education", in Benedict, C., Schmidt, P., Spruce, G., & Woodford, P. (eds.) *The Oxford Handbook of Social Justice in Music Education*. Oxford: Oxford University Press, 78–90.

Screenskills. (2018) *About the Tick*. London: Skillset [Online]. Available from: www.screenskills.com/education-training/tick/about-the-tick/ [Accessed December 2018].

Snowden, C. (2014) "Universities Are 'Not Just for Getting a Job,' Says One of Britain's Leading Academics", *The Independent*, 8 June. London [Online]. Available from: www.independent.co.uk/news/education/education-news/universities-are-not-just-for-getting-a-job-says-one-of-britains-leading-academics-9507191.html [Accessed December 2018].

Thompson, P. & McIntyre, P. (2013) "Rethinking Creative Practice in Record Production and Studio Recording Education: Addressing the Field", *Journal of the Art of Record Production*, 8.

Yorke, M. and Knight, P.T. (2006) *Embedding Employability into the Curriculum: Learning & Employability Series 1*. York: The Higher Education Academy.

12

Designing Vocational Training for Audio Engineers at a Distance

Challenges, Reflections, and Recommendations

Duncan Williams

Introduction

This chapter addresses the design of a full credit remote access module as part of an undergraduate degree course in music technology with a particular focus on sound recording technology at a public university in Texas.[1] Although the audio-visual industries, and to a certain extent the corresponding educational sector, have traditionally regarded music technology as a vocational skill, and as such, one that must be learned in practice, the client university required that all content be delivered, facilitated, and assessed entirely electronically—a challenge that necessitated a number of particular pedagogical approaches. Distance learning allows students to study technical content at any time, and at their own pace.

The method of practical assessment used would be appropriate to many other creative sections of the audio-visual engineering sector where technical knowledge and achievement of professional standards is an overarching mark of skill, and thereby an indicator of successful learning. A combination of multiple-choice assessment and virtual face-to-face reinforcement can be leveraged to maximize academic differentiation and contextualization of course content. This chapter will focus on appropriate curriculum selection and design for distance learning, illustrating the advantages and disadvantages of such a system with technical and vocational content delivery in practice, and includes an analysis of the difficulties inherent in teaching and assessing such topics remotely as well as examples of the curriculum content to give a context for assessment.

With the age of the personal computer it is no surprise to see distance learning techniques adapted to delivery electronically, often using an "in-house" *Virtual Learning Environment* (VLE) that facilitates browser-based display of reference and interactive material. A review of literature addressing this electronic mode of distance learning exists (Philips and Merisotis 1999). It is often stated that distance learning is "no new thing"—in the United Kingdom. For example, the Open University has offered distance learning courses for adults up to doctoral level for many decades, and teachers offering material for remote study via postal services can be traced back much further than this (Holmberg 2005). In fact, the application of the Edison Phonograph to music education, predicted by the inventor himself (Edison 1878), as a means for musicians to not only hear example performances from their instructors at a distance, but also, by return, submit recordings of their own performances for assessment using the postal service, is an early example of music technology informing and supporting distance learning practice, though the curriculum was not usually tailored to the delivery format *per se* (Kelleher 2012).

The pedagogical success of distance learning strategies is often judged by comparative analysis with traditional teaching (Ni 2013). In the case of a student enrolled on a vocationally based course—by which for the purposes of this chapter we mean a full undergraduate degree program which includes practical training in music technology (or indeed namesakes such as sound engineering, audio technology, recording, etc.)—traditional teaching would most likely be delivered face-to-face in classrooms, laboratories, workshops, and studio suites. Exam performance is a common method for the summative assessment of theoretical content, whereas students' practical skills are typically developed, and assessed, by a combination of formative observation and summative coursework submission.

"Value Added" data (a UK-specific term referring to a government statistic) can also be used as a tertiary metric—indicating student progress against their academic profile on entry to the program. Student satisfaction, or even "attitude" toward the content and style of program delivery are also occasionally monitored as further benchmark measures with which to compare the success of distance learning with face-to-face teaching. "Blended" learning, a system combining distance learning and traditional methods, has gained some popularity amongst the Further and Higher Education sector ("Scenarios, Strategies, Guidelines and Tools", 2012). In the UK, further education refers to study below undergraduate degree level, but beyond compulsory school-age (secondary) study.

A challenge with providing degree level programs in music technology, at least when combined with a vocational focus, is the lack of an industry standard qualification level. Anecdotally, it appears that many sound engineers working at the highest level in industry have learned their trade via unpaid internships, apprenticeships, or autodidactically. However, the advent of manufacturer certification for particular software packages does address a benchmark skill set in some respects, as does the efforts of accrediting bodies to standardize with industry (in the UK this falls under the aegis of the government funded *Creative Industries' Sector Skills Council*). It is difficult to ascertain from the job market whether or not such accreditation has become a criterion for subsequent entry to the workplace. Nonetheless, full degree programs with a focus on developing vocational skills in music technology, as well as broader, more generalized academic skills, have become part of the repertoire of many universities (both publicly funded and specialized institutions).

Motivation

There are at least two distinct advantages over traditional (face-to-face) teaching for an institution delivering a program via distance learning (Oblinger 2000). First, the field of potential student applications is no longer constrained by geographical location—a student can apply, and potentially study remotely, from anywhere. Similarly, the university human resources committee can "cherry pick" subject specialist staff to deliver the content, regardless of their location. A university or college based in an area with limited audiovisual industry might employ instructors from more prolific industry areas—London, Nashville, Los Angeles, or New York, for example—selecting instructors with firsthand industrial experience and an ongoing professional practice. This selection of adjunct faculty from industry (whom often continue to work in industry outside of their teaching duties) gives reassurances to students (and potential applicants) that relevant and current industry practice is being taught (though the technical element of the program might need to be supervised by a local technician, or by a video demonstration in the case of some content). Both of these motivating factors have an additional advantage for students enrolled in a distance learning program.

First, if the course content has been designed with care and appropriate pace, students should be able to navigate the content at their own speed, allowing for technically complex material to be covered. Practicable formats for delivery of technically complex material depend on the context—for example, DAW techniques might be feasibly delivered remotely by screen capture video sequencing, or fundamentals of sound by more traditional reading and interactive exercises. Physical audio engineering related to signal flow is more difficult to recreate virtually or remotely without appropriate setups at both ends of delivery—e.g. a duplicate patch-bay and outboard equipment (although some virtual instruments, such as the Reason platform, offer virtual simulations of patch-bays with similar signal flow). This is particularly relevant to the field of music technology, where both an aesthetic and technical understanding is vital to the overall curriculum, and will be discussed further later in the chapter.

If technically complex or "dense" material was all delivered face-to-face, students would have to ensure they kept up with the speed of delivery in the classroom (or indeed, workshop, or studio), which can mean keeping pace with (or conversely, waiting for) others in the student body (Otondo 2016). Second, the practice of *active listening* or *critical listening* (Walzer 2015) can "eat" into face-to-face teaching time, similarly active practice (Digolo et al. 2011; Seddon and Biasutti 2009). Developing a critical ear is a standard part of most music technology curricula, and hence giving the students an unconstrained amount of focused listening time, and space free from classroom distractions, for example, can be a strong advantage to be exploited by distance learning. Although the technical accuracy of a home monitoring environment can be compromising, high-quality open back headphones are reasonably affordable and the previous bandwidth limitations involved in delivering audio content online are less restrictive with current internet provision.

Existing Strategies

Currently, the delivery of distance learning in the music technology field can be broadly divided into one of three "camps", based on their general pedagogic approach (Booth et al. 2003):

1. Vocational programs without formal assessment ("hobbyist" courses)
2. Blended learning, content designed to support formative assessment on higher level programs (for example, to support degree-level study at university)
3. Full online delivery, including summative assessment, of higher level programs

A fourth approach, using totally virtualized environments, for example modeled using 3-D technology (Dickey 2005) (Halvorson et al. 2011), social media platforms (Salavuo 2008), or VR/AR/XR approaches (Jordan et al. 2016), where XR is a relatively new term for *extended reality*, were at the time of developing this material considered beyond the scope of this work, though they each offer the possibility of providing interesting tools to overcome limitations such as access to specific studio equipment, and indeed some research from the broadcast audio industry is considering these techniques for use in novel production paradigms (Pike et al. 2016). However, this work focuses on material that can be used within traditional (existing) distance learning technology, with a view to wider dissemination and the focus on both broader academic and vocational skills.

Virtual and distance learning technology has now become relatively standardized—many institutions use the same, or similar, technology, and hence this chapter focuses on the broader pedagogical issues of such an approach with specific reference to the vocational training and assessment of sound engineers and music technologists. Two papers presented at conventions of the Audio Engineering Society have evaluated specific technology and selection of appropriate material (Papanikolaou et al. 2000) using database/web interfacing technology (Kalliris et al. 2001), with a third paper outlining the initial stages of design and development of this module (Williams 2012). Beyond the Audio Engineering Society community however, there has been recent scholarship on the use of e-learning particularly for music education, and to some extent in music technology, for example the interested reader may wish to see the work of Himonides (2018), Ruokonen and Ruismäki (2016) or Crawford (2017) in these contexts.

Design and Assessment

This section outlines the design and rationale behind assessment of the remote access module (MUT360, Commercial Recording and Composition) for the Sound Recording Technology program at a public university in Texas. The module was designed in conjunction with technologists and learning specialists at the university, as well as faculty from the Sound Recording Technology program.

Course Design

Research on the definition of the discipline of music technology in the UK concluded that there is no definitive job description for a music technologist (Boehm 2005). As a descriptor, music technology traverses many fields and specializations, but some might take issue with the label of music technologists (Boehm 2007). Nevertheless, designing any course for music technologists requires an immediate and encompassing consideration of technical provision. There are many possible tasks which might be deemed teaching in the delivery of this provision—and the task of course design can be discrete or integrated with the teaching (in other words, may be carried out by the same person, though not necessarily). Teaching might include technical demonstration, moderation of discussion threads, delivering video lectures or seminars, and so on.

In this case, a particular demand on the teaching aspect of the delivery is the emphasis on "hands-on" experience, without which a student would reasonably be expected to struggle in industry as many entry level positions will provide little time for practice, and in any case, without technical experience it would be difficult to contextualize much of the material involved in this type of sound recording course—a course aimed at practical production with some theoretical knowledge. In this case, the fundamental question posed must be: *can students access the technology they require to complete industry standard work remotely?*

However, distance learning is not as disadvantaged as it might initially seem when compared to face-to-face methods in this regard—real industry work often requires the engineer to be "in field" (perhaps carrying out a recording of Foley, cues, or dialogue, for example), and contemporary composers or sound designers working on commercial music seldom work in large teams in the recording studio (with notable exceptions in the world of high-budget gaming or film soundtracks).

Therefore, basing assessment briefs on such examples of industry practice fulfills both the requirement to train students for the real world of work, and negates the need for a centralized equipment "hub" in a school, college, or university (recording studio suites, workshop laboratory or the like) which might have traditionally provided "face-to-face" students with access to hardware technology. However, by the same token, the implication is that the development of interpersonal skills and teamwork are two elements of traditional teaching that are extremely difficult to replicate or promote via distance learning, particularly with this "solo" approach.

Fostering collaboration and interaction between students is a useful approach for many elements of the study of music technology, and while an online forum allows for some interaction and discussion amongst the student body, the technology in many traditional VLE systems is not currently advanced enough to readily allow for true distance collaboration on practical projects in real time. Although increasingly social media style platforms (Salavuo 2008), synchronous group video calling (Abdalrahman et al. 2018) and other cloud computing platforms (Kim et al. 2018) might address these challenges, in this case the course content was nevertheless designed with the need to foster such skills in other ways.

Course Content

The curriculum was designed with the development of two streams of skills in mind: holistic (transferable, broader educational) skills and specific vocational skills, though as might be expected there is a degree of crossover and reinforcement between the two.

The course we describe here is aimed at third-year students majoring in sound recording technology. Each cohort contains up to 30 students, with a broad range of backgrounds. The course has run successfully each spring semester since 2012, most recently at the time of writing in 2019. The interplay between the skill set elements presented in Table 12.1 is important. An understanding of the fundamentals of sound from acoustic principles can inform the students own practice of critical listening, which we might consider as a development of aural analysis (not unlike the aural analysis a musicologist might require in their field, but rather, focused on the elements of recorded sound, thereby core to music technology).

In recent years there have been many advances in critical reflection on the processes and practices of music production, especially in educational contexts. The interested reader might consider Zagorski-Thomas (2007) or indeed the journal and conferences of the Art of Record Production, who frequently discuss how to develop critical listening in the context of practical music technology, for example interviews with educators (Wells 2012) about teaching content, or technological changes as the industry evolved from analog to digital recording (Bennett 2012). These elements can, once understood in principle and *heard* in practice, be deployed, through the remaining vocational elements. These include signal processing, sequencing, and editing, thereby allowing the student to develop and demonstrate their mastery of the vocational skills required to work with music and technology, which the design team considered important when specifying the course. Thus, student work could, in context, be self-directed and single-handed, but ultimately should require collaboration (developing teamwork and interpersonal skills, juggling deadlines, and negotiating priorities), which involves working in small teams outside of scheduled contact time.

The advantages of group work are somewhat axiomatic based on the author's own praxis, but several pedagogic theories also ground this thinking, both as a dialogic practice (Skidmore and Murakami 2016) and in terms of the so-called *design ladder*, wherein knowledge is developed in response to real circumstances

Table 12.1 Example holistic and vocational skill set streams targeted by curriculum design

Holistic	*Vocational*
Collaboration	Sound
Analysis (written)	Processing
Critical listening	Sequencing

Source: For more detail on holistic approaches to distance education the interested reader may wish to see (Bozkurt 2019) and for the influence of model on vocational training (Suartini 2019).

(Wrigley and Straker 2017). If we consider the design ladder as a possible "route" students might take through these skill sets, it becomes clear that the quickest way for the student to learn some of the skills required for work with music technology is not necessarily the most advantageous in terms of creating a more balanced academic profile, in other words a set of transferable academic skills beyond the immediate requirements of the discipline, or indeed the development of material suitable for academic assessment (Barber et al. 2015). If the route the student takes through the material can influence their learning in a dramatic way, the order of curriculum presentation must be considered carefully. Figure 12.1 illustrates an idealized route through this process.

Content was based on real-world industry experience and fed directly into the design of assessment work. The "yardstick" assessment criteria applied was simple: *is the work submitted at, or close to, professional broadcast standard.* Much of the module content was therefore focused on broadcast practice and deliverables, including written and theoretical examples from professional radio and television, with technical discussion of the necessary associated standards. Publishing briefs and contracts from an anonymized major label were used to provide context as to the type of brief practitioners might expect if engaged in professional work, for student analysis, along with example session files and documentation. A clear advantage of delivering this material online was the ability to "embed" audio and video examples, and include example sessions (generally Pro Tools sessions with stem files) for practical experimentation by students, augmenting the written material. For students without access to Pro Tools alternative sessions (.wav stem files for use in any multitrack DAW) could be provided, but in the case of this course all students are given a license for Pro Tools when they enroll.

For example, when discussing the use of side-chain filtering and dynamic compression to remove vocal sibilance in a "de-essing" process, as shown in Figure 12.2, students can download and audition "before" and "after" versions of a processed vocal, read a written explanation of the signal flow, click through to further references, or download an example session file to experiment with directly. In this case students are required to complete a spoken word edit for an audio book production (also known as a "cough edit" in UK broadcast circles). The signal processing involved in de-essing has numerous applications throughout commercial recording as it involves an understanding of signal flow, dynamics and equalization processing.

Encouraging students to develop critical listening skills and a skeptical eye for proponents of "Gear Acquisition Syndrome" can, to some extent, future-proof a course like the one presented in this chapter. The design philosophy here was to teach concepts rather than specific pieces of equipment. For example, in the case of de-essing, dynamic range is always dynamic range; even if the controls on a specific unit vary in layout (or tonal characteristic). Similarly, in choice of DAW—underlying concepts are more important than operational, platform-specific details. Industry tradeshows and the popular music technology press have been shown to encourage consumer desire for new products, usually by promising quality closer and closer to professional standards, yet the baseline for professional audio performance in the commercial world has changed very little.

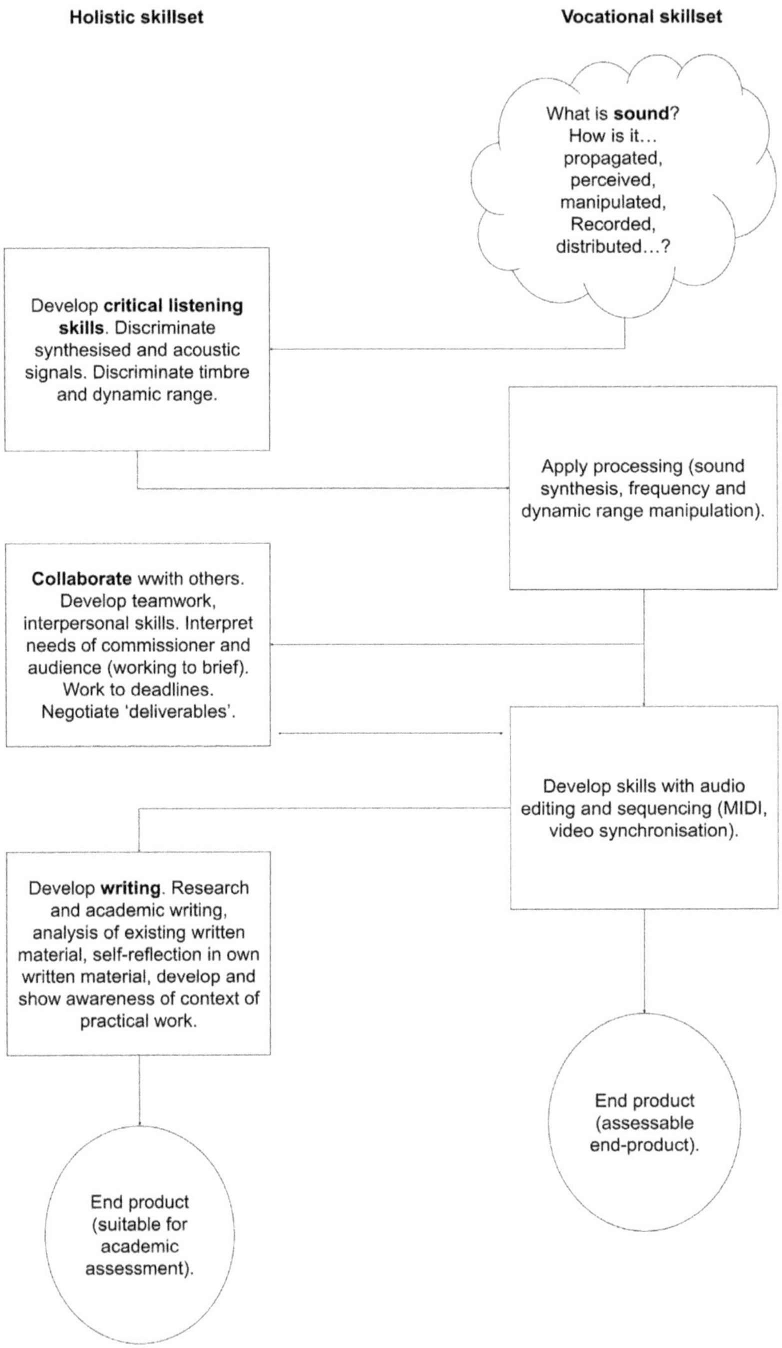

Figure 12.1 Possible routes a student might take through this course as an online-only offer in music technology studies.

Source: Teamwork and collaboration are difficult to develop in a distance based course but are essential elements of the interpersonal skill set that a music technology program to encourage. Much of the music industry relies on these skills (Bennett 2017; Cain and Cursley 2017).

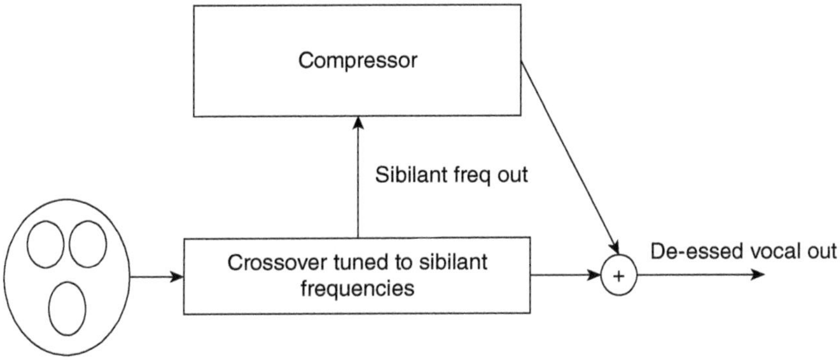

Figure 12.2 Slide illustrating de-essing technique. Sidebars allow students to audition audio files and session files.

In the view of the design team, it was important to discourage this type of thinking amongst the student body, and concentrate on the concepts and principles of sound. In order to achieve this, our course content prioritizes the principles of sound and the necessity for broadcast quality in any deliverables rather than the particular features and function of brands of equipment, software specificities, or the like. As detailed in the curriculum development section of this chapter, the design team felt that the baseline principles of sound and music technology begin with acoustics, sound synthesis, and digital audio recording. Rather than stressing, for example, the learning by rote of keyboard shortcuts which are only applicable to a particular digital audio workstation (reinforcing an inherent bias in the content which might be present either from the instructor, the institution, or market pressure from specific manufacturers), time should be spent developing skills which are common in all such workstations: gain staging, signal-to-noise ratios, appropriate use of dynamic range compression, creative and corrective use of frequency equalization or volume automation, and the like.

As audio educators, we might consider these essential competencies, a concept which in the context of DAW use the interested reader might refer to recent work by Walzer (Walzer 2016), or previous work in the context of digital literacy and how competency across domains is important for student development (Raymond 2008). Skills which might be acquired via distance learning could include, for example, production analysis (which necessitates a combination of critical listening with writing skills). The overarching curriculum selection principle is that these fundamentals will be applicable across the audio industry regardless of platform or role, serving the students in better stead when entering the world of work.

The core curriculum of the course described here can therefore be represented by three overarching themes:

1. Sound: acoustic principles and sound synthesis
2. Signal processing: frequency and dynamic range
3. Sequencing and editing: audio and MIDI, synchronization

Assessment Structure

The assessments were structured to encourage self-directed learning with reinforcement and contextualization of technical material in mind, using reframed material from the module content to achieve this purpose. *Checking* or *reinforcing learning* are techniques which can be a useful focus for the classroom based educator—allowing a reasonably experienced teacher to spot when students are not grasping the material, to potentially reframe the subject or give alternative examples (McKenzie 2003). The virtual environment does not afford much opportunity for students to ask for further detail or clarification, or for the course leader to monitor student progress in an intuitive informal manner. In order to keep challenging students, and provide an opportunity for formative assessment, (Nicol and Macfarlane-Dick 2006; Sadler 1998) multiple-choice "quizzes" were added to the assessment sections of the content. Here, 10% of the final grade was allocated to these multiple-choice tests.

To integrate some of the benefits of this "checking/reinforcing" model, a weekly virtual chat seminar was built into the material, giving the delivery a methodology somewhat akin to a "blended" model (Crawford 2017). In this case 20% of the final grade was allocated to this, in order to encourage students to make use of the time. The remaining assessments were allocated to submitted written work (in traditional research report format), with 20% of the final grade, and submitted practical work based on the industry briefs illustrated in the course content, worth 40% of the final module grade. The grade boundaries awarded were as follows:

- Notional "A": 593–660 Points
- Notional "B": 527–592 Points
- Notional "C": 461–526 Points
- Notional "D": 395–460 Points
- Notional "F": 394 Points and Below

A complete breakdown of the assessment structure and mark allocation is shown in Table 12.2.

Note that some course content did not contribute to the final grade awarded.

Table 12.2 Assessment structure and mark allocation

Assessed content	*Mark allocation*	*Contribution to final grade*
Discussions:	133 points	20%
Quizzes:	66 points	10%
Online tutorial attendance:	100 points	15%
Weekly projects:	272 points	40%
Final project:	100 points	15%
Total possible:	931 points	100%

Remote Assessment

Devising and formalizing an assessment structure based on the curriculum content highlighted in section 2 presented its own challenges. It can be difficult to re-create the highly pressurized environment of recording, broadcast, or live sound reinforcement in a classroom or simulated environment, and many established "real-world" programs offer access to studio and concert facilities similar in design and specification to the real world of work for this reason. Music technology by its nature requires access to industry standard facilities (microphones, preamplifiers, and acoustically treated spaces suitable for engaging in critical listening could be considered a bare minimum requirement). This raises questions of accessibility more generally, for example, what if potential students did not have access to the necessary level of facilities? In recent years, access to unprecedented computing power and affordable technology (with the exception of acoustically treated spaces) has made the practicality of a home recording studio more and more viable. Nevertheless, it presents a complication in the case of remote access.

In this case, the commissioning university made a commitment to provide access to such facilities at a minimal staffing cost (requiring studio technical staff only), generating a saving on Faculty teaching hours. There are legitimate concerns when it comes to cost-cutting efforts in terms of the quality of the learning experience for students. Our intention is that with carefully designed online material, the student experience can be at least maintained or possibly even improved. In other cases, work experience might be used to supplement the online delivery with hands-on training (and indeed, many "traditional" courses award credit for successful completion of work experience or internships) to encourage this practice.

A suitably equipped student, such as one already serving an apprenticeship or internship, or a working professional seeking an academic qualification to augment their experience, could perfectly reasonably produce the assessment material independently with access to the bare minimum facilities—the assessed content is not targeted at any specific platform or manufacturer and so the equipment used is of significantly less importance than the development of a working understanding of the underlying principles of sound and music, and the skill to produce work of broadcast quality. While this notion does create an intrinsic lack of parity in the equipment being used across the student body in the creation of their portfolio of practical assessment submissions, a similar lack of parity can be seen in the professional arena. It is not uncommon in industry to see successful engineers and facilities operate across a range of differing (or even multiple) platforms.

There is a further consideration in the huge technological advances being made by the open-source software community with regards to music technology. While hardware and commercial software products are undoubtedly widely used in the professional arena, the criteria for delivery of finished material to a broadcast standard can certainly be achieved with alternative resources. In fact, the ability to create multi-tracked arrangements, apply comprehensive editing techniques, and utilize a wide range of synthesis using a DAW gives most students of music technology access to more metaphorical "firepower" than most large-scale recording studios in the 1980s and earlier, where tape-based recording techniques were prohibitive in terms of track count, and physical limitations might restrict the use of numerous outboard processors. The trick, then, is often to discourage the use of too much virtual technology given the hugely expanded range of options faced by newcomers to the field.

Conclusions and Further Work

Distance learning allows students to study technical content at any time, and at their own pace. The method of practical assessment used would be appropriate to many other creative sections of the audio/visual engineering sector where technical knowledge and achievement of professional standards is an overarching mark of skill, and thereby an indicator of successful learning. The combination of multiple choice assessment and virtual face-to-face reinforcement takes advantage of, and answers traditional pedagogical demands for, academic differentiation and contextualization of course content. Theoretically, submitted reports (written work) should be pedagogically "neutral", i.e. neither advantageous nor disadvantageous if delivered via distance learning when compared to traditional teaching methods.

Current technology has become sufficiently advanced for virtual "chat" seminars to be useful for reinforcing autodidactic learning, creating a shift toward a blended learning model that can be useful for the delivery and assessment of vocational content for the trainee audio engineer. However, assessment of vocational skill in, for example, the sound reinforcement field would not be practical with the technology used here. The need to access appropriate facilities in order to maintain parity with industry for assessment purposes, and to reach a minimum standard for critical listening is in some ways insurmountable with the current technology. Virtual private network access to studio software technology (or simulated, virtualized technology) might, in the future, yield an alternative to access issues, but the requirement for development of interpersonal skills such as teamwork and communication pose more significant obstacles to such delivery. An evaluation of real-time collaborative environments would therefore make for useful further work in order to establish a platform for developing teamwork and cooperation between students undertaking a vocational program in music technology.

Acknowledgments

The author gratefully acknowledges the support of the School of Music, in particular Kevin Kelleher, director of Sound Recording Technology at Stephen F. Austin State University, Texas, USA, and the support of the Faculty of Music, University of Oxford, UK, while undertaking a portion of this work.

Note

1. The nomenclature varies between the US and UK. In the UK a module is a component of a course, assigned a certain number of credits. For example, a combination of compulsory and optional modules

conveying 120 credits would need to completed per year of a three-year undergraduate program, with 360 credits required on completion of the degree. Similarly, 480 credits for four-year programs—honors degrees generally take three to four years to complete, though the requirement can vary slightly from institution to institution, and at post-graduate study level. The notional equivalent in most US systems is units (academic modules, representing credits toward the course) and subjects (which may contain various units). In either system, modules, courses, or component units may last a single semester or span a longer duration. The term course might mean a complete degree program, or a combination of modules or units in either case, so to avoid potential dichotomy of meaning here, for the purposes of this chapter, we consider module to be interchangeable with unit, and degree to be interchangeable with course.

References

Abdalrahman, A.E.M., Altaieb, M.O., Others, (2018). "E-learning based on cloud computing technology", *Proceedings of the 1st International Conference on Computer Applications & Information Security (ICCAIS)*, IEEE, 1–6.

Barber, W., King, S., and Buchanan, S., (2015). "Problem based learning and authentic assessment in digital pedagogy: Embracing the role of collaborative communities", *Electronic Journal of e-Learning* 13, 59–67.

Bennett, J., (2017). "Towards a framework for creativity in popular music degrees", in: *The Routledge Research Companion to Popular Music Education*, edited by Gareth Dylan Smith, Zack Moir, Matt Brennan, Shara Rambarran, and Phil Kirkman. Abingdon-on-Thames, England: Routledge.

Bennett, S., (2012). "Endless analogue: Technological precursors in contemporary recording practice", *Journal on the Art of Record Production* 7.

Boehm, C., (2005). "The thing about the quotes: "Music technology" degrees in Britain", *Proceedings of the International Computer Music Conference, New Orleans. International Computer Music Association*, New Orleans, IA.

Boehm, C., (2007). "The discipline that never was: Current developments in music technology in higher education in Britain", *Journal of Music, Technology & Education* 1, 7–21.

Booth, R., Clayton, B., Hartcher, R., Hungar, S., Hyde, P., and Wilson, P., (2003). *National Centre for Vocational Education Research, 252 Kensington Road, Leabrook, South Australia 5068, Australia.* Volumes 1 and 2. ERIC.

Bozkurt, A., (2019). "From distance education to open and distance learning: A holistic evaluation of history, definitions, and theories", in: *Handbook of Research on Learning in the Age of Transhumanism*. Hershey, Pennsylvania, USA: IGI Global, 252–273.

Cain, T., and Cursley, J., (2017). "Researching pedagogy: Questions, methods, conceptual framework and analysis", in: *Teaching Music Differently*. Hershey, Pennsylvania, USA: Routledge, 33–41.

Crawford, R., (2017). "Rethinking teaching and learning pedagogy for education in the twenty-first century: Blended learning in music education", *Music Education Research* 19, 195–213.

Dickey, M.D., (2005). "Three-dimensional virtual worlds and distance learning: Two case studies of active worlds as a medium for distance education", *British Journal of Educational Technology* 36, 439–451, doi:10.1111/j.1467-8535.2005.00477.x

Digolo, B.A., Andang'o, E.J., and Katuli, J., (2011). "E-learning as a strategy for enhancing access to music education", *International Journal of Business and Social Science*, 2 (11).

Edison, T.A., (1878). "The phonograph and its future", *The North American Review* 126 (262), 527–536.

Halvorson, W., Ewing, M., and Windisch, L., (2011). "Using second life to teach about marketing in second life", *Journal of Marketing Education* 33, 217–228, doi:10.1177/0273475311410854

Himonides, E., (2018). "The misunderstanding of music-technology education: A meta perspective", in: *Creativities, Technologies, and Media in Music Learning and Teaching: An Oxford Handbook of Music Education*. Oxford, UK: Oxford University Press.

Holmberg, B., (2005). *The Evolution, Principles and Practices of Distance Education*. Oldenburg, Germany: Bibliotheks- und Informationssystem der Universität Oldenburg.

Jordan, D., Müller, F., Drude, C., Reinhold, S., Schomakers, V., and Teistler, M., (2016). "Spatial audio engineering in a virtual reality environment", in: Prinz, W., Borchers, J., and Jarke, M. (Hrsg.), *Mensch und Computer 2016: Tagungsband*. Aachen, Germany: Gesellschaft für Informatik e.V., doi:10.18420/muc2016-mci-0217

Kalliris, G., Dimoulas, C., Papanikolaou, G., Rizakos, C., and Sevastiadis, C., (2001). "Development of a distance-learning environment, using database driven dynamic web pages", *Proceedings of the 110th Audio Engineering Society Convention*, Amsterdam, Netherlands.

Kelleher, K.D., (2012). *The Contributions of Thomas Edison to Music Education*. Boston, USA: Boston University Library.

Kim, M., Pickover, C.A., and Salapura, V., (2018). *Music Practice Feedback System, Method, and Recording Medium*. US Patent US9672799B1.

McKenzie, J., (2003). "Pedagogy does matter", *The Educational Technology Journal* 13, 1–6.

Ni, A.Y., (2013). "Comparing the effectiveness of classroom and online learning: Teaching research methods", *Journal of Public Affairs Education* 19, 199–215.

Nicol, D.J., and Macfarlane-Dick, D., (2006). "Formative assessment and self-regulated learning: A model and seven principles of good feedback practice", *Studies in Higher Education* 31, 199–218.

Oblinger, D.G., (2000). "The nature and purpose of distance education", *The Technology Source Archives*, URL www.technologysource.org/article/nature_and_purpose_of_distance_education/ (accessed 8.3.12).

Otondo, F., (2016). "Music technology, composition teaching and employability skills", *Journal of Music, Technology & Education* 9, 229–240.

Papanikolaou, G., Pastiadis, C., Dimoulas, C., Kaliris, G., and Galatsopoulou, F., (2000). "Deployment of internet resources for digital audio learning courses", *Proceedings of the 108th Audio Engineering Society Convention*, Paris, France.

Philips, R., and Merisotis, J., (1999). "What's the difference? A review of contemporary research on the effectiveness of distance learning in higher education", *Journal of Distance Education*, 14 (1), 102–114.

Pike, C., Taylor, R., Parnell, T., and Melchior, F., (2016). "Object-based 3D audio production for virtual reality using the audio definition model", *Proceedings of the 2016 Audio Engineering Society International Conference on Audio for Virtual and Augmented Reality*, Los Angeles, CA.

Raymond, S., (2008). *Digital Storytelling in the Classroom: New Media Pathways to Literacy, Learning, and Creativity*. Thousand Oaks, CA: Corwin Press, 228.

Ruokonen, I., and Ruismäki, H., (2016). "E-learning in music: A case study of learning group composing in a blended learning environment", *Procedia-Social and Behavioral Sciences* 217, 109–115.

Sadler, D.R., (1998). "Formative assessment: Revisiting the territory", *Assessment in Education: Principles, Policy & Practice* 5, 77–84.

Salavuo, M., (2008). "Social media as an opportunity for pedagogical change in music education", *Journal of Music, Technology & Education* 1, 121–136.

Seddon, F., and Biasutti, M., (2009). "Evaluating a music e-learning resource: The participants' perspective", *Computers & Education* 53, 541–549.

Skidmore, D., and Murakami, K., (2016). *Dialogic Pedagogy: The Importance of Dialogue in Teaching and Learning*. Bristol, UK: Multilingual Matters.

Suartini, T., (2019). "Influence application of learning model on vocational education based on quality assurance", *Sage Open* 9, doi:10.1177/2158244019851552

Walzer, D.A., (2015). "Critical listening assessment in undergraduate music technology programmes", *Journal of Music, Technology & Education* 8, 41–53.

Walzer, D.A., (2016). "Software-based scoring and sound design: An introductory guide for music technology instruction", *Music Educators Journal* 103, 19–26, doi:10.1177/0027432116653449

Wells, J., (2012). "Interview with Dave Fisher", *Journal of the Art of Record Production* 1.

Williams, D., (2012). "Distance learning strategies for sound recording technology", *Proceedings of the 133rd Audio Engineering Society Convention*, San Francisco, CA.

Wrigley, C., and Straker, K., (2017). "Design thinking pedagogy: The educational design ladder", *Innovations in Education and Teaching International* 54, 374–385.

Zagorski-Thomas, S., (2007). "The musicology of record production", *Twentieth-Century Music* 4, 189–207.

13

Accessibility

A Practical Perspective on Inclusion

Victor Zottmann

Introduction

Between May and September of 2018, an initiative based on video production became the core aspect of my final project as part of my undergraduate degree at SAE Institute in Melbourne, Australia. The project consisted of addressing the lack of audio postproduction tutorials whose content covered not only the techniques but also the artistic purpose for applying them, accessible to Brazilians. In this case, the concept of an accessible audio education was not directed toward people with disabilities. Rather, it was aimed toward including Portuguese speakers whose lack of in-depth knowledge of the English language limit their access to numerous high-quality resources available in English.

Although the project initially focused on postproduction, this chapter does not intend to limit its application only to one area; rather, the results discussed can be universally analyzed and put into practice. A total of 40 video-based audio tutorials were published on YouTube under my personal channel, *Victor Zottmann*, and while the English language was chosen in order not to lose the essence of knowledge diffusion, Portuguese subtitles were added for each video with a view to address the reality of Brazil—within a population of approximately 208 million (Silveira 2017), only 5% speak a foreign language, and 3% are fluent in English (Rodrigues 2017).

Many factors inspired the project, including comments from fellow sound designers and audio engineers across the countries I have lived in—Brazil, Canada, and Australia. They wondered why I chose such locations for my studies, as they believed those countries did not have much to offer in terms of a career in audio. Coming from Brazil, where the quality of education suffers from the lack of investment by the government, it is common not to find much support for a career in the audio industry. On the other hand, Canada and Australia are recognized for having a strong audio influence: where the former comprises both the North American motion picture and video game industries, the latter boasts a vibrant music scene. In fact, the Canadian Media Producers Association (2017) reported a growth of approximately 61,000 jobs. Similarly, the Australian non-profit society, Music Victoria (n.d.), states that Melbourne currently boasts 465 live music venues.

While some of the most common audio practices revolve around film and videogames, there are many opportunities existing in places where students would not usually expect. Networking with people who have nothing or little to do with audio, for example, is critical when the goal is to identify unique career opportunities. Pursuing a career in audio presents both challenges and prospects; indeed, opportunities can be found in business conferences, where entrepreneurs may need someone to create audio-visual material for their organization.

This chapter discusses some of the challenges in creating high-quality audio tutorials accessible to Brazilians, with the support of an interview conducted with Dr. Mauricio Fonteles, professor of the faculty of communication at the University of Brasilia, Brazil (UnB), an experienced professional in audio and video production, and course director and instructor at *OZI Escola de Audiovisual*. Formerly residing in Brasilia, OZI is an online-based audio-visual school whose facilities have expanded to make their broad curriculum available to everyone in Brazil, and those living abroad who are fluent in Portuguese via online learning platforms. While this chapter focuses on bridging an important gap related to audio education in Brazil, it also aims to help the reader develop accessible learning solutions while prioritizing instructional quality, regardless of where one lives.

Creating Accessible and Practical Tutorials

Identifying the Gap

It would be inaccurate to suggest that audio education is dependent on students being present in a physical classroom or studio. Indeed, hands-on experience with cutting-edge hardware (e.g. analog compressors and equalizers and mastering equipment) is beneficial to anyone starting out in—and already working with—audio. On the contrary, an idea worth considering is whether such contact with hardware is required for student success in the professional field.

Despite the ease of access to information, and the fact that the universal language of audio is English, it is critical to consider that non-English-speaking countries might not understand the language, thus limiting their reach to a broader array of information. In an interview conducted by David Pearce for Wired, Jason Kottke (2017), stated that writing about film is similar to dancing about architecture. He also argues that "video lends itself well to looking at how moviemakers put things together". The same can be said about audio, in that learning it through books and articles alone is not sufficient. However, by demonstrating visual and practical examples of professional applications of audio postproduction along with supplemental texts, the viewer can understand a particular message more efficiently.

Mayer (2009: 6) explains that the rationale for multimedia presentations through words and pictures is that such media takes advantage of the full capacity of humans for processing information. Furthermore, when we restrain our mode of instruction to nothing but words, we are neglecting our capacity to visually process material as well (Mayer 2009: 6). As much as audio concerns theory, which encompasses reading and writing, it also involves practice, where students learn by doing. Indeed, there is a very large array of career paths in audio, where one learning method may overtake another. This chapter emphasizes the practical learning method.

Plenty of what sets in-person education apart is not only the student-teacher relationship, but also the fact that students can interact with analog and digital mixing consoles, patch-bays and hardware processors. It is in the physical environment where students can experiment with real world professional experiences through client-based projects, as well as engage with prospective clients. On the other hand, Russo and Benson (2005) describe affective learning as the outcomes developed by students in regard to their interaction with the course and the instructor, as well as a critical indicator of distance learning success. Online courses, on the other hand, frequently fail to design interactive learning environments (LaRose and Whitten 2000), which in turn may result in students' lack of engagement with the material.

Dr. Fonteles (personal communication 2018) argued that the advantage of transforming OZI into an online school was providing opportunities for Portuguese-speakers not only residing in Brazil, but also across the world, to high quality audio-visual education. While the student-teacher relationship is prone to suffering a few compromises due to the absence of in-person interactions, "actively engaging in video conferences and social media groups with the class has shown effective results" (Fonteles, personal communication 2018).

Given the aforementioned challenges, as well as those of producing high-quality and accessible material, a few questions are worth examining:

1. What are the challenges of developing tutorials that not only add value, but also retain people's attention and are delivered in easy-to-follow and engaging ways?
2. In addition to delivering video-based educational material, what would be an effective way to supplement it with reading material without making one dependent on the other?
3. How does one provide high-quality online audio-visual education while ensuring students from different professional backgrounds are making the most out of it?
4. If the material is delivered in such a way that resonates with learners' necessities, can affective learning be possible even with isolated videos that do not belong to a course?

It is clear that inflated statistics are appealing to the eye. Views, likes, and subscribers have, for some time now, been used to indicate a creator's success. The problem with this approach is that it creates a false sense of value. In other words, when it comes to education, the statistics that matter the most belong to the value an educator brings to a learner's life. Since its inception, this project has been determined to help people get a better understanding of the audio production practice, considering the context in which to apply a certain technique, as well as prioritizing quality rather than quantity.

The early stages of my research were devoted to gathering precise information about the needs of the target audience; more specifically why people search for audio tutorials. Due to the scarcity of research available on audio education in Brazil, two surveys (Zottmann 2018) regarding the efficiency of audio tutorials (one in English and the other in Portuguese) were shared among international and Brazilian Facebook groups on filmmaking and audio. Fifty-three people responded including 43 to the English survey and ten to the one in Portuguese. While the questions for both surveys were almost identical (see Table 13.1), the Portuguese version differed with regard to the difficulties Brazilians face with the English language, as well as their preference on subtitles.

For the purposes of this project, demographic data, such as age and gender, were not prioritized as the focus was on identifying reasons why students and professionals look for audio tutorials. However, both factors have been gathered through YouTube data over time, and the results are discussed in the "YouTube Analytics" section of the chapter. Later, I refer to the survey in English as Survey #1, and to the one in Portuguese as Survey #2—or simply as the Brazilian participants.

Topics and Tools

Table 13.2 illustrates the list of audio tutorials in the chronological order of production, as well as the duration for each video, in minutes. Before entering more complicated subjects, a variety of short Pro Tools tutorials were created to practice the overall production quality and improve storytelling. The first ten tutorials did not address any common issues encountered by beginning sound editors and designers. Later topics were important and necessary to allow for creative challenges, which would significantly improve the production quality of future tutorials.

Table 13.1 Set of questions prepared for the surveys—Question 8.P and Question 9.P indicate the changes made for the survey in Portuguese

Questions on the Surveys	
Q 1.	What is your creative specialization?
Q 2.	What is your level of experience?
Q 3.	Why do you look for audio tutorials?
Q 4.	What type of learner are you?
Q 5.	Which video platform do you use the most when searching for audio tutorials?
Q 6.	Which average lengths do you prefer a tutorial to have?
Q 7.	Do you prefer tutorials that go straight to the point?
Q 8.	What is your native language?
Q 8.P	Do you have any difficulties with the English language?
Q 9.	Do you watch videos with English subtitles? If so, why?
Q 9.P	Would you prefer if foreign tutorials had Portuguese subtitles?
Q 10.	Which communication characteristics do you find the most effective?

Table 13.2 List of audio tutorials in order of production

	Tutorial name	*Duration*
1	Pro Tools HD–Automation (Part 1)	7:55
2	Pro Tools HD–Automation (Part 2)	5:32
3	Pro Tools–Removing Mouth Clicks with the Pencil Tool	6:36
4	Pro Tools–Useful Bypass Shortcuts	1:50
5	Pro Tools–User Default Plugin Settings	3:21
6	Pro Tools Quick Tip–Audiosuite Handles	2:45
7	Pro Tools Quick Tip–Cataloguing your SFX Library	1:36
8	Pro Tools–Useful Fade Shortcuts	3:36
9	PaulStretch–The Extremes of Sound Stretching	9:37
10	Film Dialogue Mix Template–Layout and Routing	6:33
11	Film Dialogue Mix Template–Inserts, Sends, & Console	8:28
12	2 Ways to Record Scrubbing in Pro Tools	7:24
13	Sound Design Tutorial: Designing Explosions with Cereal	5:45
14	The Kontakt Footstep Instrument	6:17
15	Creating Room Tone with Convolution Reverb	8:07
16	Pro Tools Quick Tip–Custom Shortcuts	2:13
17	Pro Tools Quick Tip–The Slip-Grid Mode	1:14
18	Working with Video in Pro Tools	7:55
19	Dealing with Low-Level Dialogue	5:42
20	Pro Tools–Clip Trim and Snap Shortcuts	6:21
21	Pro Tools Quick Tip–Moving Clips with Markers	2:22
22	Applying Metadata with Basehead	6:48
23	Introduction to Kontakt (Beginner Friendly)	12:07
24	Track Presets in Pro Tools 2018	5:17
25	iZotope Iris 2 – Designing Grenade Build-Ups	7:11
26	Film Sound–Editing Dialogue with Alternate Takes	9:13
27	Exploring the GRM Tools for Sound Design	10:31
28	Create a Powerful Bass Drop in Two Minutes!	4:36
29	Audio Repair–Fixing Wind and Distortion	9:04
30	Designing Tracer's Blink Ability with Everyday Objects	5:37
31	Getting Started with Reaper–Mac and PC	5:55
32	Using Audiokinetic Wwise for Linear Media	7:37
33	Designing Slow Motion Fire with Water	5:48
34	Designing Sound with Character in Mind	6:49
35	Waterproofing a Shotgun Microphone	2:05
36	Film Sound–How to Edit Ambiences	8:52
37	Creating Birds Songs with Speech	2:31
38	Smartphone Field Recording	4:33
39	Film Sound–Conforming to an Edit Decision List (EDL)	9:32
40	The Story of a Sound	5:03

The durations of the tutorials were selected from the participants' preferred suggestions given in Question 6—namely, the shorter the better; two, five, eight, or ten minutes on average; and the longer the better. Even though there were some votes for the last option, almost all tutorials were focused on lengths of up to ten minutes in order not to bore the viewer among other factors, such as ensuring that the material was delivered in a concise manner rather than delaying it with irrelevant context. In fact, these considerations originated from the responses for Question 7, where 93% and 100% of the participants from Surveys #1 and #2, respectively, preferred tutorials that went straight to the point.

The principal tools used to create all of the videos comprise the entry-level Canon EOS 200D camera (also known as Canon SL2), along with an 18–50mm lens and the Rode Videomic Pro microphone coupled with its windshield (see Figure 13.1). This was the setup required for filming either myself or supporting footage, which are referred to as B-roll in the film and television industries. In addition, the voice-overs for all tutorials were recorded with the Shure SM7B dynamic microphone, connected to the Audient iD14 audio interface (see Figure 13.2), then edited in Pro Tools with support from iZotope RX 6 for noise reduction. The voice-overs for Pro Tools tutorials were recorded through another digital audio workstation (DAW).

In cases where sound effects needed to be recorded, a kit comprising the Rode NTG-3 shotgun microphone and the Zoom H6 portable recorder was used; the Rode Blimp was chosen as the windshield and suspension system for the microphone

Figure 13.1 Rode Videomic Pro with its windshield on the left; Canon 200D on the right.

Figure 13.2 Voice recording setup: Shure SM7B connected to the Audient iD14 audio interface.

(see Figure 13.3). The screen recordings were captured through ScreenFlow, for the macOS operating system, and the overall video footage was edited in Final Cut Pro X; the subtitles for both English and Brazilian Portuguese were then added within YouTube's built-in subtitles tool.

Design Considerations

Cinelli (2003: 33) describes communication as an "interpretation of community",[1] wherein individuals seek to share ideas, feelings, as well as to influence one another's behavior. The purpose of an accessible audio postproduction YouTube channel is to effectively communicate useful and contextualized information to everyone. Despite not obtaining as many responses from Brazilians, designing an accessible audio YouTube channel does not need to be restricted to Portuguese speakers. Rather, it can be argued that accessibility is language independent: it must account for virtually everyone. In the case of video tutorials, such claim can be aided by the use of subtitles.

Khan Academy, the prestigious non-profit organization devoted to making education accessible to anyone, has been successful at utilizing an interactive learning method: in addition to interactive exercises and instructional videos, students read about the material supporting a given unit. Furthermore, being a worldwide educational portal, Khan Academy enables non-English speakers to learn the material in

Figure 13.3 Rode Blimp at the top and Rode NTG-3 shotgun microphone right below it; on the right is the Zoom H6 along with its attachable microphones and windshield on the left.

their respective languages, while also providing native English speakers with subtitles for further accessibility. Audio-visual education currently works in a similar fashion to Khan Academy's model. Whereas interactive exercises are equivalent to hands-on training with professional recording and processing equipment and client-based projects, reading is recommended as an additional requirement for taking the most out of the course.

In a study conducted by Meij and Meij (2015, cited in Gil, J.M. & Williams, V. 2017), the effectiveness of video tutorials accompanied by text-based instructions for middle school students was analyzed. It was found that while students were able to absorb multiple sets of information at the same time, the written material allowed them to dive into further detail on the topic at hand. Furthermore, Salman Khan (2015), the founder of Khan Academy, debated in his TED talk, "Let's Teach for Mastery—Not Test Scores", that the methodology applied when learning a sport or a musical instrument is that of practicing a skill as much as one can, and once one has mastered it, one then moves on to the next challenge.

In this case, sound editing and design are not subjects that can be taught through books alone. While a broad range of books on the theory of film sound and technical sound design offer valuable knowledge, truly learning its art and technique demands hard work, repetition, and the ability to think outside the box, as well as the will to challenge conventional rules. Out of the 40 tutorials created, one was specifically designed with Khan Academy's dynamics in mind in order to make the subject more engaging, in addition to providing context in which to apply the techniques.

In order for communication to be properly decoded by the audience, as proposed by Cinelli (2003), it is pivotal that the communicator be aware whether the audience would be able to receive information without any flaws: "if the audience does not manage to relate themselves with the communicator, then there was no communication" (Cinelli 2003: 33–34). The video "Sound Design Tutorial: Designing Explosions with Cereal" is considered to be one of the most helpful out of the 40 tutorials due to the way it was constructed and presented to the audience. The idea behind it was to break the myth that for the purpose of successfully creating the sound design for a project, one must have cutting-edge tools available.

The goal of the video was therefore to record at least one sound effect per day for an entire month, with the objective of designing a small sequence for a video game trailer. Since the tutorial would not only encompass visual demonstrations of the recorded sounds, but also the postproduction process, the first challenge lied upon supplementing the tutorial with text-based material, although without making them dependent on one another. The video ought to be self-sufficient in case the viewer did not want to access the reading material. The second challenge, in turn, was that of telling an effective story within a short period of time; a story whose plot revolved around overcoming common creative blocks by executing simple techniques.

Published at the non-profit website, *DesigningSound.org*, supplementary reading material comprising specific plug-in settings and sound design processes provided to the audience for further studying. Unfortunately, at the time of this writing, Designing Sound is in hibernation as neither did the lead editors have the availability required to meet the demands of maintaining the site running, nor did the contributors. Nevertheless, the website is still live; readers can access all of the content despite not being allowed to leave comments.

Comprising more than 200,000 unique new-user accesses between 2017 and 2018,[2] the advantage of the website being a well-regarded resource by the community of audio professionals is that a broader audience can have access to the tutorial. The leading audience mostly comes from American English speakers, whose demographics have accounted for 61% of all new users for the given time period; Brazilian Portuguese speakers comprise only 1.14% (2,371 new users). On the other hand, since the site does not support language translation, Portuguese speakers would therefore not completely benefit from the additional reading material. The data from Designing Sound was obtained through personal communication with the chief editors.

Other well-received tutorials, such as "Exploring the GRM Tools for Sound Design" and "Film Sound: Conforming to an Edit Decision List (EDL)", took on radically different approaches, although always seeking to relate with the audience's needs. For example, the GRM Tools tutorial announces at the very beginning that the explanation for what is demonstrated is given only at the end of the video. The reason for this choice was based on the likelihood that the audience looking for a tutorial on sound design tools is more interested in the results, rather than in the theory of what they actually do.

On the other hand, the EDL tutorial addresses a common issue faced among editors throughout the scope of the postproduction of a film or documentary project: Adjusting the time of current edits to updated versions of the film whose duration might vary, and whose footage might also be significantly modified. For beginners, this is counterproductive and time consuming as they usually do not have access to fairly expensive tools that automatically solve these problems. The tutorial clearly demonstrates when these inconveniences might happen, and what can the user do to easily rectify it using organization skills and patience.

During production I observed that tutorials whose content were limited to software shortcuts, as well as general tips and tricks, were not as helpful as those that contextualized the material being addressed. A suggestion for enhancing this study would be not to necessarily base tutorials on information that one can find easily, but rather to identify situations in which students might encounter critical obstacles whose solutions are not usually decoded by existing tutorials.

Analyzing the Data

Surveys

Recapitulating the information provided on Table 13.1, two surveys on the reason why the target audience looks for audio tutorials were conducted. The difference between both depended upon the level of English fluency from Brazilian participants, as well as the inclusion of Portuguese subtitles in foreign tutorials. Figure 13.4 displays the five specialization areas asked for participants. Despite the low amount of responses from animation, film production, and game design,

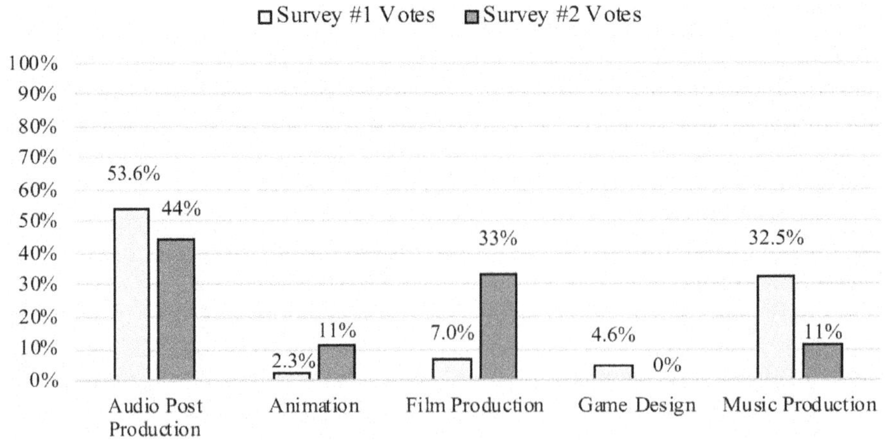

Figure 13.4 Specialization of the participants.

knowing that there could be students and professionals from visual creative areas was relevant for designing accessible audio tutorials that were engaging and easy to follow.

Figure 13.5 illustrates the level of experience of the participants. The student level must be interpreted in the literal sense of the word; that is, for those who did not have professional experience outside of school yet. While the number of Brazilian participants was significantly lower than that of Survey #1, the results indicate that despite the evident disparity between each survey, practitioners ranging from students to those with an intermediate level of experience are more inclined to engage with audio tutorials.

The survey question, "*Why do you look for audio tutorials?*", was open which may explain why there were only 33 responses for Survey #1 and six for Survey #2. While the vast majority of the answers were quite similar (i.e. related to improving a certain skill; to developing new artistic and critical perspectives, and to remaining up-to-date with current technologies), only one answer from both surveys addressed the topmost consideration for developing this project: *accessibility*. Whereas the first belong to Survey #1, the second answer was translated from English to Portuguese by the author.[3]

1. "Manuals are usually incomprehensible, written by experts for experts. I learn by seeing how it is done while having it explained in simple language, in context to the demonstration".—*Audio postproduction student (Survey #1)*.
2. "To assist in and complement presentations of design projects".—*Animation student (Survey #2)*.

From these statements, it is possible to interpret that there is, indeed, a demand for accessible audio tutorials that are easy to follow. The relevance for the response of the animation students correlates with the core aspect of this project: providing

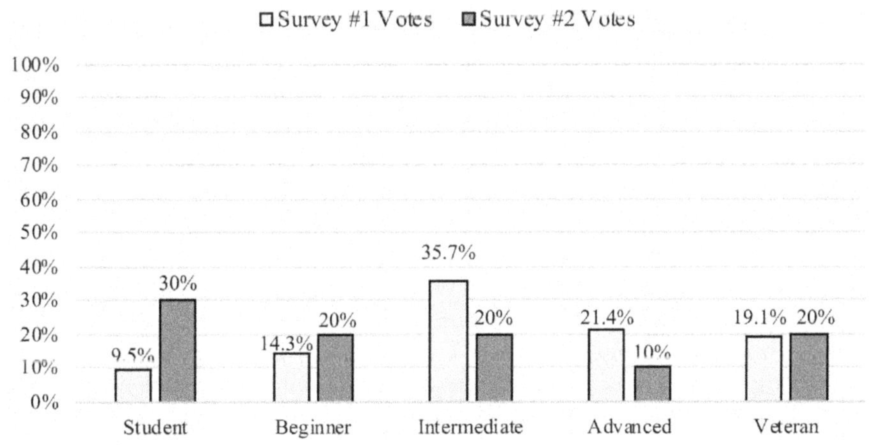

Figure 13.5 Level of experience of the participants.

accessible audio education for everyone, regardless of their creative or technical expertise.

Concerning the state of audio education in Brazil, Dr. Fonteles (personal communication 2018) stresses the importance of making audio education more accessible in Brazil not only to audio students, but for everyone interested in the subject. Indeed, there are some free resources available online, such as the non-profit website *ArtesaosDoSom.org*, which encompasses a plethora of interviews with audio professionals, books, articles, and scholarly material, as well as the mini web series available on Vimeo, "*Operário Criador*", which consists of 19 episodes on the art and technique of film sound.

Although most of the references are based on textbooks and online articles, according to Dr. Fonteles (personal communication 2018) these initiatives have already shown an evolution in accessibility, since Brazilians do not have to necessarily rely solely on English to learn about audio. However, what is currently lacking in Brazilian audio education "is that the available academic knowledge (e.g. Masters dissertations and doctoral theses) be diffused not only within the academy itself, but also in the technical and creative fields" (Fonteles, personal communication 2018). As opposed to assigning reading material to his online students at OZI as part of the curriculum, Dr. Fonteles instead incorporates the academic knowledge and research not only within the lectures, but also in his engagement with students in social media. The classes at OZI are not uniquely concentrated on practice; rather, "research is emphasized as a pivotal contribution to the acquisition of further understanding in storytelling" (Fonteles, personal communication 2019).

In 1946 (later revised in 1954 and 1969), Edgar Dale conceptualized a visual metaphor for explaining the relationship between different modes of learning: the Cone of Experience (Subramony 2002). While online search for the Cone of Experience retrieve colorful and statistical representations[4] a study conducted by Metiri Group argues that the Cone of Experience is constantly misused, as the original essentially portrays "the types of learning, from the concrete to the abstract" (2008: 6). Furthermore, it emphasizes that the inclusion of percentages to designate the efficiency of human learning abilities deviates from the true meaning of the concept originally proposed by Dale. Similarly, Subramony's study (2002) on the Cone of Experience investigates evidence of empirical attempts to prove the accuracy of Dale's concept in designing effective instructional material, arguing that many practitioners have been incorrectly examining its role in education.

In recent years the distribution of audio-visual material has sharply increased due to the popularization of smart devices. Therefore, it would not be wrong to suggest that the most popular learning modalities would indeed be a combination of both visual and auditory. In fact, when interpreting the inaccurate representation of the Cone of Experience, it can be deduced that people tend to learn much more from a combination of seeing and hearing, rather than from each modality alone.

Disregarding the debate on Dale's metaphor, and taking the underlying theory on the audio-visual modes of learning into account, Question 4 focused on asking participants what type of learner they were. Out of 43 participants for Survey #1, and ten participants for Survey #2, results showed a variety of learning preferences among both surveys, which included visual, auditory, reading and writing, and kinaesthetic (see Figure 13.6). While the project was predominantly focused on video-based audio tutorials, understanding the preferred learning types from the audience was relevant for experimenting with different types of tutorials, such as the aforementioned "Sound Design Tutorial: Designing Explosions with Cereal".

Questions 5, 6, and 7 ask participants about their preferred video platform (e.g. YouTube or Vimeo, or both); average lengths for a video tutorial, and the most appropriate way of approaching the content (i.e. by getting straight to the point or explaining the context of the tutorial right at the start). For both surveys, almost all of the participants indicated that they prefer searching for audio tutorials on YouTube—whereas four voted on Vimeo in Survey #1, none voted on the platform in Survey #2.

As illustrated on Table 13.3, the format of the inquiry on average tutorial lengths requested the participants to input their preference on a rating scheme between 1 (not desirable) and 6 (highly desirable), ranging from the shortest to the longest durations possible. The resulting sum accounts for the total amount of votes in each rating, not for the total amount of participants.

While an average of five minutes accounted for the most votes in total, it is interesting to observe the discrepancy between the extremes; the participants indicated that both the former and the latter lengths are more appealing than those between two and ten minutes. These results are rather intriguing since the data retrieved from the average watch time on YouTube, as analyzed later in the chapter, indicates

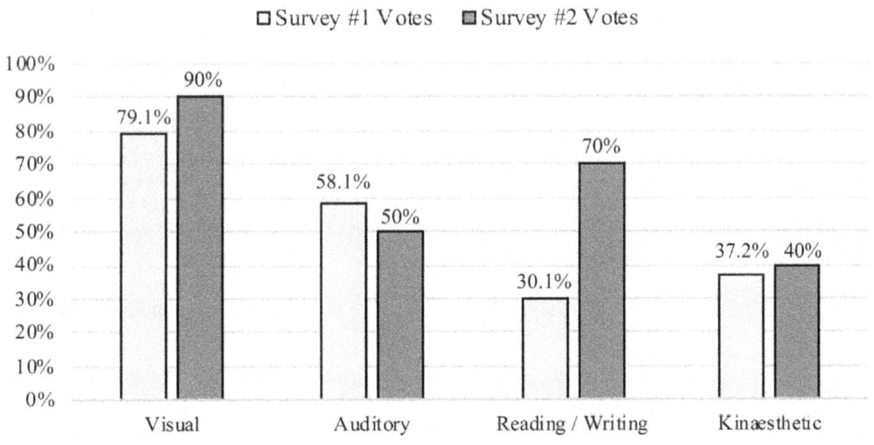

Figure 13.6 Learning type comparison between both surveys.

otherwise. By contrast, the data gathered from Survey #2 resulted in equal preference for most options; that is, Brazilian participants did not vote for the average duration of two minutes. Due to experimental purposes the layout for Question 6 from Survey #2, in particular, was different than that of Survey #1, where instead of having the participants rate their preferences, they would select only one option from those available on Table 13.3.

Table 13.4 illustrates the array of responses from the participants of Survey #1 for Questions 8 and 9. Regardless of the language being spoken (e.g. English in this case), it is always important to consider that it may be the second language of many, or all the viewers; that the audience may have hearing impairment; or simply that they could be in a situation in which watching videos on mute is the only option. While only 8 out of 43 participants (19%) from Survey #1 indicated that subtitles were valuable, this shows that there is indeed a demand for subtitles. Data from Survey #2 is omitted as all participants claimed that Portuguese subtitles would be welcomed in foreign tutorials.

As shown in Table 13.1, Brazilians were asked slightly different questions. Out of 10 participants, 8 replied they had difficulties with English; one is fluent in the

Table 13.3 Rating of lengths for audio tutorials (Survey #1)

Preferred average lengths	*Ratings from 1 to 6*						
	1	*2*	*3*	*4*	*5*	*6*	***Sum***
The shorter, the better.	6	4	4	2	6	9	**31**
2 minutes	2	10	4	7	7	3	**33**
5 minutes	13	1	10	9	3	3	**39**
8 minutes	4	7	7	10	7	2	**37**
10 minutes	4	8	7	2	4	7	**32**
The longer, the better.	7	4	1	5	4	13	**34**
	Number of votes for each rating value						

Table 13.4 Native languages from participants and their need for subtitles (Survey #1)

Language	*Number of speakers*	*How many use subtitles*
Cantonese	1	
Chinese	2	
Dutch	1	8 of 43 use subtitles.
English	30	
Latvian	1	34 do not use subtitles.
Portuguese	1	
Romanian	1	1 skipped the question.
Spanish	4	
Urdu	1	

language, and another mentioned that while it is not an issue, it is important that foreign tutorials have Portuguese subtitles so that those who are not fluent may also have access to a broader range of material. All participants agreed that the inclusion of Portuguese subtitles would enhance their learning experience.

Finally, participants from both surveys were asked to rate which communication skills they relate to the most when watching videos. The options were laid out as multiple choice. From the data presented in Table 13.5, it is possible to interpret that being clear when communicating an idea is significantly more effective than many other characteristics. Secondary to it, enthusiasm and simplicity are also attributes that should be critically considered for creating successful tutorials.

YouTube Analytics

As mentioned earlier in the chapter, in spite of age and gender always being important considerations, their influence was not emphasized when elaborating the surveys. However, data on such topics as well as others were gathered through YouTube Studio, whose analytics are accessible to the creator of the channel. The information from each tutorial was therefore collected and averaged between May 18, 2018, and August 22, 2019, where the former corresponds to the publishing date of the first video and the latter to the day in which this manuscript was last edited.

Data reveal that 44.1% of the viewers ranged between 18 and 24 years old, 22.6% between 35 and 44 years, and 33.3% between 55 and 64 years old; as for gender, whereas male users comprised 83.1% of all the views, female users consisted of 16.9% within the project's timeframe. Table 13.6 displays the total amount of watch time per country, and it is remarkable that the audio tutorials attracted many more Brazilians when compared to audiences from other nationalities.

While the total amount of watch time within the analyzed timeframe accounted for 48,531 minutes (approximately 809 hours) and 20,109 views, the data shown corresponds only to the countries whose statistics were highest. YouTube

Table 13.5 Preferred communication attributes (Surveys #1 and #2)

Options	*Survey #1 (%)*	*Survey #2 (%)*
Appropriate language	30.2	50
Clarity	67.4	100
Conciseness	34.8	40
Enthusiasm	41.9	20
Empathy	11.6	30
Simplicity	41.9	70
Tone	30.3	0
All of the above	23.3	0
	Percentage of votes for each option	

Table 13.6 Amount of watch time and views per country between May 2018 and August 2019

Country	*Watch time (minutes)*	*Views*
United States	7,879 (16.2%)	3,133 (15.6%)
Brazil	1,037 (1.7%)	472 (2.0%)
United Kingdom	688 (1.4%)	280 (1.4%)
India	261 (0.5%)	80 (0.4%)
Australia	191 (0.4%)	76 (0.4%)
Belgium	184 (0.4%)	36 (0.3%)
Russia	160 (0.3%)	97 (0.5%)
Germany	150 (0.3%)	65 (0.3%)
Ukraine	103 (0.2%)	27 (0.2%)
Canada	79 (0.2%)	38 (0.3%)
Austria	69 (0.1%)	25 (0.2%)
Total	**48,531**	**20,109**

unfortunately does not enable an option to view the demographics of all contributing countries.

Complementary to the watch time is the average view duration (AVD) among all tutorials relative to each country. The significance of this data can be interpreted as the amount of time viewers' attention is retained. It does not necessarily indicate, however, that tutorials would benefit from having smaller or longer durations, since it is plausible that viewers might skip to a certain point in the video instead of playing it back from the beginning.

On the contrary, Kim et al. (2014) observed that video tutorials with long durations are prone to having a "higher dropout rate", wherein viewers would stop watching before the conclusion. Many studies (Guo et al. 2014; Hartman-Caverly 2015; Kim et al. 2014; Meij and Meij 2015) have also recommended that video tutorials be limited for up to six minutes as the duration is an indicator of higher engagement rates—Hartman-Caverly (2015) proposed that tutorials should ideally "be capped at three minutes to ensure maximum engagement" (2015: 385).

Indeed, as highlighted on Table 13.7, tutorials with a total duration of six to seven minutes resulted in some of the highest average view durations (near or above 50%). A few outliers ranging between the last-mentioned durations were also successful in obtaining a relatively high AVD.

Despite belonging to a small fraction of the overall results, the amount of watch time with subtitles accounted for almost 19% among 19,500 views. From the data analyzed in Table 13.8, the main issue surrounding the small interaction with Portuguese subtitles lies in the discrepancy between the number of tutorial views from Brazil (see Table 13.6) and the number of views with Portuguese subtitles. However, the explanation for this could be that some Brazilian viewers did not have difficulty in following videos spoken in English.

Table 13.7 Average view duration (AVD) relative to the total duration of each audio tutorial

Tutorial Name	*Duration*	*AVD*
Pro Tools HD - Automation (Part 1)	7:55	2:58 (34%)
Pro Tools HD - Automation (Part 2)	5:32	3:30 (62%)
Pro Tools - Removing Mouth Clicks with the Pencil Tool	6:36	2:50 (39%)
Pro Tools - Useful Bypass Shortcuts	1:50	1:14 (76%)
Pro Tools - User Default Plugin Settings	3:21	1:09 (34%)
Pro Tools Quick Tip - Audiosuite Handles	2:45	1:37 (56%)
Pro Tools Quick Tip - Cataloguing your SFX Library	1:36	0:55 (40%)
Pro Tools - Useful Fade Shortcuts	3:36	1:46 (43%)
PaulStretch - The Extremes of Sound Stretching	9:37	2:33 (25%)
Film Dialogue Mix Template - Layout and Routing	6:33	2:30 (36%)
Film Dialogue Mix Template - Inserts, Sends, & Console	8:28	3:41 (41%)
2 Ways to Record Scrubbing in Pro Tools	7:24	1:46 (20%)
Sound Design Tutorial: Designing Explosions with Cereal	5:45	2:28 (42%)
The Kontakt Footstep Instrument	6:17	2:07 (34%)
Creating Room Tone with Convolution Reverb	8:07	2:47 (31%)
Pro Tools Quick Tip - Custom Shortcuts	2:13	1:26 (59%)
Pro Tools Quick Tip - The Slip-Grid Mode	1:14	0:57 (50%)
Working with Video in Pro Tools	7:55	3:26 (43%)
Dealing with Low-Level Dialogue	5:42	3:14 (58%)
Pro Tools - Clip Trim and Snap Shortcuts	6:21	2:10 (34%)
Pro Tools Quick Tip - Moving Clips with Markers	2:22	1:21 (55%)
Applying Metadata with Basehead	6:48	3:13 (48%)
Introduction to Kontakt (Beginner Friendly)	12:07	3:50 (29%)
Track Presets in Pro Tools 2018	5:17	2:24 (43%)
iZotope Iris 2 - Designing Grenade Build-Ups	7:11	2:51 (35%)
Film Sound - Editing Dialogue with Alternate Takes	9:13	2:13 (23%)
Exploring the GRM Tools for Sound Design	10:31	2:25 (22%)
Create a Powerful Bass Drop in 2 Minutes!	4:36	3:07 (70%)
Audio Repair - Fixing Wind and Distortion	9:04	2:47 (27%)
Designing Tracer's Blink Ability with Everyday Objects	5:37	2:26 (42%)
Getting Started with Reaper - Mac and PC	5:55	1:54 (28%)
Using Wwise for Linear Media	7:37	3:32 (45%)
Designing Slow Motion Fire with Water	5:48	2:48 (45%)
Designing Sound with Character in Mind	6:49	3:04 (47%)
Waterproofing a Shotgun Microphone	2:05	1:13 (55%)
Film Sound - How to Edit Ambiences	8:52	3:02 (35%)
Creating Birds Songs with Speech	2:31	0:50 (22%)
Smartphone Field Recording	4:33	1:47 (34%)
Film Sound - Conforming to an Edit Decision List (EDL)	9:32	2:52 (27%)
The Story of a Sound	5:03	1:52 (30%)

Table 13.8 Amount of watch time with subtitles/closed captions (CC)

Subtitles	*Watch time (minutes)*	*Views*	*AVD*
No Subtitles/CC	38,885 (80.1%)	16,404 (81.6%)	02:22
English	8,084 (16.7%)	3,011 (15%)	02:41
Portuguese	1,566 (3.2%)	692 (3.8%)	02:15
Total	**48,531**	**20,109**	**02:24**

Conclusion

Despite successfully attracting Brazilians to educational material that would not otherwise be as easily obtained, it is important to recognize the limitations of this study; maintaining the essence of knowledge diffusion by making the tutorials in English contributed somewhat negatively to the development of an educational resource solely designed for Brazilians. Indeed, the addition of Portuguese subtitles did prove to be a welcome feature by the target audience, although perhaps the study would have been even more beneficial toward Brazilians if the material had been designed strictly for them.

Being that accessibility is the focus of this project, it is possible to consider the research as a modest success for providing online tutorials in audio education. However, the current stage of distance education in Brazil was not taken into account during production, as the focus was instead on researching efficient and effective methodologies for designing accessible tutorials. Therefore, a more thorough assessment of the demographic demand for such content is likely needed for future research.

Nevertheless, the relationship of language-specific online tutorials with distance education may be derived from the delivery of educational material; although subtitles fail to bridge the gap in terms of establishing a strong connection with the learner. Given this issue, it can be argued that online audio education that is exclusively based on video tutorials could be classified as difficult to quantify, given that the student-teacher relationship plays a significant role in the learning process.

Gerbase argues that "the student wants to be seduced by information, to be urged to venture into the world of knowledge, not to be led by the tedious images and sounds that compose a meaningless experience" (2006: 2). Drawing from this perspective, Dr. Fonteles has found it beneficial to the distant student-teacher relationship at OZI that personalized video conferences between teacher and learner be made to supplement the online learning experience (Personal Communication 2019). If traditional education is to evolve from the physical format to the online and distance-learning format without suffering any language and aesthetical disconnections, educators and course developers must find ways to balance theoretical and practical learning such that students remain engaged.

As with many media, distance education requires a language of its own to be as efficient as in-person education—video, for example, requires its unique communication approach to successfully retain the viewer's attention. The complexity of creating effective audio tutorials for audiences with different specializations requires an understanding of, and experimenting with, a variety of communication methods.

Concerning the first two questions proposed earlier in the chapter, data suggest that tutorials that are delivered in both concise and clear manners are inclined to be more effective. In fact, the supplementary reading material for the "Sound Design Tutorial: Designing Explosions with Cereal" video contributed positively to a better distribution of content, meaning that the tutorial could be smaller and more focused. Similarly, the aforementioned well-received tutorials introduced were equally successful in addressing the first question by briefly contextualizing the way the material was presented.

Given these positive results, can affective learning be possible even with isolated videos that do not belong to a course? Affective learning represents the attitudes students develop about the course, the topic, and the instructor (Russo and Benson 2005). The current state of the YouTube channel does not adequately support the argument by LaRose and Whitten (2000), in that the interaction between the instructor and the students is pivotal for developing affection. Further, while the study focused on engaging with Brazilians, much of the research emphasized producing accessible tutorials that would appeal to both English and Portuguese speakers.

Russo and Benson comment on the relevance of immediacy to improve affective learning. According to the authors, immediacy is understood as "the examination of presence in online learning" (2005: 55). Moreover, Anderson (1979, cited in Russo and Benson 2005) defines immediacy as a function of the psychological distance that communicators put between themselves and the object of their communication. Despite not engaging with the audience earlier, it is not impossible to develop a strong personal connection with isolated educational videos. Indeed, for educators developing material without an existing course, the challenge is much higher.

Accordingly, this study sought to better understand how Brazilians respond to and connect with quality audio education derived from common issues encountered by sound editors and designers. Whereas the first few tutorials were devoted entirely to software shortcuts, the subsequent videos attempted to approach the desired outcome by working on the most efficient communication methods, as specified by the participants from both surveys. Moreover, data retrieved from YouTube gathered important information necessary to consider expanding such efforts. Future research might consider initiatives that would be devoted solely toward producing audio-visual content for Brazilians. While the addition of subtitles demonstrated to be an effective measure, perhaps creating a learning environment exclusively delivered in Portuguese would be more advantageous to the target audience.

The practical perspective on inclusion springs from the possibility of facilitating the understanding of audio and its meaning in visual media among diverse learners. Whereas the first two questions (respective to the purpose of the study) have been successfully answered, further investigation must be conducted in order to address questions three and four. Despite the simplicity in which the ideas are communicated, data revealed that the content ended up being more directed toward audio students and professionals rather than the whole creative spectrum.

Lastly, the challenge in moving forward lies in examining efficient ways to not only clearly communicate the relevance of audio as a storytelling device, but also to expand its significance in other professional areas. Devoting subsequent material exclusively to Brazilians will be essential for amending an important gap within audio education in Brazil, with a view to encourage the pursuit of a career in the field, as well as approximate them even more to knowledge that currently appears to dwell within pages of seemingly unattainable resources.

Notes

1. The term was conceptualized by Cinelli.
2. The website's data analytics were disabled from 2018 onwards to improve the server's central processing unit (CPU) usage.
3. Original text in Portuguese: "Para auxiliar e complementar apresentações de projetos de design".
4. For examples of the misused versions, please search for the Cone of Experience on Google Images. The original version can be found on Dale (1946, 1954, 1969), and on Metiri Group (2018).

References

Canadian Media Producers Association (2017). "Profile 2017: Economic Report on the Screen-Based Media Production Industry in Canada", *CMPA*, viewed 5 May 2019, www.primetimeinottawa.ca/wp-content/uploads/2018/02/Profile-2017.pdf

Cinelli, N. (2003). "A influência do Vídeo no Processo de Aprendizagem", *Masters thesis*, Universidade Federal de Santa Catarina.

Dale, E. (1946). *Audio-Visual Methods in Teaching*. 1st edn, Dryden Press, New York.

Dale, E. (1954). *Audio-Visual Methods in Teaching*. 2nd edn, Dryden Press, New York.

Dale, E. (1969). *Audio-Visual Methods in Teaching*. 3rd edn, Dryden Press, New York.

Gerbase, C. (2006). "Desafios na construção de uma estética audiovisual para educação à distancia (EAD)", *Logos 24: cinema, imagens e imaginário*, Universidade Estadual do Rio de Janeiro.

Gil, J.M., & Williams, V. (2017). "Byte-Sized Learning: A Review of Video Tutorial Engagement in a Digital Media Skills Course", *Teaching Journalism and Mass Communication*, 7 (1), 14–21.

Guo, P.J., Kim, J., & Rubin, R. (2014). "How Video Production Affects Student Engagement: A Empirical Study on MOOC Videos", *Proceedings of the First ACM Conference on Learning @ Scale (L@S '14) Conference*, Atlanta, March, 41–50, doi:https://doi.org/10.1145/2556325.2566239

Hartman-Caverly, S. (2015). "Brevity, Complexity, Availability: Extrapolating Patron Preferences from Video Analytics", *College & Research Libraries News*, 76 (7), 384–386, doi:https://doi.org/10.5860/crln.76.7.9348

Khan, S. (2015). "Let's Teach for Mastery—Not Test Scores" [online video], viewed 2 April 2019, www.ted.com/talks/sal_khan_let_s_teach_for_mastery_not_test_scores

Kim, J., Guo, P.J., Seaton, D.T., Mitros, P., Gajos, K.Z., & Miller, R.C. (2014). "Understanding In-Video Dropouts and Interaction Peaks in Online Lecture Videos", *Proceedings of the First ACM Conference on Learning @ Scale Conference (L@S '14)*, Atlanta, March, 31–40, doi:http://dx.doi.org/10.1145/2556325.2566237

Kottke, J. (2017). "The World's Best Film School If Free on YouTube", *Wired*, viewed 17 April 2019, www.wired.com/story/youtube-film-school/

LaRose, R., & Whitten, P. (2000). "Rethinking Instructional Immediacy for Web Courses: A Social Cognitive Exploration", *Communication Education*, 49 (4), 320–338, doi:https://doi.org/10.1080/03634520009379221

Mayer, R.E. (2009). *Multimedia Learning*. 2dn edn, Cambridge University Press, New York.

Metiri Group (2008). "Multimodal Learning Through Media: What the Research Says", *Cisco*, San Francisco.

Music Victoria (n.d.). "Victorian Music Fast Facts", viewed 5 May 2019, www.musicvictoria.com.au/about/victorian-contemporary-music-industry-fast-facts

Rodrigues, A. (2017). "Por que ainda não somos fluentes em inglês?" *Exame*, viewed 23 April 2019, https://exame.abril.com.br/carreira/por-que-ainda-nao-somos-fluentes-em-ingles/

Russo, T., & Benson, S. (2005). "Learning with Invisible Others: Perceptions of Online Presence and Their Relationship to Cognitive and Affective Learning", *Educational Technology & Society*, 8 (1), 54–62.

Silveira, D. (2017). "Brasil tem mais de 207 milhões de habitantes, segundo IBGE", *Globo G1*, viewed 23 April 2019, https://g1.globo.com/economia/noticia/brasil-tem-mais-de-207-milhoes-de-habitantes-segundo-ibge.ghtml

Subramony, D.P. (2002). "Dale's Cone Revisited: Critically Examining the Misapplication of a Nebulous Theory to Guide Practice", *Proceedings of the 2002 National Convention of the Association for Educational Communications and Technology*, Dallas, TX, 423–428.

van der Meij, J., & van der Meij, H. (2015). "A Test of the Design of a Video Tutorial for Software Training", *Journal of Computer Assisted Learning*, 31 (2), 116–132, doi:https://doi.org/10.1111/jcal.12082

Zottmann, V. (2018). "A Study on Accessible Audio Education", *Designing Sound*, viewed 8 May 2019, http://designingsound.org/2018/07/31/a-study-on-accessible-audio-education/

14

Future Educational Goals and Actionable Items

Teaching Communication Skills for Audio Education Institutions and Educators

Doug Bielmeier

Introduction

Over the last two decades I have been a student, professor, and program coordinator at several types of audio education programs, as well as a practitioner in industry as a staff engineer, mix engineer, producer, A1-A2 live sound engineer, and an AV/Installation consultant. When I reflect on how my education prepared me for my career, I am reminded of many occasions when I felt deficient. However, these deficiencies are not related to missing technical skills or a lack of musical knowledge. Instead, I find myself regretting occasions when I had a hard time dealing with clients/co-workers and as a result misspoke, communicated poorly, became frustrated with clients, lost a gig/client, or never got the job in the first place. My education did not explicitly teach these skills and as a result, I learned them *on-the-job* or *client-by-client*. As a result, I often wonder how my career may have been different or what opportunities I may have missed because of my communication skills. I have discovered over the years that my students also suffer from the same apprehension with these social and communication skills. As an educator and researcher, I am committed to understanding the deficiencies and strengths in educational programs to better prepare students for industry.

In my classroom, I often ask students the dreaded question: what do you do when the client is singing out of tune? For the average student, understanding how to socially navigate the situation often eludes them. Many students flounder when asked which action to choose: (a) tell the singer they are out of tune, (b) fix it without telling the artist, or (c) redo the performance knowing it cannot be fixed. This often sparks healthy conversations about how to approach the issue and the skills needed to tactfully elicit change. After a guided discussion, the students soon conclude that their musical and technical skills are useless without being able to navigate the social component of this common recording studio situation.

This chapter identifies research and developments of the last decade to identify major areas for improvement within audio education. It will not only attempt to answer the when/how to learn/teach communication skills but also addresses the myriad factors further complicating the acquisition and application of these skills. Thus, this chapter offers specific and actionable items for both educators and administrators for use at their institutions and in their classrooms.

Research and Developments

Research and developments over the last decade have identified four major areas for improvement within Audio Education: (a) fostering communication skills, (b) aligning industry and audio education expectations, (c) enabling apprenticeship skills, and (d) establishing situated-learning environments. Understanding these areas of needed improvement is the basis for the audio education innovations discussed later in this chapter.

Fostering Communication Skills

Defining what makes audio engineers good at their jobs can be described in many ways, such as technical proficiency, working well with clients, and business prowess. To better define these skills for educators, Tough's research identified 160 skills employers desired from audio students (Tough 2009, 2010). More than half of these skills focused on communication skills. The top five ranked core competencies were (a) demonstrate the ability to work hard and complete projects, (b) recognize the need for personal responsibility, (c) demonstrate the ability to be an effective listener toward co-workers and clients, (d) demonstrate the ability to communicate clearly and tactfully with clients and co-workers, and (e) develop the ability to be professional around clients. Collectively, these skills can be termed as employability skills and development of these skills in aspiring engineers is critical to their success after graduation.

Historically, communication skills are highly revered in the audio industry. Leading professional engineers identify personal skills and presence as key skills needed to be effective in the studio (Jackson 1988) and some engineers feel that people skills are of greater importance than technical skills (Lambert 1989). For example, giving coherent directions to clients can be important for a good collaboration process that keeps clients happy and working hard (Rubel 2007). Furthermore, fostering good client collaborations can be especially challenging in an industry where clients may have big egos and studios may have political hierarchy (Sanders 1994). More seasoned engineers, producers, or musicians feel that well-developed communication skills are what sets them apart from individuals who are new to the business (Lambert 1989).

In the contemporary industry, communication and employability skills are even more highly valued by employers. Employability skills include professional approach, equipment knowledge, troubleshooting, presentation/pitching, human interaction, and networking (Bragança 2015). The emphasis on good communication and employability skills is due to the increased (a) importance of client interaction, (b) responsibilities of engineers/producers, and (c) diversity of roles in the music industry. The importance of fostering positive client interaction is important as studio sizes have reduced and freelance working has increased. The primary role of the producer/engineer is to establish an environment that is conducive to creativity and allows musicians to flourish (Vandemast-Bell et al. 2015).

Engineers must give and receive clear and helpful feedback to colleagues and clients throughout creative projects (Lefford & Berg 2013). Engineers who demonstrate strong communications skills can build and retain client lists and, thus, may have an advantage in the competitive music industry (Carugo 2015). The type and amount of responsibility for engineers/producers has grown beyond the music alone. Successful engineers must excel at self-management, team working, business and customer awareness, problem solving, communication and literacy, and application of technology (Thompson & Mosley 2015). The diversity of roles in the industry has changed to include audio for film and video games. In these expanded roles, the audio engineer may be responsible for on-set sound recording, which requires communicating with (a) the producer for call times and potential location obstacles; (b) the director about the aesthetics of how the audio should be captured; and (c) lighting, wardrobe, and camera departments about practical setup and placement of microphones and recording equipment during production (Lankford & Schwartz 2018).

Despite this demand for good communication and employability skills, these topics are often overlooked in audio education as it can be complicated to recreate this industry experience in the classroom (Bragança 2015). Furthermore, assessing communication skills competencies are often the most difficult to access within the audio classroom due to their subjective and qualitative use and evaluation (Lightner 1993; Sanders 1994; Walsh 1996; Day & Koorland 1997; Paulson 2001; Patterson et al. 2002).

To develop an approach to incorporate communication skills into the classroom, it is important for educators to understand how and where these skills and core competencies are learned to design curricula that can foster mastering these skills. The New Hires Survey (NewHS), an independent research project, asked new hires (or recently graduated people) in the audio industry what skills they have and where they were learned (Bielmeier 2016). Not surprisingly, the New Hires reported that they had mastered technical skills by learning them at their audio programs. However, more than half of New Hires reported that they learned key social and communication skills (e.g. patience in the studio, professionalism with clients, and effective listening) on-the-job or on-their-own. These findings may suggest that there is room for improvement of teaching communication skills in the classroom.

Educators have several suggestions as to how to include communication skills in educational programs: (a) extracurricular activities (Carugo 2015), (b) the partnering method (Rodgers 2014), (c) the purposeful use of social media in coursework, and (d) interdisciplinary team interaction (Nixon et al. 2015; Lankford & Schwartz 2018). While these independent activities help students gain some proficiency in these communication skills, an overall educational approach has yet to be developed. This larger strategy must be conceived, developed, and implemented at the classroom and institutional level with the collaboration of stakeholders, students, teachers, and local industry.

Industry and Audio Education Relationship

Research shows that there is a feeling among music industry experts that a large gap exists between industry needs and university requirements (Vandemast-Bell et al. 2015). This feeling is supported by evidence that suggests most recent graduates do not have the specific talents required to meet the demands of the contemporary Creative Industries (Thompson & Mosley 2015). Specifically, there is a mismatch between the courses required by universities and the skills graduates need to succeed and grow in the music industry (Vandemast-Bell et al. 2015). Unfortunately, some industry professionals believe that experience with music, people, and time management cannot be learned in a laboratory or educational setting and are best learned from direct interaction with a *master craftsperson* for several years (Huws 2017). This strained relationship between industry and audio education has made aligning efforts of both parties challenging.

The inclusion of industry activities into education is not a new concept. Since the 1980s, audio education programs have tried to teach professional attitudes and increase awareness of what goes on day-to-day in the world of audio (Hirsch 1985). Some argued that audio education programs had failed to teach social skills and ways to deal with clients (Jackson 1988; Lambert 1989). Furthermore, many recording programs were not addressing these skills and consequently, were failing to fully prepare their students to meet the needs and desires of industry (Lambert 1989). These assertions and complaints spawned research into audio education and curriculum in the early 1990s. Research found employers complaining that many graduates had unrealistic expectations for interaction, weak customer service skills, and lacked communication skills (Lightner 1993). Similar research indicated that the ability to work well under stress, be an astute observer, be easy to work with, and have a sense of humor are paramount for aspiring engineers (Sanders 1994). Yet another study found that educators and employers agreed that customer relations and studio protocol were most important (Walsh 1996).

Research in the last decade examines the industry and audio education relationship and ways to incorporate industry needs into the classroom. Tough (2009) surveyed a panel of recording engineering experts to identify skills that students graduating from a four-year audio programs should have. The results mirrored the past research because the top ranked skills were communication skills. Bielmeier (2013a) created the Audio and Recording Employer Survey (ARES), which asked employers to indicate their New Hire's level of proficiency of the core competencies identified by Tough (2009) (Bielmeier 2013b). The ARES results suggested that employers were critical of skills obtained by New Hires and lacked confidence in their audio program training. Furthermore, employers reported that the studio did not resemble the classroom and suggested internships or on-the-job training are the best ways to develop and master these skills. Thompson and Mosley (2015) interviewed industry professionals in the UK, who emphasized the importance of a more industry-focused curriculum to help prepare graduates for work in the Creative Industries and argued as to the necessity of earning a degree to gain employment in the industry. Similarly, Bielmeier (2017) conducted a that surveyed audio education students to ascertain which components of their education were missing (Bielmeier 2017). The results suggested that the students wanted more focus on business and interpersonal skills dealing with clients and employers in the industry.

Fortunately, this decade's worth of research has led to a stronger relationship among research, education and industry, and resulted in a more fully developed conceptual basis for audio engineering practice than ever before. However, audio educators deal with myriad issues when attempting to educate aspiring engineers of the future (Lefford & Berg 2013).

The challenges facing audio engineering education parallel problems encountered across engineering disciplines (Lefford & Berg 2013). Specifically, industry professionals feel that engineering students are technically adept, but unable to respond to real-world engineering problems (Lefford & Berg 2013). In response to this criticism, several collaborative academic initiatives have been implemented to foster problem solving in authentic real-world environments. However, some educators feel that the constraints of university education in terms of timetabling, modular structures, teacher-led learning, and a focus on tools has weakened the focus on real-world problem solving (Lefford & Berg 2013). Furthermore, as university programs replaced apprenticeships used by previous generations, the focus moved away from developing social, aesthetic, and interpersonal skills (Davis & Parkers 2013; Patryk 2012). A realignment with industry goals may better prepare students for industry after graduation.

Another aspect of audio education to consider is the desires of the student. Current research suggests that the contemporary student has different learning expectations from previous generations. The contemporary student processes information differently and expects learning to be meaningful, diverse, and not based around lectures or slide presentations (Kirschner et al. 2011; Prensky 2010). Furthermore, contemporary college students demand learning that provides real-world skills that are significant and prepares them for future employment (Rodgers 2014).

Audio education of future graduates must keep up with changes in the industry, technology, and employment needs, and thus, audio students can benefit from a wide, multidisciplined curricular approach (Lankford & Schwartz 2018). Research conducted by Tough (2010) suggests that future audio graduates should have a multidisciplinary education that focuses on communication, leadership, troubleshooting and electronics, and be balanced or take priority over theoretical, business, and MIDI content (Tough 2009, 2010). Today's colleges and universities have the unique opportunity to close the gap between audio industry and education by providing students with sound technical understanding while creating an authentic real-world experience by working with artists, engineers, producers, and record labels in their community (Vandemast-Bell et al. 2015).

Apprenticeship Skills

In the current music industry, the roles for audio engineers are diversifying to include audio for games, apps, and plug-in development. However, at the same time the modes and contexts in which aspiring engineers can learn is decreasing. Specifically, long-term, formal and informal apprenticeships are disappearing university audio educational programs in the UK and US are growing (Lefford & Berg 2013; Thompson & Mosley 2015).

Prior to the 1970s, most recording engineers learned skills solely by apprenticeship or on-the-job training (Ashley 1973) and apprenticeships provided a direct gateway to employment in the recording industry (Davis & Parkers 2013). Apprenticeships taught important technical and interpersonal skills in a studio setting. Apprenticeship programs became less common as formal audio programs were established at universities and specialized training programs in the 1990s. The reduction of apprenticeships was caused by (a) miniaturization of the recording studio by digital audio workstations and computer advancements, (b) cost for studios to train engineers, and (c) streamlined recording process (Liston 1975; Strawn 1997; Tough 2009).

During the transition from apprenticeship to university education, the hope was that the apprenticeship system would not be lost, but rather transformed via internship (Kenny 1990). Unfortunately, a 2014 survey of audio engineering educators in the US found only 38% of audio programs required internships (Bielmeier 2014). Furthermore, a 2013 survey of audio industry employers suggested that many employers felt recent graduates are unprepared socially and often are poorly suited to work in the studio environment (Bielmeier 2013a). These findings suggest internships alone are not preparing students for employment in the audio industry and better understanding the value of apprenticeships may be helpful.

The contemporary view of apprenticeships and internships asserts the importance of allowing aspiring engineers to learn studio operations (social and technical) by observation, application, and then problem solving (Davis & Parkers 2013; Lefford & Berg 2013). While the apprenticeship experience may only expose aspiring engineers to a limited number of skills, this system can provide mentees with a foundation for learning more complex skills later on (Davis & Parkers 2013; Lefford & Berg 2013). Unfortunately, data does not confirm that apprenticeships offer better retention of audio skills as compared to classroom work nor provide improved employment rates after graduation (Jewson 2013). In fact, only half of apprentices achieve the prescribed qualifications to become audio engineers (Jewson 2013).

Furthermore, little research exists on how audio education programs prepare students for apprenticeships or internships. The Educator Survey (ES) aimed to determine whether students were being taught the ability to be mentored (Bielmeier 2017). The ES surveyed audio educators as to the skills focused on in their classrooms, departments/institutions, and internship programs. Educators of the survey reported that the ability to work well under the authority of a mentor was a skill not focused on in their classroom or institution. This deficiency is compounded by a lack of institutions requiring internships.

Contemporary educators maintain that the apprenticeship system has been transplanted into the classroom (Lefford & Berg 2013; Carugo 2015; Nixon et al. 2015; Fick 2015). These classrooms reflect the apprenticeship system by employing (a) cognitive apprenticeships (Lefford & Berg 2013), (b) problem-based learning (Fick 2015), (c) industry mentorship (Carugo 2015), and (d) situated classrooms (Nixon et al. 2015). These methods provide structure, coaching, and mentorship found in the traditional apprenticeship model. Also, the contemporary classroom emphasizes peripheral participation and situated learning, which were an essential part of the apprenticeship process (Davis & Parkers 2013).

Situated-Learning Environments

The concept of creating situated-learning environments in which students learn and work in a classroom that is authentic to industry is relatively old (Lave et al. 1991). Situated-learning environments are the foundational theoretical framework for instructional methods of (a) self-directed learning (Hiemstra 1994), (b) problem-based learning (Savery & Duffy 1995), (c) inquiry-based learning (Oliver 2008), and (d) project-based learning (Bell 2010). These instructional methods include authentic scenarios, communication, and outcomes of coursework to mirror that of industry (Herrington 2006).

Contemporary thought in audio education concurs that situated learning is an essential part of the apprenticeship process and learners immersed in situations benefit from the inherent social complexity (Davis & Parkers 2013; Lefford & Berg 2013; Snyder 2016). Situated learners can see application of theory in real-world environments while exposing them to the pitfalls and triumphs of studio work (Carugo 2015; Vandemast-Bell et al. 2015). Nevertheless, incorporation of situated learning environments remains a challenge in many classrooms.

The incorporation of situated learning (that is authentic to industry) into classrooms and coursework is highly debated. Some educators challenge the traditional assertion that the creation of state-of-the-art facilities alone will create authenticity (Snyder 2016). Institutions with this traditional belief tend to tout multi-million-dollar studio spaces. Conversely, other educators posit that authentic learning should incorporate (a) collaborative partnerships, (b) relevant unpaid work experience, and (c) real-world audio problems early in the degree process (Bragança 2015; Carugo 2015; Fick 2015). Institutions with this focus tend to showcase cooperative educational programs within the local audio industry.

Innovations for Audio Education and Actionable Items

Several innovations in audio education programs, curricula, and culture can be suggested from the review of this decade's research and developments into the acquisition and development of communications skills for graduating students entering the industry. These innovations should be fostered at the institution and classroom level and be supported by ongoing research. The following sections provide a framework of actionable items for innovation in institutions, classrooms, and research.

Institutional Innovation

Audio educational institutions have the largest impact on the pace and direction of audio education. Thus, it is important for institutions to discuss how their audio programs are preparing students for work in the industry. Specifically, institutions should (a) form curriculum committees, (b) integrate local and regional industry, (c) establish student mentorship, and (d) require internships. Table 14.1 describes the actionable items for innovation at audio education institutions and summarizes the reason for implementation and tasks required to achieve the innovation.

Formation of Curriculum Committee

A curriculum committee is a group of audio educators and administrators that work to continually update the audio education curriculum for an institution and ensure students receive the most effective and relevant curriculum for the practice of audio recording. Therefore, audio engineering curriculum committees should assess their institution's focus on communication skills to address deficiencies found in research of audio educational institutions (Bielmeier 2013a, 2013b, 2016; Tough 2009). Specifically, audio educators should address communication skills emphasized in their classroom and administrators should identify communication skill goals for their program. Furthermore, the entire committee should consider the inclusion of other work place skills, such as social interaction, business, and marketing.

Table 14.1 Actionable items for innovation at audio education institutions

Actionable item	*Reason*	*Task*
Form curriculum committees	Address communication and employability skills taught	• Review accreditation standards and audio education curricula • Align communication skills with specific courses • Update learning objectives/outcomes with communication and employability skills
Integrate local and regional industry	Align expectations of industry and audio education institutions	• Identify local industry that align with courses and curricula • Leverage educator and student current relationships • Student involvement in national groups, institutions, and contests
Establish student mentorship	Engender apprenticeship skills	• Identify mentoring experts: educator, employers, and graduates • Leverage existing partnerships with audio education institutions

Another role of the curriculum committee is to perform a complete review of accreditation standards, program of study, and individual course content and learning objectives/outcomes. This institutional review should result in meeting all accreditation standards. Accreditation bodies act as an intermediary between formal education and industry aiming to ensure that graduates learn the necessary skills and knowledge in their education based on the needs of the industry (Thompson & Mosley 2015). Unfortunately, accreditation standards are not universal.

Accreditation standard approaches vary by country. In the UK, there are two main audio educational accreditation bodies: Joint Audio Media Services (JAMES) and Institute of Engineering and Technology (IET). The JAMES accreditation is supported by a large consortium of professional associations that include (a) Association of Professional Recording Services; (b) Music Producers Guild; (c) Music Managers' Forum; (d) Professional Lighting and Sound Association; (e) UK Screen Association; (f) British Academy of Songwriters, Composers and Authors; and (g) Interactive Audio Special Interest Group (JAMES 2019). The IET accreditation covers wide range engineering accreditations beyond the music industry (IET 2019). Alternatively, the US has a wide variety of accrediting bodies at the reginal and national level, which vary in scope (e.g. engineering and music) or educational rigor (e.g. professional, certificate, two-year programs, or four-year programs). The National Association of Schools of Music (NASM) is the accrediting body for some four-year music-school based audio education programs. NASM's standards suggest that students work in a situated learning environment and the acquisition of communication skills is learned as a result of their implementation (NASM 2019).

Reviewing accreditation standards and comparing them to an institution's curriculum has several benefits, such as identifying bias and deficiencies. Identifying technical bias in an audio program may be the most important aspect of the review (Bielmeier 2013a, 2013b, 2016, 2017). Change in curriculum can be large institutional changes or small changes to classroom projects or procedures. To address curricular deficiencies, learning objectives can be met by creating new courses or modifying existing course objectives. For existing courses, modifying what is evaluated in current projects may be warranted. For example, communication skills acquisition is inherent in technical or technology-based learning objectives, but are often not focused on or graded. Therefore, a recording project can be created that requires students to find recording talent for a vocal overdub, and the emphasis (i.e. points or credit of the final project grade) should be given to the process by which the student locates the vocal talent and the student's interaction with talent. This can be evaluated by using client surveys, employing student reflection journals, and/or observing student client interactions during production meetings (Herrington 2006; Hiemstra 1994; Lankford & Schwartz 2018).

Local and Regional Industry Integration

Integrating institutions with the local and regional industry is paramount to address the misalignment of expectations and perspectives of employers, audio institutions, and their graduates as indicated in this decade's research (Bielmeier 2013a,

2013b, 2014, 2016, 2017; Tough 2010). The ease and prevalence of such integration is more likely in urban areas or audio departments within music industry programs. However, an important factor is the quality of these local collaborating businesses. Access to industry standard recording studios and other audio facilities may be difficult for students attending a rural vocational audio program. Therefore, these programs should concentrate on aligning themselves with regional or even national associations so that these industry partnerships are representative of their graduates' future job market.

When identifying local, regional, and national collaborators, institutions can focus on the relationships and associations already held by educators and students. Educators can be a great source of industry exposure and when institutions hire new full-time and/or adjunct faculty, a candidate should be evaluated on their potential for industry collaboration or tangential industry activities to engage students. Schools can foster this industry connection by incentivizing and prioritizing this collaboration into the promotion and tenure process of their employees. Industry participation may include (a) attending industry conventions, (b) sitting in on local studio sessions, and (c) facilitating entire courses that are held off campus in the real world.

Students already working in the industry may also provide a source of collaboration, and institutions should leverage their connections and relationships to aid in coursework and departmental events. It is quite common in many institutions to bring successful alumni to teach sessions for current students and give master classes. However, this requires universities to have a robust alumni association and a specific point person for alumni relations and development. Furthermore, institutions should encourage students' involvement in industry-based groups or associations. The Audio Engineering Society (AES) allows students to form local chapters at their local institution and attend meetings/events locally, regionally, or nationally (Audio Engineering Society 2019). Unlike the university or faculty created industry connections, these groups allow students to form connections with their specific career goals in mind. The nature of these groups is that they can be independent from the institution's own agenda. In addition, AES provides funding, design contests, and opportunities for distinction in both technical productions, mix contest, and research presentations at their international conventions (Audio Engineering Society 2019). AES conventions provide students with opportunities to network with peers, industry experts, educational authorities, and employers from current hardware manufactures/designers. Clubs, fraternities, and other student groups should also be encouraged; however, the institution should encourage these groups to engage in purposeful interaction with the local, regional, and nation industry.

Establish Student Mentorship

To engender mentorship and develop apprenticeship skills within students, institutions should consider the formation of a mentorship system. This is imperative as the research of this decade indicated a lack of focus by educators and institutions on the apprenticeship skills employers desired of their new hires (Bielmeier 2013a, 2013b, 2014, 2016, 2017; Tough 2010). Most institutions have a student advising system, which is different from a mentorship system. Advisors are often assigned many students and do not have audio industry experience. The focus of these advisors is to ensure students complete all courses in their majors and graduate.

Mentors, instead, are experts in the field of audio with similar career aspirations as their mentee and may have industry experience within a specific field of audio. Mentors can be professors or staff at the institution, graduate students, higher-level students, and local or regional industry members (Davis & Parkers 2013). Mentors should be assigned only a few students/mentees as the focus is on meaningful interaction and intervention in the mentees' education and career development (Lankford & Schwartz 2018). The purpose of these mentorships is to develop students apprentice skills, while preparing them for their role in an internship or co-op.

Contemporary educators tout the merits of mentorship citing its ability to allow students to apply practical skills while working in positions of responsibility with confidence, knowing there is support (Carugo 2015; Lefford & Berg 2013). Furthermore, research suggests teachers and industry experts feel mentorship is the foundation of problem-based learning and is imperative for recreating situational learning environments (Fick 2015). However, research from Huws (2017) warns that teaching and mentoring should not solely focus on technical elements and should include employability skills.

Require Internships

Internships, co-ops, and/or work study should be required by audio institutions for fulfilment of the degree. Research since the 1970s strongly suggests that employers desire new hires to be able to work under the supervision of a mentor, understand daily operation of a studio, and work within the hierarchy of a studio (Ashley 1973; Gadhoke 1978; Gander 1978; Liston 1975; Lodge 1978; Manquen 1978; Plunkett & Fink 1979; Stockham 1977; Strawn 1997; Tough 2009). Some institutions resist mandating internship programs for several reasons: (a) large number of students, (b) poor quality of student, and (c) small local industry size (Bielmeier 2014).

However, these challenges can be overcome by the creation of an internship coordinator position within an institution's audio program. This coordinator should be a current member of the audio or entertainment industry, and thus, able to work with local businesses and associations to create or develop internship programs (Vandemast-Bell et al. 2015). Furthermore, the coordinator is responsible for nurturing these business relationships and identifying needs/wants of each business. Moreover, the coordinator should have a strong understanding of each student and pair them with appropriate businesses. For institutions that cannot incur the cost of hiring a full-time internship coordinator, they can incentivize and prioritize mentorship systems for faculty and staff. These mentors can act as liaisons for students and this coordinator. Part of the mentor's role is to help mentees identify their strengths, interests, and develop professional documents to apply for internships and future employment.

Classroom Modifications

In the classroom, how projects are assessed can have a major impact on which skills are learned. Therefore, educators should examine the following areas of their curricula: (a) balance the acquisition of technical skills, (b) use of recursive feedback, and (c) design situated-learning classrooms (Bragança 2015; Carugo 2015; Davis & Parkers 2013; Huws 2017; Vandemast-Bell et al. 2015) Table 14.2 summarizes actionable items for innovation in audio education classrooms and summarizes the reason for implementation and tasks required to achieve the innovation.

Balancing Acquisition of Technical Skills

At the classroom level, new or additional learning objectives/outcomes should be implemented based on the reviews conducted by curriculum committees of audio institutions. Instructors should review courses to assess the balance of technical skills acquisition with social, communication and business skills. What may seem a tremendous undertaking is often implemented by the addition of few extra points or project tasks within an existing course. Outcome-based assignments should be graded via process-based assessment (Herrington 2006; Rodgers 2014; Savery & Duffy 1995).

For example, grading the output or artistic quality of a novice or basic-level project is not purposeful. Such a project may require students to hook up an audio interface and record via a microphone. Therefore, credit/points should focus on the process, both the technical and social skills pertinent to the assignment. Table 14.3 shows an example of outcome- and process-based assessments for a

Table 14.2 Actionable items for innovation within audio education classrooms

Actionable Item	*Reason*	*Task*
Balance of acquisition of technical skills	Balance technical, social, communication, and employability skills	• Review coursework • Balance evaluation of technical, communication, and employability skills
Use of recursive feedback	Foster communication and listening skills among clients/peers	• Identify projects with inherent feedback aspect • Implement feedback loops within project
Design situated-learning classrooms	Better prepare students for internships and employment	• Assess audio education facilities that align with local industry and student desired employment opportunities • Implement project specifically for work in these spaces

Table 14.3 Example of outcome- and process-based assessments for a novice audio recording project for students to learn to use and record with audio interface and microphone

Skills evaluated	*Outcome-based assessments*	*Process-based assessments*
Ability of student to wire up audio interface	X	X
Ability of student to record via a microphone	X	X
Time management		X
Follow procedures		X
Student reflection		X

novice audio recording project. Typically, outcome-based assessments should be replaced by assessments that focus on (a) time management (e.g. hard project due dates), (b) following procedures (e.g. specify project format and documentation), and (c) student reflection (e.g. enable ability to self-correct behaviour).

A process-based assessment approach borrows from the cognitive apprenticeships model, which suggests that future engineers need strong procedural knowledge of technology (Lefford & Berg 2013). This strong foundation in audio practice and production is especially needed in the audio for game industry because employees are typically from disparate audio backgrounds and specializations (Bragança 2015). Furthermore, research shows that some educators feel a multidisciplinary approach to audio education helps students obtain an understanding of their own workflow and how it applies to other roles in the creative fields of film and media (Lankford & Schwartz 2018).

Use of Recursive Feedback Loop

Educators should review projects for the integration of a recursive feedback loop, which fosters the ability of students to work under a mentor (Bielmeier 2014; Tough 2010). A recursive feedback loop requires students to submit projects/coursework, receive feedback from educators, and then make alterations in response. Potentially, students could rework a specific project several times until they have reconciled all feedback to the client's satisfaction. This recursive feedback loop integrates one-on-one or mentor/mentee opportunities within the classroom and hones a student's ability to communicate with clients/co-workers and respond to feedback (Douds 1985; Dunn 1988; Grundy 1984; Hirsch 1985; Kenny 1990; Pritts 1985; Jackson 1988; Merchant 2011).

During this feedback process, it is important for the educator to determine the number of iterations and an appropriate grading scheme for each reiteration to create incentive and meaning for each round of feedback. For example, a simple vocal overdub project could contain a first draft of a project followed by a round of feedback from the vocalist/client and educator. Furthermore, this project could be peer-reviewed, thus creating collaborative learning from fellow classmates and audio majors (Lankford & Schwartz 2018). In addition, these projects can be posted on the internet to receive feedback from audiences outside the institution and create a dialogue about how to deal with both constructive and non-constructive criticism (Kirschner et al. 2011).

A recursive feedback loop may be implemented into many types of classroom assignments: consider a project that involves a basic recording session in studio where students are asked to record a vocalist or single instrument. Students will submit a *first draft* to enable receiving and implementing feedback into the project prior to the final submission. Students will be graded primarily on their ability to update all related session documentation and make feedback-based alterations to the recording or mix. Table 14.4 shows an example of process-based assessment for a recording project of a single vocalist/musician in a recording space. Note that in the table, *Blackboard* refers to an internet-based classroom interface that facilitates electronic submission and assessment of work, and online discussion within the classroom (Blackboard Inc. 2019).

Another approach to infusing feedback into the classroom is to include one-on-one mixing lessons in addition to class meetings and coursework (Merchant 2011). Associate Professor John Merchant at Middle Tennessee State University has used mix lessons to help students learn how to receive feedback from a mentor/client/

Table 14.4 Example of process-based assessment for recording project of a single vocalist/musician in a recording space

Project objective	*Student activity*
Pre-session plan	• Locate musician • Reserve studio time to record with musician • Coordinate with musician to create recording approach • Create/define session documentation (e.g. info. sheet, track sheet, and recording plan with diagrams) • Locate/reserve microphones and equipment to record
Recording session	• Record musician using one/few microphones • Apply proper studio signal flow to record in Pro Tools
Submission of first draft	• Submit first draft by uploading Pro Tools session files and docs to Blackboard
First draft feedback	• Receive in-class feedback from classmates and educator • Receive feedback via Blackboard from classmates • Give feedback to other students via Blackboard
Implement feedback	• Make changes to first draft recording • Update documentation based on feedback
Submission of final draft	• Submit final draft by uploading Pro Tools session files and docs to Blackboard • Submit final draft to musician
Final draft feedback	• Receive feedback from musician • Receive in-class feedback via Blackboard from classmates and educator • Give feedback to other students via Blackboard

professor and learn to adjust based on this feedback. A balance of how often these lessons occur should be determined by the educator to meet the needs of each course.

Design Situated Classrooms

Creating situated learning environments is also a part of preparing students for internship and employment in the industry (Bielmeier 2014; Carugo 2015; Davis & Parkers 2013; Herrington 2006; Paulson 2001; Rodgers 2014; Thompson & Mosley 2015; Tough 2010; Vandemast-Bell et al. 2015). Learning facilities may include recording studio, critical listening room, or audio design/maintenance/repair laboratories. The type of learning facility at a specific audio institution should reflect the future industry environments in which students seek to be employed. However, it is difficult and financially challenging for universities to keep pace with advancements in technology and burgeoning fields of employment within the recording industry.

State-of-the-art facilities built only a decade ago, may no longer be useful in teaching employable skills for the industry of today. For example, understanding the signal flow of large format analog consoles for live-sound may no longer be relevant to the common employment opportunity for recent graduates. Large format analog consoles are not used in small clubs and corporate events, and have been replaced by iPads and computers to adjust volume levels. Therefore, an assessment of facilities should be conducted by the faculty and staff working in those facilities with deference to the audio program's mentors, internship coordinator, and student's employment goals.

A situated classroom is used by the Blackbird Academy, which is a nine-month formal audio training program under the aegis of veteran recording engineer/educator Mark Rubel. Blackbird students are working with real clients at a business located on Nashville's historic Music Row. Unfortunately, this type of classroom may not possible at many locations where the recording industry is not an active community.

Research Innovation

Research of this decade suggests a misalignment of expectations between audio education institutions and industry research (Bielmeier 2013a, 2013b, 2014, 2016, 2017; Tough 2010). To align expectations between industry needs and curricular design, audio education institutions and industry should agree on (a) what skills are most needed by audio students at graduation; (b) where/when students should acquire specific skills related to technology and employability; and (c) terms of partnerships/interactions related to mentorship, internships, merit-based scholarships competitions/opportunities, and audio conventions. Future research into each of these areas can provide insight into the perspectives of both parties.

Table 14.5 summarizes the areas of research that should be conducted to improve audio programs and the skills of graduates. First, to understand the alignment

Table 14.5 Actionable items for innovation within audio education research

Research question	*Tasks*	*Objective*
Alignment of educational and accreditation	• Identify perspectives of institutional administrators and accreditation bodies • Determine skills most important to students	• Learning outcomes of audio education programs meet industry needs • Prepare students for common employment in audio industry
Alignment between education and industry	• Explore deficiencies of new hires described by employers in ARES	• Define employability skills students should have
Standardization of audio educational curriculum	• Aggregate data collected from alignment studies between education, accreditation, and industry • Propose new guidelines for audio education for entire US	• Tailor guidelines to specializations in audio industry levels of audio education • Guidelines suggest modes for balancing technology/employability skills, and timetables for implementing • Suggest plan for standardization of audio education

between educational institutions and accreditation bodies, research should be conducted to identify the perspectives of institutional administrators and accreditation bodies to determine which are the most important skills for students. This issue seems to be most prevalent in the US were accreditation bodies vary greatly across the country. This research could lead to a better understanding of what formal audio programs are teaching and how this fits with industry needs. Second, to understand the alignment between educational institutions and industry, a follow-on study of the ARES Study should be conducted to further investigate the deficiencies of new hires described by employers (Bielmeier 2017).

The ARES study should be used as a baseline for talking points and qualitative interviews. Furthermore, the 150 optional clarification comments in the ARES study should be further investigated as these comments were extremely critical of the new hires' skills, personalities, work ethics, and expectations for a career in the audio industry. Third, research should investigate the possibility of standardization of audio education curriculum across the US specific to both specialization and graduates desired field of employment. The research should aggregate data collected from alignment studies between educational institutions, accreditation bodies, and industry, and propose new guidelines/standards for audio education to be applied across the US. These new guidelines could be tailored to specializations within the audio industry and/or different levels of formal education training programs (e.g. certificate programs, two-year programs, or four-year programs). Furthermore, these standardizations should suggest modes for incorporating and balancing technology and employability skills. Also, guidelines may suggest timetables for implementation of specific skills throughout the curricula.

Summary of Actionable Items

This section discussed methods to improve audio education to better prepare students for the industry. Educational institutions can support audio educators by forming curriculum committees, integrating local and regional industry, establishing student mentorship, and requiring internships. In the classroom, educators can focus on balancing communication and technical skills, employing the recursive feedback loop, and establishing a situated classroom. Future research should focus on alignment among educational institutions, accrediting bodies, and industry. In addition, future research should study the standardization of audio education curricula to meet the needs of the contemporary music industries.

Conclusion

The recording and entertainment industries are dynamic with constantly evolving technology and ever-changing modes of consumption and content creation. To keep pace with the changing demands of the industry, educational institutions need to continuously improve and modify educational outcomes. Such modifications are required to provide students with the foundational skills and theory to begin a career in industry. However, in application, providing a strong recording industry foundation has proven to be challenging for educators and institutions, and at times near impossible.

Understanding the research and developments of the past decade helps to illuminate the contemporary perspectives of employers, employees, educators, and students. The areas of improvement identified provide insight and actionable items for continued innovations in audio institutions and the maintenance of a healthy relationship between industry and audio programs. Ultimately, as members of the audio community, it is vital to support and maintain the collective mandate of audio programs that seeks to provide students with the skills and education necessary to begin purposeful careers in the audio and entertainment industry.

References

Ashley, J.R. (1973) "What is the need for university level education in audio", *Journal of the Audio Engineering Society*, 21(1), 98.

Audio Engineering Society (2019) "Student competitions and awards" [Online], www.aes.org/students/awards/ (accessed: 4 May 2019)

Bell, S. (2010) "Project-based learning for the 21st century: Skills for the future", *The Clearing House*, 83(2), 39–43, doi:10.1080/00098650903505415

Bielmeier, D. (2013a) "What skills new recording engineers have and where they learn them: A survey of new recording engineers' perceived skill sets and those observed by their employers", Doctoral dissertation, Argosy University, Washington, DC.

Bielmeier, D. (2013b) "Why didn't you learn this at recording school: Critical comments by employer", Proceedings of the Audio Engineering Society 50th International Conference: Audio Education, Murfreesboro, July, 1–10, www.aes.org/e-lib/browse.cfm?elib=16836

Bielmeier, D. (2014) "Apprenticeship skills in audio education: A comparison of classroom and institutional focus as reported by educators", Proceedings of the Audio Engineering Society 137th Convention, Los Angeles, October, 1–10, www.aes.org/e-lib/browse.cfm?elib=17424
Bielmeier, D. (2016) "Audio recording and production education: Skills new hires have and where they reported learning them", *Journal of the Audio Engineer Society*, 64(4), 218–228, www.aes.org/e-lib/browse.cfm?elib=18130
Bielmeier, D. (2017) "Audio recording production students report skills learned or focused on in their programs", Proceedings of the Audio Engineering Society 143rd Convention, New York, NY, October, 1–10, www.aes.org/e-lib/browse.cfm?elib=19211
Blackboard Inc. (2019) "BlackBoard" [Online], www.blackboard.com/index.html (accessed: May 2019)
Bragança, R. (2015) "Modding game audio for education", Proceedings of the Audio Engineering Society 138th Convention, Warsaw, Poland, May, 1–5, www.aes.org/e-lib/browse.cfm?elib=17655
Carugo, D. (2015) "Fasten your seatbelts! Accelerating students on work experience into positions of responsibility", Proceedings from the Audio Engineering Society 26th UK Conference: Audio Education, Glasgow, UK, March, 1–4, www.aes.org/e-lib/browse.cfm?elib=17848
Davis, R. & Parkers, S. (2013) "Collaboration, creativity, and communities of practice: Music technology courses as a gateway to the industry", Proceedings of the Audio Engineering Society 50th International Conference: Audio Education, Murfreesboro, July, 7–4, www.aes.org/e-lib/browse.cfm?elib=16851
Day, S. & Koorland, M. (1997) "The future isn't what it used to be: Student competencies for the 21st century", *Contemporary Education*, 69(1), 34–40.
Douds, T. (1985) "The importance of audio education", *SPARS Papers*, 2(Spring), p. 1.
Dunn, K. (1988) "School studios: Budget or bust", *Mix*, July, pp. 57–58, 61–62, 168–170.
Fick, J. (2015) "From creativity to science and back again: Supporting audio students through active teaching approaches", Proceedings of the Audio Engineering Society 139th Convention, New York, NY, October, 1–6, www.aes.org/e-lib/browse.cfm?elib=17921
Gadhoke, R. (1978) "Curriculum in recording engineering", Proceedings of the Audio Engineering Society 60th Convention: Trends in Audio Education, A Symposium, Los Angeles, May, 1–35, www.aes.org/e-lib/browse.cfm?elib=10254
Gander, M.R. (1978) "Balancing theory and practice in audio education: Experience of a recent graduate", Proceedings of the Audio Engineering Society 60th Convention: Trends in Audio Education, A Symposium, Los Angeles, May, 1–35, www.aes.org/e-lib/browse.cfm?elib=10254
Grundy, A. (1984) "Institute of audio research", *SPARS Papers*, 2(Fall), p. 1.
Herrington, J. (2006) *Authentic learning environments in higher education*. Hersey, PA: IGI Global.
Hiemstra, R. (1994) *Self-directed learning: The sourcebook for self-directed learning*. Amherst, MA: HRD Press.
Hirsch, H. (1985) "Education for the audio renaissance", *SPARS Papers*, 2(Spring), p. 1.
Huws, C. (2017) "The education of the next generation of pro-audio professionals", Proceedings of the Audio Engineering Society 143rd Convention, New York, NY, October, 1–11, www.aes.org/e-lib/browse.cfm?elib=19213
IET (2019) "The Insitution of Engineering and Technology" [Online], www.theiet.org/about/ (accessed: 27 May 2019)
Jackson, L. (1988) "Studios speak to the schools", *Mix*, July, pp. 72–88.
JAMES (2019) "JAMES HE & FE course accreditation overview" [Online], www.jamesonline.org.uk/accreditation/accred_overview/ (accessed: 27 May 2019)
Jewson, N. (2013) "Communities of practice in their place: Some implications of changes in the spatial location of work", In *Communities of Practice* (pp. 166–180). London: Routledge.
Kenny, T. (1990) "Digital diplomas", *Mix*, July, p. 64.
Kirschner, F., Paas, F. & Kirschner, P. (2011) "Superiority of collaborative learning with complex tasks: A research note on an alternative affective explanation", *Computers in Human Behavior*, 27(1), 53–57, doi:10.1016/j.chb.2010.05.012
Lambert, M. (1989) "Education in the school of hard knocks", *Mix*, July, p. 23.
Lankford, E. & Schwartz, A. (2018) "Case study: An interdisciplinary audio curriculum", Proceedings of the Audio Engineering Society 145th Convention, New York, NY, October, 1–6, www.aes.org/e-lib/browse.cfm?elib=19844

Lave, J. & Wenger, E. (1991) *Situated learning: Legitimate peripheral participation*. Cambridge: Cambridge University Press.

Lefford, M. & Berg, J. (2013) "From practice to research and back again: Research skills in audio engineering education", Proceedings of the Audio Engineering Society 50th International Conference: Audio Education, Murfreesboro, July 2013, 7–20, www.aes.org/e-lib/browse.cfm?elib=16841

Lightner, J.W. (1993) "A survey of the professional audio industry in an eight-state region to assess employers' perceived value of formal audio education and their perceived training needs for entry-level employee", Doctoral Disseratation, Ferris State University, Big Rapids, MI.

Liston, F.E. (1975) "An analysis of management personnel within the music industry", Dissertation, Abstracts International.

Lodge, T. (1978) "A curriculum in music industry arts", *Journal of the Audio Engineer Society*, 15(1), 7–9.

Manquen, D. (1978) "An audio design engineering certificate program for BS students in electrical engineering", *Journal of the Audio Engineer Society*, 15(1), 17–18.

Merchant, J. (2011) "A revised approach to teaching audio mixing techniques: Applying the deliberate practice model", Proceedings of the Audio Engineering Society 131st Convention, New York, NY, October, www.aes.org/e-lib/browse.cfm?elib=16069

NASM (2019) "National Association of Schools of Music" [Online], https://nasm.arts-accredit.org/accreditation/standards-guidelines/handbook/ (accessed: 27 May 2019)

Nixon, P., Young, T., Brown, E. & Wiltshire, L. (2015) "Interdisciplinary benefits: Encouraging creative collaboration", Proceedings from the Audio Engineering Society 26th UK Conference: Audio Education, Glasgow, UK, March, 1–7, www.aes.org/e-lib/browse.cfm?elib=17863

Oliver, R. (2008) "Engaging first year students using a web-supported inquiry-based learning setting", *Higher Education*, 55(3), 285, doi:10.1007/s10734-007-9055-7

Patryk, G. (2012) "Netlabels and democratization of the recording industry" [Online], http://firstmonday.org/ojs/index.php/fm/article/vi ew/3770/3278 (accessed: 25 April 2012)

Patterson, C., Crooks, D. & Lunyk-Child, O. (2002) "A new perspective on competencies for self-directed learning", *Journal of Nursing Education*, 41(1), 25–31, doi:10.3928/0148-4834-20020101-06

Paulson, K. (2001) "Using competencies to connect the workplace and postsecondary education", *New Directions for Institutional Research*, 110(1), 41–54, doi:10.1002/ir.10

Plunkett, B. & Fink, D.G. (1979) "Hiring in the audio industry: You may be an engineer, but can you drive the train", *Journal of the Audio Engineer Society*, 15(1), 22–31.

Prensky, M. (2010) *Teaching digital natives: Partnering for real learning*. Thousand Oaks, CA: Corwin.

Pritts, R. (1985) "To school or not to school", *SPARS Papers*, 2(Fall), p. 2.

Rubel, M. (2007) "Notes from the P&E wing: The art of listening". *MIX* [Online], http://mixonline.com/recording/applications/audio_n otes_pc_wing_2/index.html

Rodgers, J. (2014) "Partnering approaches for teaching music technology", Proceedings of the Audio Engineering Society 137th Convention, Los Angeles, October, 1–6, www.aes.org/e-lib/browse.cfm?elib=17425

Sanders, D.H. (1994) "The professional preparation of the audio engineers: A survey of studio personnel and recommendations for school curricula design", Dissertation, Abstracts International.

Savery, J. & Duffy, T. (1995) "Problem based learning: An instructional model and its constructivist framework", *Educational Technology*, 35(5), 31–38.

Snyder, K. (2016) "Understanding project-based learning in the audio classroom: Using PBL to facilitate audio storytelling", Proceedings of the Audio Engineering Society 141st Convention, Los Angeles, October, 1–6, www.aes.org/e-lib/browse.cfm?elib=18405

Stockham, T.G. (1977) "Records of the future", *Journal of the Audio Engineering Society*, 25(11), 892–895, www.aes.org/e-lib/browse.cfm?elib=3308

Strawn, J. (1997) "Technological change: The challenge to the audio and music industries", *Journal of the Audio Engineering Society*, 45(3), 170–184, www.aes.org/e-lib/browse.cfm?elib=7863

Thompson, P. & Mosley, B. (2015) "Audio education: Its purpose and place in formal education and the creative industries", Proceedings from the Audio Engineering Society 26th UK Conference: Audio Education, Glasgow, UK, March, 19–25, www.aes.org/e-lib/browse.cfm?elib=17866

Tough, D. (2009) *Developing a consensus-driven, core competency model to shape future audio engineering technology curriculum: A web-based modified Delphi study*. Nashville, TN: Middle Tennessee State University.

Tough, D. (2010) "Shaping audio engineering curriculum: An expert panel's view of the future", Proceedings of the Audio Engineering Society 129th Convention, San Francisco, October, 1–12, www.aes.org/e-lib/browse.cfm?elib=15726

Vandemast-Bell, P., Werner, D. & Crossley, J. (2015) "Seeking best practice for education and training in the recording studio", Proceedings from the Audio Engineering Society 26th UK Conference: Audio Education, Glasgow, UK, March, 21–29, www.aes.org/e-lib/browse.cfm?elib=17868

Walsh, E.J. (1996) "Important occupational skills and knowledge needed in the preparation of the recording engineer: A survey of faculty perceptions", Dissertation, Abstracts International.

15

Addressing the Gender Imbalance in the Theater Sound Industry

A Conservatoire Approach

Clare Hibberd

Introduction

In his recently published book *Sound Design for the Stage*, theater sound designer Gareth Fry acknowledges that while success in the industry involves ability and luck, middle-class, white males are more likely to succeed. He notes that acknowledgment of sexism, racism, and privilege in the industry is rare and that work is required to implement change (2019).

This chapter compares research into gender imbalance in theater with the growing research on gender imbalance in audio education. A distinctive subfield within audio education, theater sound in higher education in the UK is regularly taught alongside other theater disciplines. Undergraduates are often asked to make a choice of which of these disciplines (stage management, sound, lighting, technical stage) they wish to specialize in. Given that the education of theater sound is directed toward technology, sound production, and listening skills (Bennett 2019), it is clear that it demands a large array of abilities. Brown (2010) argues for theater sound to be understood as an art form, whereas Deiorio (2019) maintains that theater sound design is related to the people that collaborate together to create a production.

The undergraduate BA Production, Technology, Management course at the Royal Conservatoire of Scotland will be used to demonstrate the implementation of how institute-wide intervention is working to drive change in gender stereotypes. The work carried out on gender imbalance in theater by Michelle MacArthur (2015) for Canadian Equity in Theater can be likened to trends in the UK and US. Several smaller inquiries have been carried out in the UK and Scotland with comparable results. This will be explored in more detail later in the chapter. In response to an abundance of research into gender in relation to music education over the last 20 years, Dr. Elizabeth Dobson's focus on collaboration and creativity in the arts has recently led to her paper about all-female communities and the impact this has had on creativity in audio education. Digital Audio Ecofeminism (DA'EF) is described as the set of values that female collectives, networks, and organizations demonstrate in order to prioritize equality and reject capitalism

(Dobson 2018). This draws on the work of global audio groups, which will also be examined throughout the chapter.

The following review of literature will illustrate the complex nature of the issues that women in theater face, from a point of view of the context and the educational structures that ultimately serve those roles.

Gender Imbalance in Theater

Michelle MacArthur (2015), lead researcher for the Canadian Equity in Theatre (EIT) initiative, states that women occupy a significantly lower number of positions in creative theatrer roles.[1] Comparisons with this study can be made with an analysis of Scottish theater carried out by Christine Hamilton (2016), who found that between 2014 and 2015, women made up just 11% of sound designers and composers in Scottish publicly funded theaters. In further support to this, Elizabeth Freestone researching for *The Guardian* newspaper in 2012 found that of the top ten subsidized theaters in the UK, female sound designers made up only 5 of 436 employment opportunities in sound (Freestone 2012).

As part of MacArthur's *Best Practice Recommendations* (2015), she recommends that there should be an increase in the production of plays by women and other marginalized groups recognizing that intersectionality (how social categorizations of race, class, and gender overlap creating interdependent systems of discrimination and disadvantage) is prevailing throughout western society. Kimberlé Crenshaw, in her TED talk "The Urgency of Intersectionality" uses the case of DeGraffenreid vs General Motors in 1975 to explain this social injustice. DeGraffenreid sued General Motors for discrimination on the basis that black people were employed for one type of job, while women were employed for another (Crenshaw 2018). This is comparable to a theater environment where black women occupy even less roles in the workforce.

In support of this, Barnes (2015) fights to unravel barriers experienced by women, giving a voice to performers, directors, and choreographers in musical theater. Her account is raw and unapologetic in tone, calling out producers and directors. Despite the predominance of female audiences in musical theater, Barnes questions why shows are mainly produced by men. Interviewing performers such as Mel La Barrie,[2] the underrepresentation of women as part of the creative team is explored. The link between female characters being portrayed as victims and the place of women in society is explored with the suggestion that the authentic female voice is often overlooked in this genre, compared to other areas of the arts.

Kerbel (2017) also addresses discrimination issues in the workplace, offering advice though her experience in trying to achieve gender equality in theater, encouraging an open and caring approach that is inclusive of all creative input. As well as addressing unconscious bias, looking at artistic and administration roles and the theatrical process, part of Kerbel' s guide includes practical ways to implement change. First, she recommends gathering a combination of qualitative and quantitative data in order to gain evidence, both independently and as an organization. Second, listening to individual experiences and accepting the positive and negative impact of any action (such as outreach groups). Kerbel similarly clarifies that data collection can highlight unconscious bias and allow for interventions to take place. For example, she suggests making a list of all freelancers hired over a period of time in order to explore any gender gaps and addressing imbalances moving forward.

Gender Imbalance in Audio Education

To understand the clear lack of female representation in theater sound, it is useful to explore the pathway that university students undertake into the industry. Most students studying theater sound at the Royal Conservatoire of Scotland have a musical education background. Some have had opportunities to assist with school productions and some have shadowed friends or family in the professional theater industry. In all cases, students have engaged with music, whether that be in an academic setting or not. Although not a requisite on the *BA Production Technology Management* course, music is a desirable subject in order to demonstrate a student's listening, collaborative, and creative skills. Interestingly, research has shown that despite girls achieving highly in music compared to boys, this is unlikely to be mirrored in the numbers of women studying music technology in higher education (Born and Devine 2015).

Hallam et al. (2008) found that although female students perform better in school music exams, no consistent gender differences in musical skill can be found. In fact, data gathered in 1994 showed that boys were underrepresented in learning to play an instrument and 60% of students learning an instrument were girls. The reasons that were given for these statistics include stereotypical expectations (including those from parents), peer pressure, and boys preferring instruments that require less practice. The question that arises from these findings is that if girls are studying music at school, at what point does this translate to fewer females studying music technology in higher education?

A possible explanation is that boys demonstrate greater interest in music linked to technology compared to girls (Hallam et al. 2008). Armstrong (2003) suggests that the compositional process mediated by music technology is more conducive to the working styles of boys such as nonverbal communication. In interviews, girls communicated their dissatisfaction with the lack of encouragement to use live instruments to compose during lessons. Armstrong recommends that music teaching should allow for a pluralist approach, to include opportunities for all types of learner and to be flexible around creative output. Theater sound careers encompass a huge variety of skills including understanding complex technology, collaborating with other creatives, and artistic skill. No two theater sound engineers have taken the same path and this could be understood as critical in developing new ideas and productions.

Brilliance Bias

Brilliance bias, as conceived by Caroline Criado Perez, is a concept where an academic subject is culturally recognized by society to require a "brilliance" (p. 100) that can't be easily learned. She suggests that culturally we don't naturally assume women to be naturally talented. This includes areas such as maths, physics, music, and computer science—all of which can be attributed to audio education. All of these subjects are less likely to include women as a majority workforce (Criado Perez 2019). Her book, *Invisible Women*, goes further to suggest other reasons for obstructions to subjects that could be compared to the audio industry. Female students, are significantly less likely to receive funding or to be offered mentorship in higher education (Criado Perez 2019). Additionally, letters of recommendation for women are less likely to include stand-out adjectives such as "remarkable" or "outstanding" (p. 102).

The combination of factors described here would suggest that both the students and the educators are affected by cultural expectations, leading to the scenario of low numbers of female students and low numbers of female educators acting as role models in the field. As a female audio educator, I receive far less bias in teaching compared to working in the industry. As a theater sound engineer, a common greeting by visiting staff would be to ask where the "sound guy" was. As a lecturer these comments are rarely heard. I attribute this to the societal gender bias of having more women in teaching roles (so therefore more acceptable whatever the subject) and inclusive environment nurtured at the Royal Conservatoire of Scotland.

Gender Stereotypes

Susan McClary (1991) makes the link between musical elements and the cultural context from which they derive. Her work as a musicologist in interdisciplinary contexts examines the political, social, and cultural meaning of music conventions. Combining musicology with feminism, McClary makes a clear argument for the patriarchal characterizations of musical structure and performance. To put this into context, Thompson and Stevenson (2015) looked at perceptions of popular electronic musicians at UK higher education institutions, finding that formal music education has a significant impact on the development of musicians practice, suggesting that there is a need to develop a much more comprehensive music curricula that would address the needs of a far wider student population.

Lucy Green (1997) in her work on informal music making, suggests that musicians pick up skills and knowledge through watching and listening to other musicians, in a very similar way that theater sound practitioners pick up expertise. If the work of McClary and Green were both to be applied to theater sound, it could be argued that if the structure and performance of music is essentially "male" in form, the skills passed down from sound operator to operator are also "male". In order to mix a piece of music, an understanding of musical structure is vital. Knowing when to push a solo instrument or allow a build to swell naturally is developed through hours of practice, listening, and observing. This suggests an entirely different obstruction to success that female identifying theater sound practitioners could face, further from "mansplaining", a term widely used since the publication of Rebecca Solnit's pivotal essay, "Men Explain Things to Me", where the author argues that abuse of power, be it online, in the home, or in the workplace should be addressed together by men and women (Solnit 2014).

Studies that have explored the link between gender stereotypes and assumptions about expertise include Hollingshead and Fraidin (2003). They found that males and females showed similar gender-stereotyped behavior. Tasks at work could be directly linked to males or females, for example, women taking up organizational roles whereas men are more likely to be problem solvers. In a theatrical setting, this could be compared to the higher likelihood that males progress to A1 (Audio 1) positions faster and more readily than women. A2 (Audio 2) tasks tend to involve a greater amount of working with performers compared to sound desk operation that an A1 would perform. Although no data exists currently about these specific theater roles, this is a widely recognized occurrence in the UK. Julia Pascal, writing for *The Guardian*, after the male dominated Olivier Awards of 2018, calls for a quota system to be employed in British theater in order for equality to be taken seriously. She highlights that both the National Theatre and the Royal Shakespeare Company have never had a female artistic director and this lack of acknowledgment marginalizes women into a minority workforce (Pascal 2018).

The pedagogy of music technology has close links to theater sound design. Often the equipment is similar although the execution of production can involve far greater numbers in a theatrical environment. Collaboration is inevitable in theater, with sound designers often needing to work closely with all departments, including set, costume, and music to ensure that the needs of performers are met (Deiorio 2019). Studies into gender bias in music technology education have been carried out in the UK since the formation of the National Curriculum (for schools) in 1992. Born and Devine (2015) suggest that although music technology is now found extensively throughout higher education, little has changed to mobilize gender, class, or race stereotypes in the field. Music technology degrees are overwhelmingly male and lower in social class profiles compared to traditional music degrees. Following this pattern of comparison, music technology commands a 90% male dominance compared to 55% female dominance in the learning of traditional music. Furthermore, BME (Black and Ethnic Minority) students make up just 6% of students compared to the 11% across all degree subjects. Carola Boehm (2006) argues that changes need to be made to the way that music technology degrees in Britain are taught, allowing for more opportunities for interdisciplinary learning in an environment that does not need to adhere to strict methodology. Students should not be held back by the structure of how a subject is taught—instead universities should be encouraging students to build new knowledge through free experimentation. Boehm labels this concept a postmodernist approach, where fragmentational knowledge and self-organization are the foundations of the subject.

From the research, this collaborative and interdisciplinary approach should be an opening for greater opportunities for a wider demographic of students to enter the profession. More data would be required to prove this and track any increase in female participation as music pedagogy at school level becomes more inclusive.

Audio Equity Pledge

Dobson's (2018) Audio Equity Pledge is a resource for people working in education and audio-related industry who are interested in social equality. Three actions are recommended to those wishing to pledge allegiance. The first focuses on collaboration. This involves supporting groups currently engaging with women and girls in sound, including understanding and addressing unconscious bias. The second action is about perspective, understanding intersectionality and how this affects women in sound. The final action of the pledge is to change environments to create learning spaces that are fit for purpose.

The following sections of this chapter examine the close links between the teaching of audio education within the BA Production Technology Management course at the Royal Conservatoire of Scotland and collaborative work with local groups and individuals who share the principles of the Audio Equity Pledge. As Dobson (2018) makes clear in her research, the action on gender imbalance within the industry needs to be addressed on multiple levels. Support needs to be given to female groups already active. Unconscious bias should be checked and environments need to be welcoming to everyone. For each section of the Audio Equity Pledge—collaboration, perspective, and changing environment—references will be made to demonstrate what is being achieved.

Collaboration

Soundgirls is a US non-profit organization that provides a support network for women and nonbinary people working in professional audio.[3] Opportunities are sought in order to inspire and empower young women in the field and to expand opportunities through cooperation, collaboration, and diversity. Chapters exist all over the world,[4] including the UK. Soundgirls Glasgow has been in operation for just over a year and local support is steadily growing. The chapter meets regularly for drop-in sessions, and audio-based workshops have been held to promote new learning. As an educator, I find myself in a prime position to promote the work of Soundgirls and to lead drop-in sessions and workshops. Although currently in its early stages, plans are in process for a greater number of events to expand participation in the near future.

Stellar Quines is an award-winning Scottish Theater Company that promotes the diversity of women and girls by nurturing,[5] empowering, and creating award-winning theater. The current "M*****classes" organized by the company are specific artist development opportunities to support women in technical theater roles (Stellar Quines 2018). In May 2019, a free collaborative event between the Audio Engineering Society (AES), Soundgirls, and Stellar Quines took place at the Royal Conservatoire of Scotland. Over 50 participants attended with workshop leaders delivering four sessions based on theater sound. Positive feedback at the end of the sessions included comments that the opportunity to network with other women (especially those starting out in the industry) was highly useful, with established female audio specialists wanting to offer their support for future events.

In light of the #MeToo campaign, several theater societies published statements of support. The Federation of Scottish Theatre and Society of London Theatre both responded with a commitment to ensure that theater is a safe space for all. Theaters such as the Royal Court and the Young Vic have developed comprehensive staff training programs as well as workshops (#takeupspace) to support women in all aspects of creativity.

By setting a standard of safe collaboration in audio education, the platform is set for emerging professionals, with the hope that more positive outcomes will seep into the industry.

Perspective

As the only conservatoire in Scotland, the Royal Conservatoire of Scotland is in a unique position, receiving pedagogical direction and guidance from several perspectives. It is useful to understand why teaching practices have been adopted in audio education. The Equality Act of 2010 underpins education policy and law in Scotland. Nine protected characteristics[6] ensure that equal access and protection for those groups are enforced by law. Under the specific chapter relating to higher education, The Equality Act states that institutions must not discriminate against a person in the arrangements it makes for deciding admissions, the terms on which admissions are offered or by not admitting a student. Many institutes in the UK use the definition of the nine characteristics to further inform their policies and practices beyond admission and actively seek ways of providing further support for these groups.

Further to this, education in Scotland is supported by Education Scotland, a Scottish government executive agency that funds quality and improvement. Operating independently from the government allows impartiality and accountability. The "Embedding Inclusion, Equity and Empowerment" program aims explicitly to promote inclusive practice within the education system drawing upon several legislative duties of local authorities. These include the Children and Young People Act, Additional Support for Learning Act, Equalities duties, the "Getting It Right for Every Child" (GIRFEC) policy, and Safeguarding (Embedding Inclusion, Equity and Empowerment 2018b).

Supporting children and young people's learning is the key aim of a range of national policies, legislation, strategies, and frameworks, from early years to positive, sustained destinations, known as the Supporting Learners Framework. The framework addresses three types of support. First, *entitlement to support* states that every child and young person is entitled to support with their learning. Second, *universal support* ensures the responsibility of staff to promote well-being, inclusion, equality, and fairness by the creation of personal learning plans. This also includes the need to provide opportunities for students to discuss their learning and development. Third, *targeted support* focuses on overcoming barriers to learning, whether these are long or short term, through staged intervention.

"Developing the Young Workforce" (DYW) is a strategy that aims to better prepare young people for the workplace and to reduce youth unemployment by 40% by 2021. Of ten key aims, one is directly relatable to impacting the number of females applying to courses. It states that opportunities and experiences for all learners will be improved, with a focus on reducing gender imbalance on course take-up (Developing the Young Workforce 2018a).

The Scottish Funding Council (SFC) is a non-departmental public body of the Scottish Government. It was established by the Further and Higher Education Act of 2005 and invests around £1.8 billion of public money each year with the ambition that Scotland will be the best place in the world to learn, educate, research, and innovate. Investment includes supporting research in Scottish universities and widening access to provide more people with more routes into learning and skills (Scottish Funding Council 2018). Aligning with the DYW strategy, the SFC created the Gender Action Plan (GAP). The GAP was created as a way to support institutions to address gender imbalance and inequality. The main ambition of the GAP is that no college or university subject will have a gender imbalance of greater than 75% of one gender.

Given the strong leadership currently offered by the Scottish government in meeting gender and diversity targets, it will be of great interest to witness the expected changes in education and the effect that this will have on the theater sound industry as a whole. Indeed, if targets are met, the gender diversity landscape of Scotland will become far more equal.

For audio education, this support at government level, should provide the impetus that is required for change to occur in institutions. Policy now dictates clear targets that need to be met, linked to funding. In other words, Scottish institutions that wish to continue to receive high levels of funding for their courses will need to comply. In theater sound this should see an increase of 20% of female students.

Changing Environments

As one of the four modules of the BA Production Technology Management course, *Personal and Professional Development* is emphasized as equally important to student development as stagecraft, management, and communication. Certainly, at the outset, students can find this weighting challenging and unusual compared to previous learning experiences. A female educator across several audio-based modules ensures that female sound designers and technicians are included in the teaching as mentors and workshop leaders. Activities are also designed to be inclusive to a variety of learning styles, giving the students the choice to decide how they work most effectively. There is no pressure on students to follow predetermined procedures or processes to reach their creative goal. A student who wants to compose music for a production is supported in their decision as much as a student who wants to concentrate on the system design. The RCS safe space policy goes someway to address this, citing that everyone in the room should have their opinions valued.

Personal reflection is carried out regularly by teaching staff, and fortnightly team meetings allow good practice to be shared between colleagues. Peer learning is an integral part of the learning process and delivered as part of all modules, including personal and professional development. All students by their third year are expected to lead other students, ensuring that no team members are left behind in their understanding. Where students are less confident, staff intervention is "on the job". For example, a student struggling to communicate with a director about their sound design concept could be offered a tutorial to go through their ideas before they need to present their work. At all times students are encouraged to dig deep and find their own problem solving methods. This may appear harsh or unusual for an undergraduate course but the rigour that is demanded at this level is known by the staff to equate to the working world of theater audio.

Conclusion

It is evident that within Scotland there is currently a great emphasis from government to drive forwards gender equality. The liberal government setting in Scotland allows conservatoire-style teaching and learning to thrive and grow in the current challenging political environment. Programs implemented by the Royal Conservatoire of Scotland to address the Gender Action Plan have initially shown that the gender gap is slowly closing, evidenced by the most recent report (Caplan 2017). This has been achieved with no direct "female only" initiatives, but rather a comprehensive approach that addresses a multifaceted approach to inclusivity.

Summarizing, transferable teaching methods from the Royal Conservatoire of Scotland approach to teaching theater sound draws two conclusions that can be demonstrated by the institute as a whole and the educator. From the point of view of the educator, every learning opportunity should be approached with the mindset that everyone present (staff and students) have a valid opinion. Learners should be involved in goal setting, content, and style of delivery and there should be a universal understanding that learning to complete a task or solve a problem is a key motivator for students. Above all, a pluralistic approach to teaching and learning needs to be adopted that is inclusive of all people.

Notes

1. Little progress has been made since reports carried out in the 1980s and then again in the early 2000s that indicated that women are accountable for a significantly lower number of positions (Macarthur 2015).
2. Mel La Barrie is a prolific UK musical theater performer. Credits include principal roles in Wicked (Universal Stage Productions), Matilda (Royal Shakespeare Company), Les Miserables (Cameron Mackintosh Ltd) and Mary Poppins (Cameron Mackintosh Ltd).
3. https://soundgirls.org/
4. A chapter is a geographically local group of members that are able to meet regularly for formal and informal meetings.

5. www.stellarquines.com/
6. The nine protected characteristics as identified by the Equality Act 2010 are age, disability, gender reassignment, marriage and civil partnership, pregnancy and maternity, race, religion and belief, sex and sexual orientation.

References

Armstrong, V. (2003) "Girls and Boys" Technological Toys: Music Composition in the Computerized Classroom. *GEMS (Gender, Education, Music, & Society), the on-line journal of GRIME (Gender Research in Music Education)*, 7(5).

Audio Engineering Society. (2019) *Theatre Sound Assembly* [Online]. Available from: www.aes-uk.org/forthcoming-meetings/theatre-sound-assembly/ [Accessed 29 April 2018]

Barnes, G. (2015) *Her Turn on Stage: The Role of Women in Musical Theatre*. McFarland, Carolina, United States of America.

Bennett, S. (2019) *Theory for Theatre Studies: Sound*. Methuen Drama, London.

Boehm, C. (2006) *The Thing About the Quotes: 'Music Technology' Degrees in Britain* [Online]. Available from: http://eprints.staffs.ac.uk/5064/1/Betweening_Ucas_ICMC_7.0.pdf [Accessed 10 August 2019]

Born, G. and Devine, K. (2015) Music Technology, Gender, and Class: Digitization, Educational and Social Change in Britain. *Twentieth-Century Music*, 12(2), pp. 135–172.

Brown, R. (2010) *Sound: A Reader in Theatre Practice*. Palgrave Macmillan, London.

Caplan, R. (2017) *RCS Mainstreaming Equality 2017* [Online]. Available from: www.rcs.ac.uk/wp-content/uploads/2017/04/Mainstreaming-Equality-Report-2017.pdf [Accessed 27 December 2018]

Crenshaw, K. (2018) Film. *The Urgency of Intersectionality* [Online]. Available from: www.ted.com/talks/kimberle_crenshaw_the_urgency_of_intersectionality [Accessed 28 December 2018]

Criado Perez, C. (2019) *Invisible Women*. Chatto & Windus, London.

Deiorio, V. (2019) *The Art of Theatrical Sound Design: A Practical Guide*. Methuen Drama, London.

Dobson, E. (2018) Digital Audio Ecofeminism: The Glocal Impact of All-Female Communities on Learning and Sound Creativities. In L. De Bruin, P. Burnard and S. Davis (Eds.) *Creativities in Arts Education, Research and Practice: Glocalised Perspectives for the Future of Learning and Teaching*, pp. 201–220. Brill Sense, Leiden.

Education Scotland. (2018a). *Developing the Young Workforce: Scotland's Youth Employment Strategy* [Online]. Available from: www.gov.scot/publications/developing-young-workforce-scotlands-youth-employment-strategy/ [Accessed 7 November 2018]

Education Scotland. (2018b) *Embedding Inclusion, Equity and Empowerment* [Online]. Available from: https://education.gov.scot/what-we-do/Embedding%20inclusion,%20equity%20and%20empowerment [Accessed 26 December 2018]

Education Scotland. (2018c) *Supporting Learners—From Early Years to Positive, Sustained Destinations* [Online]. Available from: https://education.gov.scot/Documents/supporting-learners-legislation-policies.pdf [Accessed 27 December 2018]

Federation of Scottish Theatre. (2017) *Sexual Harassment and Abuse Statement* [Online]. Available from: www.scottishtheatre.org/news/sexual-harassment-and-abuse [Accessed 7 August 2019]

Fry, G. (2019) *Sound Design for the Stage*. The Crowood Press, Wiltshire.

Green, L. (1997) *Music, Gender and Education* [Online]. Available from: https://books.google.co.uk/books?hl=en&lr=&id=pZgBNUqIbhQC&oi=fnd&pg=PR11&dq=lucy+green+music+gender+and+education&ots=51-1jLQTG7&sig=61EgrXpHnCU5JgbZNtE72lI1n4Q&redir_esc=y#v=onepage&q=lucy%20green%20music%20gender%20and%20education&f=false [Accessed 9 August 2019]

Freestone, E. (2012) The Guardian. *Women in Theatre: How the "2:1 Problem" Breaks Down* [Online]. Available from: www.theguardian.com/news/datablog/2012/dec/10/women-in-theatre-research-full-results [Accessed 6 November 2018]

Hallam, S., Rogers, L. and Creech, A. (2008) Gender Differences in Musical Instrument Choice. *International Journal of Music Education*, 26(1), pp. 7–19.

Hamilton, C. (2016) *Where Are the Women? Analysis of Creative Roles by Gender in Scottish Theatre 2014/15* [Online]. Available from: www.stellarquines.com/provocation/gender-in-scottish-theatre/where-are-the-women/ [Accessed 29 December 2018]

Hollingshead, A.B. and Fraidin, S.N. (2003) Gender Stereotypes and Assumptions About Expertise in Transactive Memory. *Journal of Experimental Social Psychology*, 39, pp. 355–363.

Kerbel, L. (2017) *All Change Please: A Practical Guide for Achieving Gender Equality in Theatre*. Nick Hern Books, London.

Macarthur, M. (2015) Achieving Equity in Canadian Theatre: A Report with Best Practice Recommendations. Equity in Theatre. *Playwrights Guild of Canada*, April [Online]. Available from: https://playwrightsguild.ca/wp-content/uploads/2018/04/FINAL-EIT-Report_4-22-15.pdf [Accessed 29 April 2018]

McClary, S. (1991) *Feminine Endings: Music, Gender & Sexuality*. 2nd Ed. University of Minnesota Press, Minneapolis.

Pascal, J. (2018) *Women Are Being Excluded from the Stage. It's Time for Quotas. The Guardian*, 24 April [Online]. Available from: www.theguardian.com/commentisfree/2018/apr/24/women-theatre-quotas-stage-gender [Accessed 11 August 2019]

Royal Conservatoire of Scotland. (2018a) *BA Production Technology & Management Undergraduate Introduction* [Online]. Available from: www.rcs.ac.uk/courses/ba-production-technology-management/ [Accessed 7 November 2018]

Royal Conservatoire of Scotland. (2018b) *Gender Action Plan* [Online]. Available from: www.rcs.ac.uk/why_rcs/vision-and-governance/equalityanddiversity/gender-action-plan/ [Accessed 7 November 2018]

Scottish Funding Council. (2016) *Gender Action Plan* [Online]. Available from: www.sfc.ac.uk/web/FILES/Corporate_publications_SFCCP052016_GenderActionPlan/SFCCP052016_Gender_Action_Plan.pdf [Accessed 28 December 2018]

Scottish Funding Council. (2018) *SFC's Role in Scotland's Future* [Online]. Available from: www.sfc.ac.uk/about-sfc/about-us/about-us.aspx [Accessed 28 December 2018]

Society of London Theatre. (2017) *Joint Statement from the Theatre Industry* [Online]. Available from: https://solt.co.uk/about-london-theatre/press-office/joint-statement-from-the-theatre-industry/ [Accessed 7 August 2019]

Solnit, Rebecca. (2014). *Men Explain Things to Me: And Other Essays*. Granta Publications, London.

Spotlight. (2019) *Melanie La Barrie* [Online]. Available from: www.spotlight.com/interactive/cv/5734-3423-8863 [Accessed 6 August 2019]

Stellar Quines. (2018) *Artist Development M*****Classes* [Online]. Available from: www.stellarquines.com/artist-development/masterclasses/ [Accessed 29 April 2018]

Thompson, P. and Stevenson, A. (2015) Exploring the Experiences, Perceptions and Reflections of Popular Electronic Musicians at UK Higher Education Institutions. *Journal of Music, Technology & Education*, 8(2), pp. 199–217.

UK Government. (2018) *Equality Act 2010* [Online]. Available from: www.legislation.gov.uk/ukpga/2010/15/contents [Accessed 27 December 2018]

16

Enhancing Student Employability Through Innovative Program Design

A Case Study

Jude Brereton, Helena Daffern, Marc Green, Frank Stevens, and Andy Hunt

Introduction

For more than 30 years the University of York has been at the forefront of audio education, establishing a pioneering MA/MSc Music Technology program in the mid 1980s that shaped the nascent discipline of music technology. Originally taught jointly across the departments of Music and Electronic Engineering, it was a truly interdisciplinary subject where knowledge, concepts, methods of investigation, and outcomes were combined and contrasted, and often difficulties in finding a common language and working methodology were encountered, reflecting that interdisciplinarity is an inherently complex endeavor (Stember 1991).

Students originally enrolled in the master's program held undergraduate degrees in either music or STEM (science, technology, engineering, and math); the MA/MSc gave these students an opportunity to work across disciplines, taking classes in a variety of modules (learning units) provided across the two departments. Nevertheless, the interdisciplinary nature of this early master's degree program still yielded to externally enforced disciplinary divides in requiring students to make a choice at the culmination of their studies; whether to focus on "music", gaining an MA (Master of Arts) qualification or "technology", gaining an MSc (Master of Science) qualification. This severed the two disciplines again on graduation where they had in fact been integrated throughout the program.

In the mid 1990s undergraduate Music Technology System programs (MEng and BEng)[1] were established in the Department of Electronic Engineering, heavily influenced by the interdisciplinary ethos of the successful MA/MSc program. As these programs have grown, developed, and transformed over the last 30 years, a more multidisciplinary nature has become established, reflecting the general evolution of the discipline of "music technology" articulated by Boehm (2007, 2014), who locates music technology between the three knowledge domains of art, science, and technology.

Originally concerned mainly with the use of technology in the musical arts, "music technology" in the Department of Electronic Engineering at York has developed to encompass a variety of scientific and technical specialist fields, including acoustic science, audio software programming, music psychology, sound design, and sound recording. For the most part, students now on our undergraduate programs have obtained level 3[2] (mostly GCE A-Level qualifications) in maths and physics, usually alongside one or two more science subjects. Some also have A-Level music or music technology, although this is not a requirement, and many have kept music making as a hobby alongside their math and science studies at school.

When music technology was first established at the University of York, it did not appear as a subject taught in high schools, although technology was beginning to find its place in the music classroom (Finney and Burnard 2010). Music technology was introduced as a GCE A-Level subject in the mid-1990s, more or less coinciding with the introduction of our undergraduate programs. Indeed academic colleagues working in the Department of Electronic Engineering at the time were involved in validating the first syllabus for music technology A Level and writing the teacher's guide. However, it is interesting to note that in the most recent A-Level music technology subject specification the exam board places an overall emphasis on "making music with technology" placing it within the "art" and "technology" knowledge domains—"technology is the servant of music, not an end in itself" (Edexcel 2013, p. 19).

Both the departments of Music, and Theatre, Film, Television and Interactive Media (in the Faculty of Arts) also offer degree programs that emphasize the art and technology domains with a focus on the creative use of technology, music composition, sound design, and sound for visual media. In contrast, our undergraduate programs combine a solid grounding in analog and digital electronic engineering with specialist teaching modules in areas such as audio software programming, room acoustic modeling, digital music synthesis, psychoacoustics, musical instrument interfacing, music perception, and emerging audio technologies. For us the emphasis is on "making technology for music making" placing it more closely toward the science and engineering knowledge domain.

MSc Program Design

The aspiration for the new MSc in Audio and Music Technology, designed in 2013 and admitting its first students in 2015, centered on regaining a distinctive interdisciplinary ethos, influenced by some of the latest scholarship on inclusive engineering education (Heywood 2005; Peters 2018). It was steered by a desire to foreground active and collaborative learning (Kilgo et al. 2015) and incorporate elements of music, art, and creativity within the curriculum (Madden et al. 2013). The vision was not only to achieve better outcomes for students but also to provide a program that was flexible enough to admit graduate students from a variety of disciplinary backgrounds, in music, creative arts, music technology, audio engineering, and other branches of engineering, math, technology, and science, reflecting the diverse teaching and research backgrounds of the staff involved.

Given that UK school pupils are expected to specialize early after age 16, most taking only three or four subjects to Level 3, there is a sense of needing to choose either a science or an arts route at this stage. Data on exam entries show only one-third of 18-year-olds in England take a combination of fine arts (including music) and science subjects (Sutch 2014).

Those who have opted for an arts specialism route at Level 3 are potentially unable to access our more science-based type of music technology at undergraduate level. We did not want to exclude the many capable graduate students who had chosen undergraduate degrees in the creative arts or music and therefore admit students who can demonstrate an aptitude for the technical aspects of music technology, for example through technology-based hobby projects, or a demonstrable interest in the science of sound.

For those who have taken degrees in related disciplines such as engineering, math, or sciences, a key requirement for admission to our program is a demonstrable interest and experience in either making technology for music making, for example building microprocessor-based synthesizer; or making music with technology, for example experience in using studio technology and/or a digital audio workstation.

Design Vision and Ethos

The key objectives of the MSc center around active learning through guided work, and practical learning outcomes (Biggs 1996; Allen and Tanner 2007; Robinson 2016) alongside a desire to promote accessibility to a wider group of potential students (Bosman et al. 2017).

The focus on "what students do" and "what students can do when they graduate" aligns closely with the constructivist approach of the York Pedagogy (Robinson 2016) developed at the University of York, UK. This campus-wide framework ensures consistent academic program development across all departments and subjects. It requires a set of Program Learning Outcomes (PLOs) or statements that summarize what graduates will be able *to do* after taking the program (Table 16.1).

The PLOs for our MSc in Audio and Music Technology outlined here have been specifically developed not only to fulfill the York pedagogy but also to closely match the categories of our accrediting body—the Institution of Engineering and Technology (IET) (IET 2018), which uses the Accreditation of Higher Education Programmes (AHEP) framework of engineering competences to assess technical courses across the UK (Engineering Council 2014). Our MSc is weighted toward the science/engineering knowledge domains in terms of program learning outcomes, in order to satisfy the AHEP, the expectations of an MSc qualification, and alignment with other engineering programs taught within the department. For this reason our MSc, unlike some other music technology programs, is not accredited by JAMES (Joint Audio Media Education Support), the professional body who accredits courses that prepare students for work in the music and entertainment industries (JAMES 2019): Our focus is on scientific analysis, design, and implementation of new technologies rather than the use of technology to create music.

PLO	*On successful completion of the programme, graduates will be able to:*
1	**Subject Knowledge:** Conduct research into digital audio signal processing, audio software programming or acoustic analysis, advancing the state of knowledge by applying specialist engineering techniques techniques and research methodology.
2	**Engineering Analysis:** Extract and critically evaluate literature and other data about complex audio, acoustic and music technology systems through analytical and computational methods and modelling.
3	**Engineering Design:** Design innovative industry relevant engineering solutions for research-based problems in audio software and/or hardware, music technology and acoustics.
4	**Practical Skills:** Apply professional skills in engineering, audio analysis, programming, critical listening, acoustic modelling, construction and measurement, to independently solve technically challenging research-based problems.
5	**Technical Communication:** Communicate, debate, and contextualise information in a succinct, professional and technically accurate manner to key stakeholders which might include event managers, engineers, musicians and members of the public. This will include the ability to interpret and write technical documentation to a professional standard.
6	**Management & Personal Development:** Proficiently manage themselves, teams and complex projects in preparation for future careers in the field of audio and music technology as an individual practitioner or in industry.

Table 16.1 Program learning outcomes for the MSc in audio and music technology.

When setting entry/admissions requirements this desire to widen access to students from different disciplines required the means to assess the aptitude of potential students rather than relying on typical qualification assessment. The protocol to achieve this includes asking students with no formal science or engineering qualifications for evidence of working with the technology-related aspects of audio and music, perhaps through home studio recordings, assisting with sound control at live events, audio-related multimedia projects, computer programming for games, hobby electronics projects or tinkering with technology such as hacking existing hardware, or mini-projects based on readily available electronic microprocessors.

In order to design a program that is accessible but challenging, the content needed to share and celebrate the strengths, skills, and knowledge of the students from diverse and differing disciplinary backgrounds (Salminen-Karlsson 2002; Bernardini and de Poli 2007; Peters 2018; Bosman et al. 2017). Once we had identified the essential core content required in our program in terms of both subject knowledge and professional and transferable skills, we identified areas where students might be less confident and provided optional additional content to support those students. For example, students who have progressed through an educational background in arts and humanities with experience in music making and music theory may be less confident in their skills relating to the mathematical and computing aspects of the program (see zero-to-hero workshops later in the chapter for more detail).

Indeed, the emphasis on required expertise and fluency in mathematics, which is common to most engineering education, is less prioritized in our program. The math required is taught as needed, and placed in a relevant context. We believe our practical approach to incorporating mathematical content helps to break down some of the potential barriers to studying audio engineering.

We were also conscious that different teaching styles might have been experienced by students with different backgrounds. Therefore, we incorporated a mix of teaching approaches so that students could learn from each other and not be intimidated by unfamiliar formats. Presently, content is delivered through lectures, seminar-style sessions, workshops, and practical activities in the computer or electronics laboratory. To enhance learning, the formats and the role of the students are communicated at the start of each module. This includes encouraging students to build dialogue with the seminar leader. Tutors encourage students to actively engage with and debate different concepts, in order to foster critical thinking, rather than expecting passive involvement via a lecture-only style of pedagogy.

Overview of the Program

Master's level study in the UK generally takes the form of a 12-month intensive program, with learning and teaching activities broken down into credit bearing "modules", with 180 credits completed over the year. Ten credits represent 100 hours of student work, which includes face-to-face contact time at lectures, lab activities or workshops, as well as time spent on individual study and preparation for assessments.[3] A selection of modules from the 2018/2019 iteration of the program is described in more detail as follows.

- *Audio Signals and Psychoacoustics*

This module aims to enhance understanding and increase students' knowledge of acoustics and psychoacoustics in the context of audio, music production, and listening, informed by recent developments in the field.

Fundamental concepts, vital to understanding audio as a signal, are established before moving on to application specific areas. An introduction to the basics of sound waves and audio signals builds to complex digital signal processing (DSP) techniques and a sophisticated understanding of psychoacoustics and auditory concepts.

- *Voice Acoustics and Applications*

This module introduces students to the analysis and synthesis of the human voice. They then inform, manipulate, recognize, and synthesize acoustic signals for specific speech and voice performance applications.

The content is informed by research taking place at the AudioLab across voice and spatial audio, which includes analysis of classical singing techniques (e.g. Vos et al. 2018), speech synthesis (e.g. Gully et al. 2018), and assessing the benefits of singing through our interactive virtual choir (Daffern et al. 2019). Embedding this research into the module maintains the engagement of the students and staff, but also contextualizes their learning for real-world applications.

- *iOS Programming*

The public's response to the iPhone's introduction in 2007 is widely seen as the turning point in the adoption of smartphones. By opening up their development kit and programming environment to anyone, Apple effectively launched a new industry sector.

In the autumn term we introduce students to iOS[4] programming skills, but using the niche application area of interactive musical and audio processing. Students learn the basics of coding, quickly develop their own apps on iPhones and iPads, then work toward creating their own musical instruments or sound-processing apps.

- *Virtual Acoustics and Auralization*

Enhanced by strong links to current research within the teaching team, this module focuses on the creation of virtual acoustic environments using acoustic measurement and room modeling techniques as well as the use of immersive surround sound technologies over a wide range of applications, including music production, film and television postproduction, gaming, and virtual reality systems.

- *Interfacing for Audio and Music*

Students design and build fully working music/audio systems with physical human control interfaces. This includes connecting sensors to a microprocessor via electronic circuits, using an international standard to connect the microprocessor to an external computer, and implementing a custom synthesis or sound-processing system on the computer. The final system allows users to control audio/musical algorithms in real time using physical gestures and/or a combination of hardware and software that students have designed and built themselves.

- *Personal Professional Practitioner*

This module prepares students for future roles in industry, research, or self-employment, by developing key employability skills such as self-reflection for personal effectiveness, teamwork, leadership, self-promotion and marketing, public engagement, and skills in communicating science/technology to diverse audiences.

The module allows room for the explicit development of employability skills firmly in the context of audio and music technology by academic staff, with an awareness of the multifaceted nature of the audio industry. It is designed to bring together key strands of employability, which might be characterized as knowledge, skills, and social capital (Rich 2015). The CareerEDGE model of employability (Pool and Sewell 2007) has been a key influence here, with this module structured to combine subject specific skills, career information, real-world experience, generic skills, and emotional intelligence, all underpinned by self-reflection and evaluation in class and via the e-learning log—an online learning journal that students complete weekly (Baruah et al. 2017).

Flexible and Inclusive Learning

The idea of differentiated instruction is widely accepted in primary and secondary school teaching. Indeed it is embedded in the UK government's Teachers' Standards that set the minimum requirements for (school) teachers' conduct: "A teacher must: . . . Adapt teaching to respond to the strengths and needs of all pupils: to know when and how to differentiate appropriately, using approaches that enable pupils to be taught effectively" (Department for Education 2011).

Differentiation in the higher education classroom or lecture theater is of growing interest to university educators with much discussion and consideration being given to diverse learning styles, collaborative learning opportunities, and flexibility to enhance student engagement and inclusion (Santangelo and Tomlinson 2009; Zepke and Leach 2010; Peters 2018; Marcos-Jorquera et al. 2017)

Meeting Students "Where They Are"

As the MSc is intended for students from a diverse range of academic backgrounds, the students begin the course with varying experience of the subject areas taught on the course. Previous qualifications of students on the course to date include BSc physics, BSc mathematics, BSc statistics, BSc economics, BSc audio and recording technology, BSc sound, light and live event technology, BA interdepartmental humanities and science, BEng electronic and communication engineering, BA music production, and BA theatre technical arts.

One of the first tasks in the e-learning log is a skills audit, asking students to consider which skills (generic and discipline specific) they are confident they possess, and which they would like to develop during the program, aligning these with their overall career goals and aspirations.

In the first term, three different programming languages are learned, whereas the second term has a more applied focus, including considerations of recording and microphone techniques. Students tend to have greater experience in either programming or studio skills, depending on their disciplinary background. Student supervisors—academic staff who guide a small group of students and have access to the e-learning logs—work with other teaching staff to point more experienced students toward resources that will push their abilities wherever possible. The MSc teaching is undertaken by a small team of academic teaching specialist and research-active staff, who are assisted by a number of graduate teaching assistants (who are generally doctoral candidates in the AudioLab). The small number of staff involved, and the physical proximity of our offices mean that for the most part student support is provided through informal group discussions, alongside more formal mechanisms of formative (helps to monitor student progress, but does not contribute to overall degree marks) and summative (used to evaluate student learning and contributes to overall degree marks) assessment.

The autumn term covers such a wide range of areas that the majority of material should be new even if a student has experience in a particular area. Feedback from the first MSc cohort led us to instigate a number of introductory four-hour intensive "zero-to-hero" workshops in iOS programming, MATLAB, Pure Data, and recording studio techniques.

"Zero-to-Hero" Workshops

"Zero-to-hero" workshops provide a baseline platform for the students in each of these areas to build upon, and are written assuming no prior knowledge. These sessions are also valuable for the teaching staff, who can get an initial idea of the experience and aptitudes of the new students. The relatively small cohort of students means that it is feasible to monitor progress and to provide support on an individual basis. All students are encouraged to attend, even those who feel they have well-developed skills in the area, so that a collaborative "community of learning" can be established from the start, with students supporting each other, effectively "trading" knowledge and expertise (Lenning and Ebbers 1999).

The studio "zero-to-hero" workshop, for example, starts by assuming no knowledge of recording technology at all and this baseline is shared with all students by the workshop leader. The basics of the mixing desk and digital audio workstation are explained and the workshop leader works to establish a non-judgmental environment for learning new skills, making sure that all students in the cohort are able to contribute to discussions as they wish. In small groups students are tasked to perform elementary tasks, such as routing a microphone feed from the live room to the desk. More experienced students can offer help and advice, and reflect on how they learned to use the studio themselves, especially if the skills were initially self-taught. An overall ethos of self-reflection on personal learning and the use of active learning techniques helps to establish a non-threatening learning environment. The workshop leader's role here is vital, as students who regard themselves as novices in the studio need to rely on a supportive and collaborative code of conduct underpinning the learning space.

Personal Experiences of Music

Music Perception and Critical Listening is one of two modules in the MSc that are co-taught with students on the MA in music production programme (Music Department). The module encourages students to draw on their own knowledge and experience to contemplate and investigate the perception of music, critical listening, and the technical elements of audio and music technology within a musical context. They are asked to challenge their own conceptions about music perception, drawing on their own experiences and debating their own opinions from the perspective of their practical understanding of audio and music technology as well as scholarly literature regarding music perception. For instance, one session focuses on rhythm perception, from the musical and perceptual definitions to the application of rhythmic devices within different genres with which they are familiar, and how these theories fit within the nature/nurture debate of music perception (Palmer 1997; Trehub 2003).

Active Learning and Creativity

Approaches to active learning are becoming popular in engineering curriculum developments; they are seen to be beneficial in developing enhanced problem-solving skills, social skills, creativity, and criticality (Singer and Smith 2013). For engineering educators the challenge is often to embed creativity into learning activities and assessments (Biggs 1996; Heywood 2005; Peters 2018). Since music technology is so strongly linked to music making, our students are often confident in their creative abilities and respond positively to opportunities to work and learn creatively.

In *Audio Signals and Psychoacoustics*, lectures on some of the more technical digital signal processing topics always include a variety of audio examples and often combine a discussion of the "pure" mathematical theory alongside MATLAB code examples and visualizations. This prevents the more complicated digital signal processing theories (such as the z-transform, or the Discrete Fourier Transform) seeming too abstract by relating to real-world examples and practical implementation. Examples of this include listening to the aliasing effects resulting from the sampling theorem, and visualization of the helical spiral representing the behavior of the Fourier Transform in the complex plane.

The lab work here is presented in 14 sessions, the first four including directed tasks in MATLAB, and the final ones reserved for individual work toward the assignment. These sessions allow for individual teaching relationships to be formed with each student in order to address individual needs, which have included support with individual assignment planning, working through example MATLAB scripts, and revisiting previously covered material as required. This is made possible thanks to the relatively high ratio of teaching staff to students (for some practical lab activities the small cohort of around 16 is supported by one academic staff member and one graduate teaching assistant) and the high quantity of staff-student contact time. In addition, students' entries into the e-learning log allow academic staff to be more fully aware of the different academic backgrounds, skills, and experiences of students in the cohort.

Assignments for this module include the design and implementation of a creative binaural demonstration in MATLAB and a demonstration of a psychoacoustic phenomenon in Pure Data. Both of these tasks are designed to encourage students to demonstrate their understanding of the technical material by applying their knowledge creatively.

The *iOS programming* module incorporates a variety of teaching methods, including lectures with live coding demonstrations, backed up by a series of computer-lab exercises, links to online tutorials, and back-up reading material. In this module and *Interfacing for Audio and Music* the assessment tasks seek to reward not only the success of the "product" (the final system produced) but also the "process", i.e. the approach taken by the student considering their starting point, in terms of previous experience and skills, using student's self-reflection on the learning undertaken in order to produce a viable product.

Allowing for Flexibility

The spring term allows students to choose modules in the Electronic Engineering Department, or to join MA music production students in the Music Department in *Production Techniques, Technologies and Aesthetics*, which develops theoretical and practical knowledge of audio production for pop, rock, and dance music. We work closely with students to make sure that the choice they make matches not only their previous experience but also their potential career goals.

Since students come from different backgrounds and have different aspirations, we incorporate levels of choice into assessment design where possible. In the *Interfacing for Audio and Music* module, for example, we allow students to select one of three application areas within which to build their "product" for assessment:

- An Interactive Sonification (Degara et al. 2015) device—allowing users to explore data as sound by interactively controlling how that data is converted to audio
- A music performance analysis system—allowing users to gather and process data in real time about musical performance and gestures
- A new electronic/computer musical instrument

The design and implementation forms one part of the assessment, with the student also required to produce a website to include full technical documentation as well as a video aimed at marketing the "product". See Figure 16.1 for a recent example of a visual feedback device designed for use by audio engineers mixing in ambisonic surround sound.

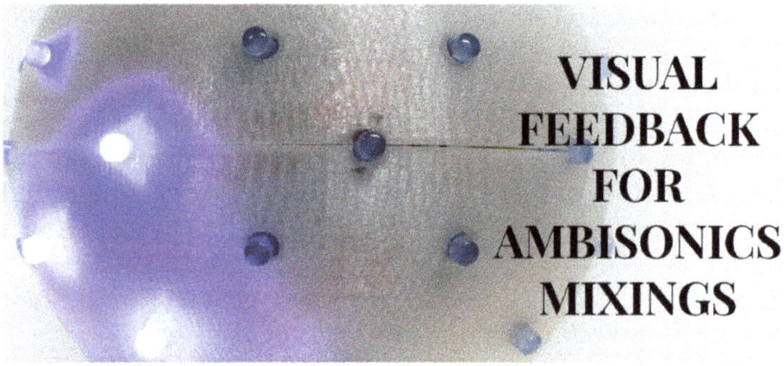

Figure 16.1 An electronic visual feedback device, designed and built by a current student as part of the assessment for the *Interfacing for Audio and Music* module.[5]

Self-Reflection on Learning

A number of assessments in the program incorporate some form of student self-reflection on learning. In *Audio Signals and Psychoacoustics*, students are required to write a reflective critique of their assignment submission, allowing them to demonstrate an ability to analyze the quality of their own work and make specific suggestions for how it could be improved (Baruah et al. 2017; MacCallum and Casey 2017; Scott 2017)

A similar approach is used in *Music Perception and Critical Listening*, where a key part of the module assessment is a group presentation on a chosen topic of music or audio perception. Students are free to choose their own topic (e.g. music in horror films, the McGurk Effect, perception of rhythm) for their individual submission comprising a critical evaluation of a portfolio of five or six 30-second-long audio examples, two of which they have created themselves. Alongside their critical evaluation, students submit a short personal reflection about their reasons for choosing the topic and what surprised, interested, or engaged them most in the module learning activities.

The e-learning log asks students not only to reflect on subject-specific skills development but also to consider wider aspects of their learning and engagement such as their initial expectations, career goals, a reflection on teaching styles encountered, their approach to note taking and reading, alongside more personal concerns such as their living, social, and cultural environment (see Figure 16.2).

In the last week of the autumn term students are encouraged to assess the impact of the program so far in terms of their career aspirations. Example e-learning log entries quoted in Table 16.2 serve to illustrate the variety of career goals and aspirations amongst the cohort ranging from further research to founding a business within the Creative Industries or becoming a music producer.

Enhancing Employability through Innovative Assessment

Since the MSc modules are shaped by problem-based learning design (dos Santos 2016) and the CDIO (conceive, design, implement, operate) framework (Clark and Andrews 2012), which has been shown to improve student engagement in engineering programs, learning is assessed through coursework rather than formal examinations. Most modules are explicitly designed to support development of student autonomy in acquiring knowledge and developing understanding through structured work, which leads toward a small practical design-and-build project or a portfolio of work submitted for assessment (McDowell et al. 2006). These coursework assessments increase in technical complexity and the amount of independent work required, at the same time assessing the acquisition of subject specific knowledge and professional skills and competencies for increased employability.

"Real-World" Assessments

A core element in the design of the MSc was working in a number of "authentic" or "real-world" assessments. We define real-world assessments as coursework that

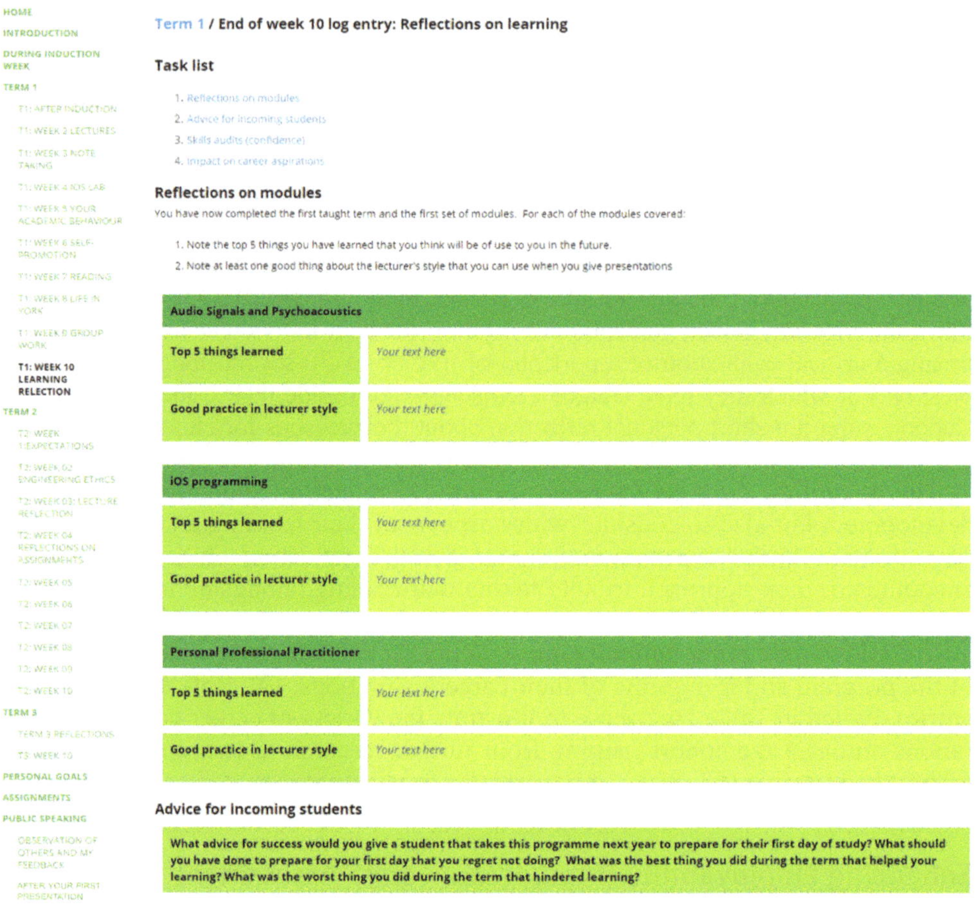

Figure 16.2 Example of an e-learning log entry template for autumn term, week 10.

exists outside of the campus and has life beyond the program. They allow us to combine the benefits of active learning—seen as increased engagement, enjoyment, and subsequently improved retention and employability (Zepke and Leach 2010; Kilgo et al. 2015; Scott 2017)—with the development and promotion of professional skills.

Three of the six MSc program learning outcomes (see Table 16.1) center around the development of such professional, or employability, skills including subject-specific engineering skills (PLO4), communication skills (PLO5), and personal development skills specifically for future careers in the audio industry, or as an individual practitioner (PLO6).

Real-world assessment tasks help foster innovation and creative skills and reinforce critical thinking and problem-solving skills by expecting students to work not just on authentic simulations but to really get their "hands dirty" in real-world tasks (Mueller 2005; Brereton 2016; dos Santos 2016).

The real-world assessment tasks are showcased on the students' personal website and LinkedIn profile that they are required to set up as part of the *Personal*

Reflecting on this term as a whole has your experience changed in any way your future career aspirations or goals, comment below? If so, go back to your Personal Goals statement and update the relevant sections.

. . . I have always considered my future pretty open. I would still love to found and run my own business, and I wish to continue writing, producing and releasing my own music. One of my goals has always been to gain a PhD and over the course of the first term I have come to realise that PhD study is almost in my grasp. Through the term and from conversations with my supervisor and lecturers I have become more focused on this goal and am currently looking into Bioacoustics as a field to research. . . .

Because I didn't do a BSc I think I should be aiming to look for opportunities to grow my skill set instead of good pay/bad progression. ...Studio work is my end goal however I want to be a successful music producer and not a bedroom recorder, this will take a lot of time and money and I would rather grow that from a hobby alongside a career and hopefully one day it will be realistically sustainable. (Long story short I would like to be a music producer but I would also like to be able to secure a mortgage! such is the music industry!).

So far, I am still determined to not think about the grade but to enjoy and be confident on my abilities and find a way to be a part of the audio industry.

Table 16.2 Example entries to the e-learning log around career aspirations.

Professional Practitioner module in the autumn term. This online presence is formally and summatively assessed, since skills in self-promotion and communication have been explicitly developed in this module.

During the *Project Development* module students produce a series of published blog posts (see Figure 16.3) detailing their progress and ideas. We encourage students to make these publicly available and join a wider online community of audio enthusiasts, industry collaborators and researchers, being aware that much academic networking and sharing of insights happens through online communication (Mewburn and Thomson 2013). The website and blog help to develop students' personal brand, showcase their digital skills and employability skills, demonstrate self-efficacy, increase their self-confidence and articulate their skills to potential employers (Thorley 2013). In a survey of recent MSc audio and music technology graduates, 8 out of 13 who responded were still using the blog they had established during the program as part of their personal brand and curated online presence.

Of course, an assessed student portfolio of work to showcase skills is quite commonplace in creative arts such as art, music composition, or dance, but almost unheard of in science or engineering degrees. More commonly engineering degree programs culminate in a "capstone" project (Miller and Olds 1994; Clark and Andrews 2012; Kilgo et al. 2015) in the final year of study.

The innovative assessments included in the program, together with the specific development of personal professional skills, provide a variety of opportunities to

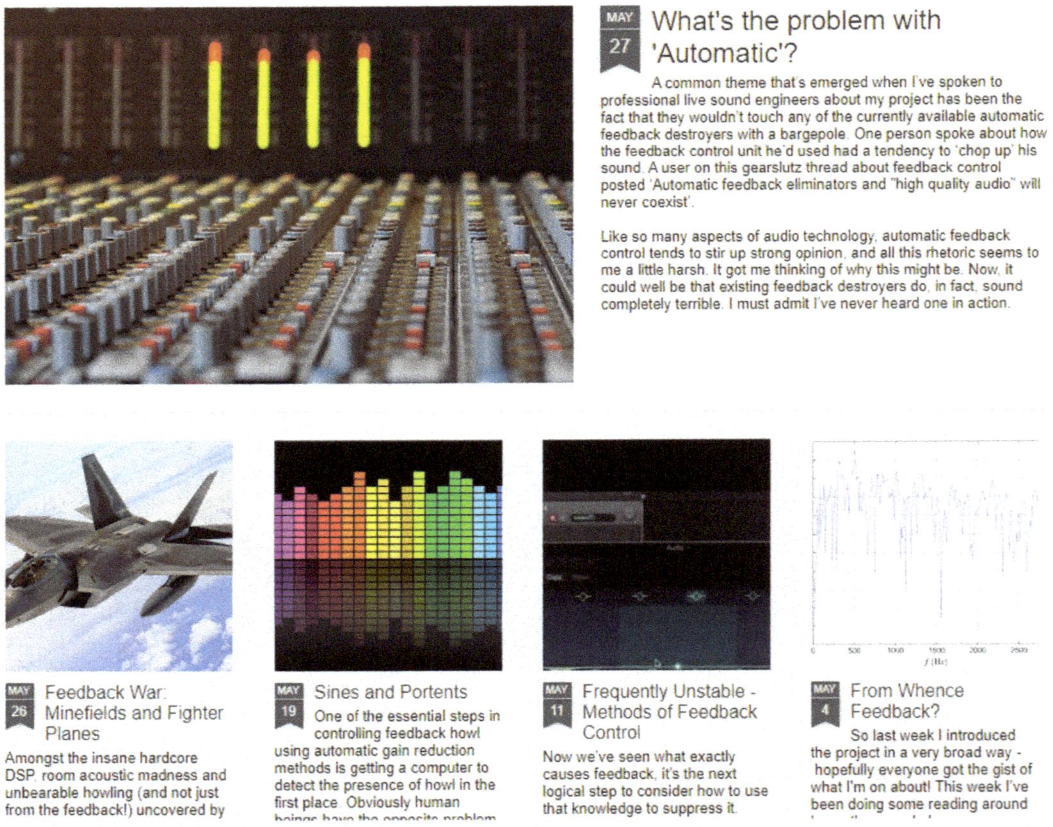

Figure 16.3 Example of an MSc student research project blog *Feedback Destroyer*.

place student learning firmly in the context of future employment goals that have been established through self-reflection in the e-learning log. Nevertheless, the use of reflective practice and the dissemination of student work (portfolio) in science and engineering education is not yet widespread (Baruah et al. 2017; MacCallum and Casey 2017; Mello et al. 2017) and some students and staff are resistant to this as a teaching and learning tool. Indeed our own electronic engineering undergraduates at York undertake a final year "capstone" project, yet the work produced by the student and the skills developed are only described between student and academic staff, and not explicitly articulated to the outside world and potential employers (Thorley 2013).

In contrast, many of the "real-world" artifacts produced by our MSc students for assessment can be showcased to potential employers, allowing them to articulate the skills they have developed via the blog and website (Brereton 2016). For example, the iPhone app developed as part of the *iOS programming* module can be highlighted here, and eventually published on the Apple App Store. Another recent graduate made good use of this opportunity to demonstrate their skills and has since secured a role as an iOS developer. We recognize there is scope for us to extend our successful use of real-world assessments into our existing undergraduate programs.

Individual Projects

The MSc program culminates in an individual research project that focuses on a relevant and up-to-date area of research or technology development and provides a strong focus on employability, especially the provision of industry-related projects.

The *Project Development* module provides students with a guided introduction to effective researching, but also has an emphasis on developing a high level of communication skills, including writing for a lay audience (through a weekly blog), for academics (via a literature survey) and oral presentation skills (through seminar presentation to peers and academics).

Final projects can extend the kinds of work done in taught modules, or combine skills from several. Previous projects have included using MATLAB and signal processing skills learned in *Audio Signals and Psychoacoustics* together with virtual acoustics techniques to investigate acoustic feedback, and creation of iOS apps for teaching and conducting listening tests.

Industry Experience

Although traditionally engineering degrees have been seen as highly vocational, with good employability inherent in the nature of the discipline, the engineering industry and professional associations are concerned that engineering graduates are entering the job market lacking the "soft" skills needed for success in the workplace and unprepared to meet the demands of working life (*Skills & Demand in Industry: 2017 IET 2017*).

Our industry-related projects give the opportunity for students to engage with industry professionals from a variety of sectors including audio engineering, technology development, app development, electronic engineering. Each year a number of industry projects are offered to the students; industry collaborators who have offered to sponsor projects include Google, AECOM, Abbey Road, Music Tribe, Allen and Heath, and Apollo Ensemble. Project sponsorship entails the industry collaborator suggesting and scoping the project. Primary supervision of the student is undertaken by an academic staff member at York, with the industrial collaborator providing secondary supervision and assisting with the final marking (grading).

Industry projects continue to be a popular choice and are a draw for potential students to the course. While there is additional work involved for staff in making the connections and setting up the projects, they prove valuable beyond the immediate benefits to the student, laying the foundations for new long-term research collaborations.

Challenges and Opportunities

Students' starting points: The biggest challenge faced by our program remains the different skill levels and experience of incoming students. We have attempted to mitigate possible gaps in knowledge as far as possible using "zero-to-hero" workshops, which receive good feedback from students; nevertheless catering for students with diverse interests, previous experience, and career goals, as well as different personal expectations and learning styles requires the use a variety of appropriate pedagogical strategies. University assessment regulations and criteria can often seemingly constrain the flexibility required to cater for such diversity. However, we have worked within the program design to include a number of different assessment methods, which suit a variety of student learning styles and personal goals.

An intensive program: The intensity of the program offers another challenge since some students will naturally adapt to the pace of the academic work better than others. Being able to effectively monitor students' progress and offer extra resources and support where possible for those who may fall behind forms part of our flexible approach. The level of intensity is, however, also a strength of the program, as committed students quickly become skilled in a variety of areas. Ultimately, this is useful, especially when a student is unsure what path to take after graduation, as they can choose what they enjoy the most and where to focus on their individual research project.

Bridging the gap between art and technology: Another challenge is balancing the relative weightings of art and technology. For example the *Music Perception and Critical Listening* module and parts of the *Project Development* module are shared between two different cohorts of students (MSc audio and music technology and MA music production) who may have different agendas and priorities that focus on different areas of music technology.

The challenge here is to judge how to pitch some of the material, especially that which is more focused on music theory and psychoacoustics, in which the two cohorts will naturally have different levels of experience. In practice, this is typically gauged and altered as the modules progress, relying on the module leaders to differentiate between students with different levels of ability in a variety of topics. The apparent different learning styles of the students can sometimes cause problems in their collaborative assignments, but this can also be a good experience for the students working in teams with a variety of points of view.

Improving diversity in audio: There is of course still much work to do around gender equality in audio engineering and improving inclusion to audio engineering and music technology degrees (Armstrong 2011; Born and Devine 2015; Young et al. 2018).

A recent search for taught MSc degrees in music technology on UCAS (the Universities and Colleges Admissions Service—a centralized service for university admissions in the UK) returns five programs centered on the technology domain of music technology, and who admit graduates with first degrees in a variety of subjects. Another five MSc degrees that are more focused on acoustics and audio technology, require a science or engineering first degree.

Since the proportion of female students in undergraduate engineering and technology degrees is still stubbornly low it is difficult to improve the gender balance at the master's level since most science/engineering based postgraduate programs in audio and acoustics require a previous engineering qualification. Our program seeks to bridge this gap with its combination of technology, music analysis, and acoustic and audio engineering content, yet admitting applicants from a variety of disciplinary backgrounds to help break down some of the real and perceived barriers to master's level study in music technology and audio engineering.

The MSc program was not designed specifically to tackle gender inequality in the field of audio engineering. Nevertheless female school pupils are twice as likely than male to study only arts and humanities subjects at age 18, and subsequently engineering at undergraduate level in the UK is still a predominantly male domain. We hope that our desire to cater for different academic journeys prior to entry will, at least in some small way, help the discipline eventually to achieve a better gender balance. Indeed, in the first four years of the program female applications and entrants have averaged between 21% and 22%, which is much higher than the current balance in the department on our undergraduate M/BEng tusic Technology systems programs and higher than the reported 6% to 8% female participation rate in the audio industry (Mathew et al. 2016). Opening up admissions to engineering-based master's level study in music technology to a wider pool of potential applicants, through careful consideration of skills and previous experience in this interdisciplinary field, has the potential to improve diversity and inclusion in audio engineering more widely.

Conclusion

Thirty years since our first music technology master's degree, and against an educational landscape where STEM disciplines are seemingly more highly valued (or at least better funded) than arts, the University of York now offers a number of degree programs in audio education across three departments and two faculties (both arts and science). All have technology, creativity, and scholarship at their heart, but differ in subject content focus, potential student recruitment, and post-program routes into employment, reflecting that music technology as a discipline is beginning to fragment along the traditional arts/science divide (Boehm 2005).

Nevertheless the MSc program in audio and music technology illustrates the potential for effective provision of a master's taught course for the audio and music industries, which could be replicated or adapted by other institutions. Of particular note are a strong pedagogical design with carefully crafted Program Learning Outcomes, flexibility in entry qualifications, rapid-progress introductory courses alongside real-world practical assessments that build up a student's portfolio together with links to (and involvement in) industry and research institutions. These features, combined with a range of teaching styles, tools, and techniques, result in a popular program that prepares students from a variety of academic and disciplinary backgrounds for a leading role in the audio and music technology workplace.

Notes

1. BEng is a three-year bachelor of engineering course; MEng is a four-year program resulting in a master of engineering qualification.
2. Level 3 qualifications are taken predominantly by pupils at age 18 before they leave high school/sixth form college and generally form part of the entry requirements for undergraduate degree courses at university.
3. The current program specification can be found at www.york.ac.uk/study/postgraduate-taught/courses/msc-audio-music-technology/.
4. The name for Apple's operating system that can run apps.
5. Further information on this device is available at https://ambfeed.weebly.com/

References

Allen, D. & Tanner, K. (2007). "Putting the horse back in front of the cart: Using visions and decisions about high-quality learning experiences to drive course design", *CBE Life Sciences Education, 6* (2), 85–89, doi:http://dx.doi.org/10.1187/cbe.07-03-0017

Armstrong, V. (2011). *Technology and the Gendering of Music Education*. Farnham, Surrey, England; Burlington, VT: Routledge.

Baruah, B., Ward, T. & Brereton, J. (2017). "An e-learning tool for reflective practice and enhancing employability among engineering students", *Proceedings of the European Association for Education in Electrical and Information Engineering (EAEEIE) Annual Conference*. Grenoble, France, 7–9 June, doi:10.1109/EAEEIE.2017.8768718

Bernardini, N. & de Poli, G. (2007). "The sound and music computing field: Present and future", *Journal of New Music Research, 36* (3), 143–148, doi:http://dx.doi.org/10.1080/09298210701862432

Biggs, J. (1996). "Enhancing teaching through constructive alignment", *Higher Education, 32* (3), 347–364, doi:http://dx.doi.org/10.1007/BF00138871

Boehm, C. (2005). "Music technology in higher education", in F. McMahon & T. Claes (Eds.), *Probing the Boundaries of Higher Education* (Vol. 12, pp. 85–90). Oxford, UK: Inter-Disciplinary Press.

Boehm, C. (2007). "The discipline that never was: Current developments in music technology in higher education in Britain", *Journal of Music, Technology and Education, 1* (1), 7–21, doi:https://doi.org/10.1386/jmte.1.1.7_1

Boehm, C. (2014). "A brittle discipline: Music technology and third culture thinking", *Proceedings of the SEMPRE MET2014 Conference: Researching Music Technology in Education: Critical Insights*, London, UK, April.

Born, G. & Devine, K. (2015). "Music technology, gender, and class: Digitization, educational and social change in Britain", *Twentieth-Century Music, 12* (2), 135–172, doi:https://doi.org/10.1017/S1478572215000018

Bosman, L., Chelberg, K. & Fernhaber, S. (2017). "Introduction to engineering: A constructivist-based approach to encourage engagement and promote accessibility", *Global Journal of Engineering Education, 19* (3), 237–242.

Brereton, J. (2016). "Making learning authentic: Real-world assessments for masters level study", *Forum Magazine, 2*, 14–15, University of York, York, UK.

Clark, R. & Andrews, J. (2012). "Engineering the future: CDIO as a tool for combating retention difficulties", in M. Rasul (Ed.), *Developments in Engineering Education Standards: Advanced Curriculum Innovations* (pp. 143–155). Hershey, PA: IGI Global, doi:10.4018/978-1-4666-0951-8.ch008

Daffern, H., Camlin, D.A., Egermann, H., Gully, A.J., Kearney, G., Neale, C. & Rees-Jones, J. (2019). "Exploring the potential of virtual reality technology to investigate the health and wellbeing benefits of group singing", *International Journal of Performance Arts and Digital Media, 15* (1), 1–22, doi:http://dx.doi.org/10.1080/14794713.2018.1558807

Degara, N., Hunt, A. & Hermann, T. (2015). "Interactive sonification [guest editors' introduction]", *IEEE Multimedia, 22* (1), 20–23, doi:http://dx.doi.org/10.1109/MMUL.2015.8

Department for Education (2011). *Teachers' Standards: Guidance for School Leaders, School Staff and Governing Bodies*. Retrieved from: www.gov.uk/government/publications/teachers-standards [Accessed 21 December 2018]

dos Santos, S.C. (2016). "PBL-SEE: An authentic assessment model for PBL-based software engineering education", *IEEE Transactions on Education, 60* (2) 120–126, doi:http://dx.doi.org/10.1109/TE.2016.2604227

Edexcel (2013). *Specification: GCE Music Technology*. Retrieved from Pearson Education Limited website: https://qualifications.pearson.com/en/qualifications/edexcel-a-levels/music-technology-2008.coursematerials.html [Accessed 21 December 2018]

Engineering Council (2014). *The Accreditation of Higher Education Programmes: UK Standard for Professional Engineering Competence* (3rd Ed.). Retrieved from: www.engc.org.uk/education-skills/accreditation-of-higher-education-programmes/

Finney, J. & Burnard, P. (2010). *Music Education with Digital Technology*. London, UK: Continuum International Publishing Group.

Gully, A.J., Daffern, H. & Murphy, D.T. (2018). "Diphthong synthesis using the dynamic 3D digital waveguide mesh", *IEEE/ACM Transactions on Audio, Speech, and Language Processing, 26* (2), 243–255, doi:http://dx.doi.org/10.1109/TASLP.2017.2774921

Heywood, J. (2005). *Engineering Education: Research and Development in Curriculum and Instruction*. Hoboken, NJ: John Wiley and Sons.

IET. (2017). *Skills & Demand in Industry:2017 IET Skills Survey*. Retrieved from IET website: www.theiet.org

IET. (2018). *Working to Engineer a Better World*. Retrieved from IET website: www.theiet.org/ [Accessed 15 December 2018]

JAMES. (2019). Retrieved from: www.jamesonline.org.uk/ [Accessed 14 May 2019]

Kilgo, C.A., Ezell Sheets, J.K. & Pascarella, E.T. (2015). "The link between high-impact practices and student learning: Some longitudinal evidence", *Higher Education, 69* (4), 509–525, doi:https://doi.org/10.1007/s10734-014-9788-z

Lenning, O.T. & Ebbers, L.H. (1999). "The powerful potential of learning communities: Improving education for the future", *ASHE-ERIC Higher Education Report, 26* (6), 2–173.

MacCallum, J. & Casey, S.C. (2017). "Enhancing skills development and reflective practise in students during their programme of study", *New Directions in the Teaching of Physical Sciences, 12* (1), doi:https://doi.org/10.29311/ndtps.v0i12.2368

Madden, M.E., Baxter, M., Beauchamp, H., Bouchard, K., Habermas, D., Huff, M., Ladd, B., Pearon, J. & Plague, G. (2013). "Rethinking STEM education: An interdisciplinary STEAM curriculum", *Procedia Computer Science, 20*, 541–546, doi:http://dx.doi.org/10.1016/j.procs.2013.09.316

Marcos-Jorquera, D., Pertegal-Felices, M.L., Jimeno-Morenilla, A. & Gilar-Corbí, R. (2017). "An interdisciplinary practical for multimedia engineering students", *IEEE Transactions on Education, 60* (1), 8–15, doi:http://dx.doi.org/10.1109/TE.2016.2566606

Mathew, M., Grossman, J. & Andreopoulou, A. (2016). "Women in audio: Contributions and challenges in music technology and production", *Proceedings of the Audio Engineering Society Convention 141*, Los Angeles, October.

McDowell, L., Sambell, A. & Sambell, K. (2006). "Supporting diverse students: Developing learner autonomy via assessment", in C. Bryan & K. Clegg (Eds.), *Innovative Assessment in Higher Education* (pp. 158–168). London, UK: Routledge.

Mello, L.V., Tregilgas, L., Cowley, G., Gupta, A., Makki, F., Jhutty, A. & Shanmugasundram, A. (2017). "Students-as-partners' scheme enhances postgraduate students' employability skills while addressing gaps in bioinformatics education", *Higher Education Pedagogies, 2* (1), 43–57, doi:http://dx.doi.org/10.1080/23752696.2017.1339287

Mewburn, I. & Thomson, P. (2013). "Why do academics blog? An analysis of audiences, purposes and challenges", *Studies in Higher Education, 38* (8), 1105–1119, doi:http://dx.doi.org/10.1080/03075079.2013.835624

Miller, R.L. & Olds, B.M. (1994). "A model curriculum for a capstone course in multidisciplinary engineering design", *Journal of Engineering Education, 83* (4), 311–316, doi:http://dx.doi.org/10.1002/j.2168-9830.1994.tb00124.x

Mueller, J. (2005). "The authentic assessment toolbox: Enhancing student learning through online faculty development", *Journal of Online Learning and Teaching, 1* (1), 1–7.

Palmer, C. (1997). "Music performance", *Annual Review of Psychology, 48* (1), 115–138, doi:http://dx.doi.org/10.1146/annurev.psych.48.1.115

Peters, J. (2018). *Designing Inclusion into Engineering Education.* Retrieved from Royal Academy of Engineering website: www.raeng.org.uk/

Pool, L.D. & Sewell, P. (2007). "The key to employability: Developing a practical model of graduate employability", *Education + Training, 49* (4), 277–289, doi:http://dx.doi.org/10.1108/00400910710754435

Riaz, H., Stiles, M., Armstrong, C., Chadwick, A., Lee, H. & Kearney, G. (2017). 'Multichannel microphone array recording for popular music production in virtual reality,' *Proceedings of Audio Engineering Society Convention 143*, Audio Engineering Society, New York, October.

Rich, J. (2015). *Employability: Degrees of Value*. Oxford, UK: Higher Education Policy Institute.

Robinson, J. (2016). *The York Pedagogy*. York, UK: University of York. Retrieved from: www.york.ac.uk/staff/teaching/themes/theyorkpedagogy/ [Accessed 15 December 2018]

Salminen-Karlsson, M. (2002). "Gender-inclusive computer engineering education: Two attempts at curriculum change", *International Journal of Engineering Education, 18* (4), 430–437, doi:https://doi.org/10.1007/s10798-015-9319-y

Santangelo, T. & Tomlinson, C.A. (2009). "The application of differentiated instruction in postsecondary environments: Benefits, challenges, and future directions", *International Journal of Teaching and Learning in Higher Education, 20* (3), 307–323.

Scott, G.W. (2017). "Active engagement with assessment and feedback can improve group-work outcomes and boost student confidence", *Higher Education Pedagogies, 2* (1), 1–13, doi:http://dx.doi.org/10.1080/23752696.2017.1307692

Singer, S. & Smith, K.A. (2013). "Discipline-based education research: Understanding and improving learning in undergraduate science and engineering", *Journal of Engineering Education, 102* (4), 468–471, doi:http://dx.doi.org/10.1002/jee.20030

Stember, M. (1991). "Advancing the social sciences through the interdisciplinary enterprise", *The Social Science Journal, 28* (1), 1–14, doi:http://dx.doi.org/10.1016/0362-3319(91)90040-B

Sutch, T. (2014). *Uptake of GCE AS Level Subjects 2007–2013. Research Division Assessment, Research and Development Cambridge Assessment.* Retrieved from: https://www.cambridgeassessment.org.uk/Images/173577-uptake-of-gce-as-level-subjects-2007-2013.pdf.

Thorley, M. (2013). "Connecting learners, employers and practitioners through emergent digital technology", *Proceedings of the SEMPRE MET2014 Conference: Researching Music Technology in Education: Critical Insights*, London, UK, April.

Trehub, S.E. (2003). "The developmental origins of musicality", *Nature Neuroscience*, *6* (7), 669–673, doi:http://dx.doi.org/10.1038/nn1084

Vos, R.R., Murphy, D.T., Howard, D.M. & Daffern, H. (2018). "Determining the relevant criteria for three-dimensional vocal tract characterization", *Journal of Voice: Official Journal of the Voice Foundation*, *32* (2), 130–142, doi:http://dx.doi.org/10.1016/j.jvoice.2017.04.001

Young, K., Lovedee-Turner, M., Brereton, J. & Daffern, H. (2018). "The impact of gender on conference authorship in audio engineering: Analysis using a new data collection method", *IEEE Transactions on Education*, *61* (4), 1–8, doi:http://dx.doi.org/10.1109/TE.2018.2814613

Zepke, N. & Leach, L. (2010). "Improving student engagement: Ten proposals for action", *Active Learning in Higher Education*, *11* (3), 167–177, doi:http://dx.doi.org/10.1177/1469787410379680

Index

Note: Page numbers in *italic* indicate a figure and page numbers in **bold** indicate a table on the corresponding page. Page numbers followed by 'n' indicate a note on the corresponding page.

For Product Safety Concerns and Information please contact our EU representative GPSR@taylorandfrancis.com
Taylor & Francis Verlag GmbH, Kaufingerstraße 24, 80331 München, Germany

www.ingramcontent.com/pod-product-compliance
Ingram Content Group UK Ltd.
Pitfield, Milton Keynes, MK11 3LW, UK
UKHW051942061126
477531UK00005B/223
9780367074432